The Quark and the Jaguar

ADVENTURES IN THE SIMPLE AND THE COMPLEX

MURRAY GELL-MANN

W. H. FREEMAN AND COMPANY
NEW YORK

The poem "Cosmic Gall" by John Updike is reprinted
from *Telephone Poles and Other Poems.*
Copyright 1960 John Updike.
Reprinted by permission of Alfred A. Knopf, Inc.
and Andre Deutsch, Ltd.

Cover image: © John Giustina/*The Wildlife Collection*

Library of Congress Cataloging-in-Publication Data

Gell-Mann, Murray.
The quark and the jaguar: adventures in the simple and
the complex / Murray Gell-Mann.
p. cm.
Includes index.
ISBN 0-7167-2725-0
1. Gell-Mann, Murray 2. Quark (Elementary particle physics)
3. Particles (Elementary particle physics) 4. Elementary particle physicists—
United States—Biography I. Title
QC774.G45A3 1994
530—dc20 94-1642
CIP

Printed in the United States of America

Second printing 1996, MB

FOR MARCIA

It's good for us—
chaos and color, I mean.

Marcia Southwick
Why the River Disappears

CONTENTS

Preface ix

PART I

THE SIMPLE AND THE COMPLEX

1 · Prologue: An Encounter in the Jungle 3

2 · Early Light 11

3 · Information and Crude Complexity 23

4 · RANDomness 43

5 · A Child Learning a Language 51

6 · Bacteria Developing Drug Resistance 63

7 · The Scientific Enterprise 75

8 · The Power of Theory 89

9 · What Is Fundamental? 107

PART II
THE QUANTUM UNIVERSE

10 · Simplicity and Randomness in the Quantum Universe 123

11 · A Contemporary View of Quantum Mechanics: Quantum Mechanics and the Classical Approximation 135

12 · Quantum Mechanics and Flapdoodle 167

13 · Quarks and All That: The Standard Model 177

14 · Superstring Theory: Unification at Last? 199

15 · Time's Arrows: Forward and Backward Time 215

PART III
SELECTION AND FITNESS

16 · Selection at Work in Biological Evolution and Elsewhere 235

17 · From Learning to Creative Thinking 261

18 · Superstition and Skepticism 275

19 · Adaptive and Maladaptive Schemata 291

20 · Machines That Learn or Simulate Learning 307

PART IV
DIVERSITY AND SUSTAINABILITY

21 · Diversities Under Threat 329

22 · Transitions to a More Sustainable World 345

23 · Afterword 367

Index 377

PREFACE

The Quark and the Jaguar is not an autobiography, although it does contain some reminiscences about my childhood and a number of anecdotes about colleagues in science. Nor is it primarily concerned with my work on the quark, although a sizable chunk of the book is devoted to some observations on the fundamental laws of physics, including the behavior of quarks. I may some day write a scientific autobiography, but my aim in this volume is to set forth my views on an emerging synthesis at the cutting edge of inquiry into the character of the world around us—the study of the simple and the complex. That study has started to bring together in a new way material from a great number of different fields in the physical, biological, and behavioral sciences and even in the arts and humanities. It carries with it a point of view that facilitates the making of connections, sometimes between facts or ideas that seem at first glance very remote from each other. Moreover, it begins to answer some gnawing questions that many of us, whether working in the sciences or not, continue to ask ourselves about what simplicity and complexity really mean.

The book is divided into four parts. At the beginning of the first part, I describe some personal experiences that led me to write it. Taking long walks in tropical forests, studying birds, and planning nature conservation activities, I became excited by the idea of sharing with readers my growing awareness of the links between the fundamental laws of physics and the world we see around us. All my life I have loved exploring the realm of living things, but my professional life has been devoted mostly to research on the fundamental laws. These laws underlie all of science (in a sense that is discussed in this book) but often seem far removed from most experience, including a great deal of experience in the other sciences. Reflecting on questions of simplicity and complexity, we perceive connections that help to link together all the phenomena of nature, from the simplest to the most complex.

When my wife read me Arthur Sze's poem in which he mentions the quark and the jaguar, I was immediately struck by how well the two images fitted my subject. The quarks are basic building blocks of all matter. Every object that we see is composed, more or less, of quarks and electrons. Even the jaguar, that ancient symbol of power and ferocity, is a bundle of quarks and electrons, but what a bundle! It exhibits an enormous amount of complexity, the result of billions of years of biological evolution. What exactly does complexity mean, though, in this context and how did it arise? Such questions are typical of the ones this book tries to answer.

The remainder of the first part is devoted to the relationships among various concepts of simplicity and complexity, as well as to complex adaptive systems—those that learn or evolve in the way that living systems do. A child learning a language, bacteria developing resistance to antibiotics, and the human scientific enterprise are all discussed as examples of complex adaptive systems. The role of theory in science is discussed, as well as the issue of which sciences are more fundamental than others, along with the related question of what is meant by reductionism.

The second part deals with the fundamental laws of physics, those governing the cosmos and the elementary particles out of which all matter in the universe is composed. Here the quark comes into its own, as do superstrings, which for the first time in history offer the serious possibility of a unified theory of all the particles and forces of nature. The theory of the elementary particles is so abstract that many people find it difficult to follow even when explained, as it is here, without mathematics. Some readers may find it advisable to skim through portions of the second part, especially Chapters 11 (on the modern interpretation of quantum mechanics) and 13 (on the standard model of the elementary particles, including quarks). Skimming those chapters, or even the whole part, does not seriously interfere with following the remaining parts. It is ironic that a portion of the book intended to explain why fundamental physical theory is simple should nevertheless be difficult for many readers. Mea culpa! The second part concludes with a chapter on the arrow or arrows of time, culminating in a commentary on why more and more complex structures keep appearing, whether in complex adaptive systems like biological evolution or in nonadaptive systems like galaxies.

The third part takes up selection pressures operating in complex adaptive systems, especially in biological evolution, human creative thinking, critical and superstitious thinking, and some aspects (including economic ones) of the behavior of human societies. The approximate but convenient notions of fitness and fitness landscapes are introduced. In Chapter 20, I describe briefly the use of computers as complex adaptive systems, for instance to evolve strategies for playing games or to provide simplified simulations of natural complex adaptive systems.

The final part is rather different from the others, in that it is concerned mainly with policy matters rather than science and with advocacy as much as scholarship. Chapter 21 follows up the discussion in the earlier parts of the book about how the diversity of life on Earth represents information distilled over nearly four billion years of biological evolution, and how human cultural diversity has a similar relation to tens of thousands of years of cultural evolution of *Homo sapiens sapiens.* In Chapter 21, I argue that it is worth a great effort to preserve both biological and cultural diversity, and I take up some of the problems, paradoxes, and challenges involved. But it is not really possible to consider those issues in isolation. Today the network of relationships linking the human race to itself and to the rest of the biosphere is so complex that all aspects affect all others to an extraordinary degree. Someone should be studying the whole system, however crudely that has to be done, because no gluing together of partial studies of a complex nonlinear system can give a good idea of the behavior of the whole. Chapter 22 describes some efforts just getting under way to carry out such a crude study of world problems, including all the relevant aspects, not only environmental, demographic, and economic, but also social, political, military, diplomatic, and ideological. The object of the study is not just to speculate about the future, but to try to identify among the multiple possible future paths for the human race and the rest of the biosphere any reasonably probable ones that could lead to greater sustainability. Here the word sustainability is used in a broad sense, including not only the avoidance of environmental catastrophe, but of catastrophic war, widespread long-lasting tyranny, and other major evils as well.

In this volume the reader will find a great many references to the Santa Fe Institute (SFI), which I helped to found and where I now work, having taken early retirement from the California Institute of Technology, where I have become a professor emeritus after being a professor there for more than thirty-eight years. A good deal of the research done today on simplicity, complexity, and complex adaptive systems is carried out by members of the Institute or, more accurately, of the Institute family.

The word family is appropriate because SFI is a rather loose organization. The president, Edward Knapp, is assisted by two vice presidents and an office staff of about a dozen remarkably dedicated workers. There are only three professors, of whom I am one, all with five-year appointments. Everyone else is a visitor, staying for periods ranging from a day to a year. The visitors come from all over the world, and a number of them pay frequent visits. The Institute holds numerous workshops, lasting a few days or sometimes a week or two. In addition, several research networks have been organized on a variety of interdisciplinary topics. The far-flung members of each network communicate with one another by telephone, electronic mail, fax, and the occasional letter, and they meet from time to time in Santa Fe or sometimes elsewhere. They are experts in dozens of specialties, and they are all interested in collaborating across disciplinary boundaries. Each one has a home institution, where research can be carried out in a satisfactory manner, but each one also prizes the Santa Fe affiliation, which permits making connections that are somehow not so easy to make at home. Those home institutions may be great industrial research laboratories, universities, or national laboratories (especially the nearby one at Los Alamos, which has supplied so many brilliant and hard-working members of the Institute).

Those who study complex adaptive systems are beginning to find some general principles that underlie all such systems, and seeking out those principles requires intensive discussions and collaborations among specialists in a great many fields. Of course the careful and inspired study of each specialty remains as vital as ever. But integration of those specialties is urgently needed as well. Important contributions are made by the handful of scholars and scientists who are transforming themselves from specialists into students of simplicity and complexity or of complex adaptive systems in general.

Success in making that transition is often associated with a certain style of thought. The philosopher F. W. J. von Schelling introduced the distinction (made famous by Nietzsche) between "Apollonians," who favor logic, the analytical approach, and a dispassionate weighing of evidence, and "Dionysians," who lean more toward intuition, synthesis, and passion. These traits are sometimes described as correlating very roughly with emphasis on the use of the left and right brain respectively. But some of us seem to belong to another category: the "Odysseans," who combine the two predilections in their quest for connections among ideas. Such people often feel lonely in conventional institutions, but they find at SFI a particularly congenial environment.

The specialties represented at the Institute include mathematics, computer science, physics, chemistry, population biology, ecology, evolutionary biology, developmental biology, immunology, archaeology, linguistics, political science, economics, and history. SFI holds seminars and issues research reports on topics that include the spread of the AIDS epidemic, the waves of large-scale abandonment of prehistoric pueblos in the southwestern United States, the foraging strategies of ant colonies, whether money can be made by using the nonrandom aspects of price fluctuations in financial markets, what happens to ecological communities when an important species is removed, how to program computers to imitate biological evolution, and how quantum mechanics leads to the familiar world we see around us.

SFI is even cooperating with other organizations in the attempt, described in Chapter 22, to model ways in which human society on our planet might evolve toward more sustainable patterns of interaction with itself and with the rest of the biosphere. Here especially we need to overcome the idea, so prevalent in both academic and bureaucratic circles, that the only work worth taking seriously is highly detailed research in a specialty. We need to celebrate the equally vital contribution of those who dare to take what I call "a crude look at the whole."

Although SFI is one of very few research centers in the world devoted exclusively to the study of simplicity and complexity across a wide variety of fields, it is by no means the only place—or even the principal place—where important research is being carried out on the various topics involved. Many of the individual projects of the Institute have parallels elsewhere in the world, and in many cases the relevant research was begun earlier in other institutions, often even before SFI was founded in 1984. In some cases, those institutions are the home bases of key members of the SFI family.

I should like to apologize for what must seem like advertising for SFI, especially since the nature of the relationship between the Institute and other research and teaching organizations has been somewhat distorted in certain books published by science writers during the last few years. What amounts to a glorification of Santa Fe at the expense of other places has angered many of our colleagues at those places, especially in Europe. I am sorry if my book gives a similarly misleading impression. The reason for my emphasis on Santa Fe is merely that I am familiar with some of the work carried on here, or by scholars and scientists who visit here, and much less familiar with research, even prior research, carried out elsewhere.

In any case, I shall mention at this point (in no particular order) a few of the leading institutions where significant research on subjects related to simplicity, complexity, and complex adaptive systems is going on and, in most instances, has been going on for many years. Of course, in doing so I risk exacerbating the wrath of the scientists and scholars at those places that I fail to include in this partial list:

The Ecole Normale Supérieure in Paris; the Max Planck Institute for Biophysical Chemistry in Göttingen, of which Manfred Eigen is the director; the Institute for Theoretical Chemistry in Vienna, where Peter Schuster has been the director (he is now engaged in starting a new institute in Jena); the University of Michigan, where Arthur Burks, Robert Axelrod, Michael Cohen, and John Holland form the "BACH group," an interdisciplinary junta that has been conversing about problems of complex systems for a long time—all of them are connected to some extent with SFI, especially John Holland, who is cochair, along with me, of the Science Board; the University of Stuttgart, where

Hermann Haken and his associates have long studied complex systems in the physical sciences under the rubric of "synergetics"; the Free University of Brussels, where some interesting work has been carried out for many years; the University of Utrecht; the Department of Pure and Applied Sciences at the University of Tokyo; ATR near Kyoto, where Thomas Ray has moved from the University of Delaware; the centers for nonlinear studies at several campuses of the University of California, including those at Santa Cruz, Berkeley, and Davis; the University of Arizona; the Center for Complex Systems Research at the Beckman Institute of the University of Illinois in Urbana; the program in Computation and Neural Systems at the Beckman Institute of the California Institute of Technology; Chalmers University in Göteborg; NORDITA in Copenhagen; the International Institute for Applied Systems Analyses in Vienna; and the Institute for Scientific Interchange in Turin.

Several friends and colleagues whose work I greatly respect have been gracious enough to look over the entire manuscript at various stages of completion. I am very grateful for their help, which has been immensely valuable, even though, because of the pressure of time, I have been able to use only a fraction of their excellent suggestions. They include Charles Bennett, John Casti, George Johnson, Rick Lipkin, Seth Lloyd, Cormac McCarthy, Harold Morowitz, and Carl Sagan. In addition, a number of distinguished experts in various fields have been generous with their time in checking on particular passages in the manuscript, including Brian Arthur, James Brown, James Crutchfield, Marcus Feldman, John Fitzpatrick, Walter Gilbert, James Hartle, Joseph Kirschvink, Christopher Langton, Benoit Mandelbrot, Charles A. Munn III, Thomas Ray, J. William Schopf, John Schwarz, and Roger Shepard. Of course, the errors that no doubt remain are my sole responsibility and not that of any of these kind and learned people.

Anyone who knows me is aware of my intolerance of mistakes, as manifested for example in my ceaseless editing of French, Italian, and Spanish words on American restaurant menus. When I come across an inaccuracy in a book written by somebody else, I become discouraged, wondering whether I can really learn something from an author who has already been proved wrong on at least one point. When the errors concern me or my work, I become furious. The reader of this volume can therefore readily imagine the agonies of embarrassment I am already enduring just through imagining dozens of serious mistakes being found by my friends and colleagues after publication and pointed out, whether gleefully or sorrowfully, to the perfectionist author. In addition, I keep thinking of the legendary figure described to me by Robert Fox (who writes about the human population problem)—a Norwegian lighthouse keeper who has nothing to do on long nights throughout the winter but read our books, searching for mistakes.

I should like to express my special thanks to my skilled and devoted assistant, Diane Lams, for all the help she has given in the process of completing and editing the book, for managing my affairs so competently that I could devote sufficient time and energy to the project, and especially for putting up with the bad temper that I frequently exhibit in the face of deadlines.

The publishers, W. H. Freeman and Company, have been very understanding of my difficulty in dealing with schedules, and they supplied me with a wonderful editor, Jerry Lyons (now at Springer-Verlag), with whom it has been a delight to work. I should like to thank him not only for his efforts but also for his humor and affability and for the many good times Marcia and I have had with him and his wonderful wife, Lucky. My gratitude is extended also to Sara Yoo, who labored tirelessly in distributing countless copies and revisions to anxious editors around the world. Liesl Gibson deserves my thanks for her gracious and very efficient assistance with last-minute demands of manuscript preparation.

It is a pleasure to acknowledge the hospitality extended by the four institutions with which I have been associated during the writing of this book: Caltech, SFI, the Aspen Center for Physics, and the Los Alamos National Laboratory. I should also like to thank Alfred P. Sloan Foundation and the U.S. government agencies that have supported my research during recent years: the Department of Energy and the Air Force Office of Scientific Research. (It may surprise a few readers to know that both of these agencies finance research, such as mine, that is neither classified nor connected with weapons. The help given to pure science by such organizations is a tribute to their farsightedness.) Support of my work through donations to SFI by Jeffrey Epstein and by Gideon and Ruth Gartner is also gratefully acknowledged.

At Los Alamos, I have been especially well treated by the director of the laboratory, Sig Hecker, by the director of the theoretical division,

Richard Slansky, and by the secretary of the division, Stevie Wilds. At the Santa Fe Institute, every single member of the administration and the staff has been most helpful. At Caltech, the president, the provost, and the outgoing and incoming chairs of the division of physics, mathematics, and astronomy have all been very kind, as has John Schwarz, as well as that wonderful lady who has been secretary to the elementary particle theory group for more than twenty years—Helen Tuck. At the Aspen Center for Physics, since its foundation more than thirty years ago, everything has always revolved around Sally Mencimer, and I should like to thank her too for her many kindnesses.

Writing has never come easily to me, probably because my father criticized so vigorously anything I wrote as a child. That I was able to complete this project at all is a tribute to my beloved wife, Marcia, who somehow inspired and goaded me into keeping up the work. Her contribution was indispensable in several other ways as well. As a poet and an English professor, she was able to cure me of some of my worst habits as a writer, although very many infelicities of style unfortunately remain and should of course not be blamed on her. She persuaded me to work on a computer, to which I have become addicted; it now seems odd that I could ever have thought of doing without one. In addition, as someone with little training in science or mathematics who nevertheless has a profound interest in both, she has been an ideal practice target for the book.

As a teacher and lecturer, I have often been advised to pick some particular person in the audience and direct the talk to that individual, even trying to establish repeated eye contact with him or her. In a certain sense, that is what I have done with this volume. It is intended for Marcia, who has tirelessly pointed out places where the explanations are insufficient or the discussions too abstract. I have changed parts of it over and over until she understood and approved. As in so many other respects, more time would have helped. There are, alas, still a number of passages where she would have preferred more clarity.

As I write the finishing touches that deadlines permit, I realize that I have never worked so hard on anything in my life. Research on theoretical physics is entirely different. Of course, a theorist does a great deal of thinking and worrying at odd times, inside or outside of conscious awareness. But a few hours' thought or calculation every day or every few days, plus a good deal of arguing with colleagues and students, have usually sufficed in the way of explicit work—time put in at the desk or the blackboard. Writing, on the contrary, means spending a huge number of hours at the keyboard nearly every day. For a fundamentally lazy person like me, it has come as quite a shock.

The most exciting part of writing this book is being constantly reminded that the project itself is a complex adaptive system. At every stage of composition I have a mental model (or schema) for the book, a concise summary of what it is going to be. That summary needs to be fleshed out with a huge number of details in order to yield a chapter or a part. Then, after my editor, my friends and colleagues, and Marcia and I have had a chance to look over a chapter, the resulting comments and criticisms on the text affect not only the text of that chapter but the mental model itself, often allowing some variant model to take over. When that new one is equipped with details in order to produce more text, the same process is repeated. In that way the concept of the entire work keeps evolving.

The result of that evolutionary development is the book you are about to read. I hope it succeeds in conveying some of the thrill that all of us experience who think about the chain of relationships linking the quark to the jaguar and to human beings as well.

PART I

THE SIMPLE AND THE COMPLEX

CHAPTER

1

PROLOGUE: AN ENCOUNTER IN THE JUNGLE

I have never really seen a jaguar in the wild. In the course of many long walks through the forests of tropical America and many boat trips on Central and South American rivers, I never experienced that heart-stopping moment when the powerful spotted cat comes into full view. Several friends have told me, though, that meeting a jaguar can change one's way of looking at the world.

The closest I came was in the lowland rain forest of Eastern Ecuador, near the Napo River, a tributary of the Amazon, in 1985. Here a number of Indians from the highlands have settled, clearing small patches of forest for agriculture. They are speakers of Quechua (called Quichua in Ecuador), which was the official language of the Inca Empire, and they have given their own names to some of the features of the Amazonian landscape.

Flying over that landscape, which stretches for thousands of miles from north to south and east to west, one sees the rivers below as sinuous ribbons snaking through the forest. Often the bends in the rivers become oxbows, like the ones on the Mississippi, and the oxbows pinch off to become lakes, each one connected to the main river by a trickling stream. Local Spanish speakers call such a lake a *cocha*, using a Quechua word that applies also to highland lakes and to the sea. The aerial observer can see these cochas in all the different stages through which they pass, starting with the ordinary river bend, then the oxbow, then the newly pinched-off cocha, and then the "ecological succession" as the lake slowly dries up and is gradually reclaimed for the forest by a sequence of plant species. Eventually, it appears from the air only as a light green spot against the darker green of the surrounding forest, and finally, after a century or more, that spot becomes indistinguishable from the rest of the rain forest.

When I came near to viewing a jaguar, I was on a forest trail near Paña Cocha, which means "piranha lake." There my companions and I had caught and cooked three different species of piranha, all of which were delicious. Those fish are not quite so dangerous as one might think. True, they sometimes attack people, and it is advisable for a bather who has been bitten to leave the water so the blood will not attract more of them. Still, piranhas are more likely to be the eaten than the eaters in their contacts with humanity.

About an hour's walk from the lake, we flushed a group of peccaries, and immediately afterward we sensed the presence of another large mammal just ahead. We smelled a strong pungent odor, very different from that of the wild pigs, and heard the crackling sounds of a heavy creature moving through the underbrush. I caught sight of the tip of its tail, and then it was gone. The master of animals, the emblem of the power of priests and rulers, had passed by.

It was not a jaguar, but another and smaller jungle cat that was to make a difference in my life, by making me aware that so many of my seemingly disparate interests had come together. Four years after the incident in Ecuador, I was getting acquainted with the flora and fauna of another forested area in tropical America, far from where the Incas had ruled. This was the region where a different Pre-Columbian civilization had flourished, that of the Maya. I was in northwestern Belize, near the Guatemalan and Mexican borders, in a place called Chan Chich, which means "little bird" in the local Mayan language.

Many speakers of Mayan languages live in the area today, and traces of the Classic Maya civilization can be found everywhere in that part of Mesoamerica, most dramatically in the physical remains of the abandoned cities. One of the grandest of those cities is Tikal, with its gigantic pyramids and temples, in the northeastern corner of Guatemala, less than a hundred miles from Chan Chich.

Speculation abounds on the collapse of the Classic Maya way of life more than a thousand years ago, but the causes remain a mystery and a source of controversy to this day. Did the common people tire of laboring at the behest of the rulers and the nobility? Did they lose faith in the elaborate religious system that maintained the power of the élite and held the fabric of society together? Did the wars among the numerous city states lead to general exhaustion? Did the remarkable agricultural practices that supported such large populations in the rain forest finally fail? Archaeologists continue to search for clues to answering these and other questions. At the same time they have to consider the relation between the definitive Classic collapse in the rain forest and what happened in the more arid region of the Yucatán, where in some places the Classic civilization was succeeded by the Postclassic, under Toltec influence.

Visiting a gigantic excavated site like Tikal is, of course, unforgettable, but for those willing to go off the beaten track, the jungle affords other pleasures as well, such as coming suddenly upon an unexcavated ruin that isn't indicated on ordinary maps.

A ruin reveals itself first as a hillock in the forest, covered, like the flat ground, with trees and shrubs. Approaching, one catches the odd glimpse of old masonry covered with moss and ferns and creepers. Peering through the foliage, one can get a general idea of the size and shape of the site, especially by climbing to a high spot. There, in an instant, one's imagination clears away the jungle and excavates and restores a small Classic Maya site in all its splendor.

The forest around Chan Chich is as rich in wildlife as in ruins. Here one can see adult tapirs wrinkling their long noses as they watch over their tiny variegated offspring. One can admire the brilliant plumage of ocellated turkeys, especially the males with their bright blue heads covered with small red knobs. At night a flashlight illuminating the top of a tree may pick out wide-eyed kinkajous clinging to the branches by their prehensile tails.

As a lifelong birdwatcher I take particular delight in recording the voices of skulking forest birds, playing back their songs or calls to attract them, and then seeing them (and recording their sounds better) when they come near. In search of birds, I found myself, one day in late December, walking alone on a trail near Chan Chich.

The first part of my walk had been uneventful. I had had no luck in recording or sighting any of the bird species I was seeking. Now, after more than an hour, I was no longer concentrating on bird calls or paying close attention to movements in the foliage. My thoughts had drifted to a subject that has occupied a good part of my professional life, quantum mechanics.

For most of my career as a theoretical physicist, my research has dealt with elementary particles, the basic building blocks of all matter in the universe. Unlike the experimental particle physicist, I don't have to stay close to a giant accelerator or a laboratory deep underground in order to conduct my work. I don't make direct use of elaborate detectors and I don't need a large professional staff. At most I require only a pencil, some paper, and a wastebasket. Often, even those are not essential. Give me a good night's sleep, freedom from distractions, and time unburdened by worries and obligations, and I can work. Whether I'm standing in the shower, hovering between wakefulness and sleep on a late-night flight, or walking along a wilderness trail, my work can accompany me wherever I go.

Quantum mechanics is not itself a theory; rather, it is the framework into which all contemporary physical theory must fit. That framework, as is well known, requires the abandonment of the determinism that characterized the earlier "classical" physics, since quantum mechanics permits, even in principle, only the calculation of probabilities. Physicists know how to use it for predicting the probabilities of the various possible outcomes of an experiment. Since its discovery in 1924, the predictions of quantum mechanics have always worked perfectly, up to the accuracy of the particular experiment and the particular theory concerned. But, in spite of this uniform success, we do not yet fully understand, at the deepest level, what quantum mechanics really means, especially for the universe as a whole. For more than thirty years some of us have been taking steps to construct what I call the "modern interpretation" of quantum mechanics, which permits it to apply to the universe and also to deal with particular events involving individual objects instead of just repeatable experiments on easily reproducible bits of matter. Walking through the forest near Chan Chich, I was pondering how quantum mechanics can be used in principle to treat individuality, to describe which pieces of fruit will be eaten by parrots or the various ways in which a growing tree can shatter a piece of masonry from a ruined temple.

My train of thought was broken when a dark figure appeared on the trail about a hundred yards in front of me. I stopped short and carefully raised my binoculars to get a closer look. It was a medium-sized wild cat, a jaguarundi. It stood across the trail, its head turned toward me, allowing me to see its characteristic flattened skull, long body, and short forelegs (features that have prompted some to call it an otter cat). The creature's length—about three feet—and uniform grayish-black coat indicated that it was an adult and of the dark rather than the reddish type. For all I knew, the jaguarundi had been standing there for some time, its brownish eyes trained on me as, bewitched by the mysteries of quantum mechanics, I drew nearer. Though obviously alert, the animal seemed utterly at ease. We stared at each other, both motionless in our tracks, for what seemed like several minutes. It even remained still as I moved closer, to within thirty yards or so. Then, having seen all it cared to see of this particular human being, it faced forward, put its head down, and slowly dissolved into the trees.

Such sightings are not very common. The jaguarundi is a shy animal. Because of the destruction of its native habitat in Mexico and in Central and South America, its numbers have decreased over the years, and it is now included in the *Red List of Threatened Animals.* Adding to the threat is the creature's apparent inability to reproduce in captivity. My experience with this particular jaguarundi resonated with my thinking about the whole notion of individuality. My memory was jogged back to an earlier encounter with individuality in nature.

One day in 1956, when I was a very young professor at Caltech, my first wife Margaret and I were returning to Pasadena from the University of California at Berkeley, where I had given some lectures on theoretical physics. We were in our Hillman Minx convertible with the top down. In those days, academics dressed a little more formally than we do today—I was wearing a gray flannel suit and Margaret had on a skirt and sweater, with stockings and high heels. We were traveling on Route 99 (not yet converted into a freeway) near Tejon Pass, between Bakersfield and Los Angeles. When passing through that area, I often scanned the sky, hoping for a glimpse of a California condor. This time, I caught sight of a large form flying low overhead and then rapidly disappearing behind the hill on our right. I was not sure what it was, but I was determined to find out. I pulled the car over to the side of the road, grabbed my field glasses, jumped out, and ran up the hill. I was deep in thick red mud most of the way. Part way up, I looked back and there was Margaret, not far behind, her elegant clothes covered with mud just like mine. We reached the ridge together and looked down on a field where a dead calf was lying. Feasting on it were eleven California condors. They constituted a large fraction of the total population of the species at that time. We watched them for a long while as they fed, flew off for short distances, landed, walked around, and fed again. I was prepared for their gigantic size (their wing spread is around ten feet), their brightly colored bare heads, and their black and white plumage. What surprised me was how easily we could tell one of them from another by their lost feathers. One had a couple of flight feathers missing from the left wing. Another had a wedge-shaped gap in its tail. None was completely intact. The effect was dramatic. Each bird was an easily identifiable individual, and the observable individuality was a direct result of historical accidents. I wondered whether these losses of plumage were permanent consequences of the condors' long and eventful lives, or simply the temporary effect of a yearly molt. (I learned later that condors change all their feathers every year.) We are all accustomed to thinking of human beings (and pets) as individuals. But the sight of those distinguishable condors strengthened powerfully my appreciation of how much of the world we perceive as composed of individual objects, animate or inanimate, with their own particular histories.

Standing, a third of a century later, in the Central American forest, staring at the place where the jaguarundi had disappeared, remembering the ragged condors, and recalling that I had just been thinking about history and individuality in quantum mechanics, it struck me that my two worlds, that of fundamental physics and that of condors, jaguarundis, and Maya ruins, had finally come together.

For decades I have lived with these two intellectual passions, one for my professional work in which I try to understand the universal laws governing the ultimate constituents of all matter and the other for my avocation of amateur student of the evolution of terrestrial life and of human culture. I always felt that in some way the two were deeply connected, but for a long time I didn't really know how (except for the common theme of the beauty of nature).

There would seem to be an enormous gap between fundamental physics and these other pursuits. In elementary particle theory we deal with objects like the electron and the photon, each of which behaves exactly the same wherever it occurs in the universe. In fact, all electrons are rigorously interchangeable with one another, and so are all photons. Elementary particles have no individuality.

The laws of elementary particle physics are thought to be exact, universal, and immutable (apart from possible cosmological considerations), even though we scientists may approach them by successive approximations. By contrast, subjects like archaeology, linguistics, and natural history are concerned with individual empires, languages, and species, and at a more detailed level with individual artifacts, words, and organisms, including human beings like ourselves. In these subjects the laws are approximate; moreover, they deal with history and with the kind of evolution undergone by biological species or human languages or cultures.

But the fundamental quantum-mechanical laws of physics really do give rise to individuality. The physical evolution of the universe, operating in accordance with those laws, has produced, scattered through the cosmos, particular objects such as our planet Earth. Then, through processes like biological evolution on Earth, the same laws have yielded particular objects such as the jaguarundi and the condors, capable of adaptation and learning, and eventually other particular objects such as human beings, capable of language and civilization and of discovering those fundamental physical laws.

For some years, my work had been concerned as much with this chain of relationships as with the laws themselves. I had been thinking, for example, about what distinguishes complex adaptive systems, which undergo processes like learning and biological evolution, from evolving systems (such as galaxies and stars) that are nonadaptive. Complex adaptive systems include a human child learning his or her native language, a strain of bacteria becoming resistant to an antibiotic, the scientific community testing out new theories, an artist getting a creative idea, a society developing new customs or adopting a new set of superstitions, a computer programmed to evolve new strategies for winning at chess, and the human race evolving ways of living in greater harmony with itself and with the other organisms that share the planet Earth.

Research on complex adaptive systems and their common properties, as well as work on the modern interpretation of quantum mechanics and on the meaning of simplicity and complexity, had been making steady progress. To further the interdisciplinary study of such issues, I had helped to found the Santa Fe Institute in Santa Fe, New Mexico.

Meeting the jaguarundi in Belize somehow strengthened my awareness of the progress my colleagues and I had made in understanding better the relation between the simple and the complex, between the universal and the individual, between the basic laws of nature and the particular, earthly subjects I had always loved.

The more I learned about the character of that relation, the more I wanted to communicate it to others. For the first time in my life, I felt the urge to write a book.

CHAPTER

2

EARLY LIGHT

The title of this book comes from a line in a poem by my friend Arthur Sze, a splendid Chinese-American poet who lives in Santa Fe and whom I met through his wife, the talented Hopi weaver Ramona Sakiestewa. The line reads, "The world of the quark has everything to do with a jaguar circling in the night."

Quarks are elementary particles, building blocks of the atomic nucleus. I am one of the two theorists who predicted their existence and it was I who gave them their name. In the title, the quark symbolizes the simple basic physical laws that govern the universe and all the matter in it. It may seem to many people that the word "simple" doesn't apply to contemporary physics; indeed, explaining how it does apply is one of the aims of this book.

The jaguar stands for the complexity of the world around us, especially as manifested in complex adaptive systems. Together, Arthur's images of the quark and the jaguar seem to me to convey perfectly the two aspects of nature that I call the simple and the complex: on the one hand, the underlying physical laws of matter and the universe and, on the other, the rich fabric of the world that we perceive directly and of which we are a part. Moreover, just as the quark is a symbol of the physical laws that, once discovered, come into full view before the mind's analytical eye, so the jaguar is, for me at least, a possible metaphor for the elusive complex adaptive system, which continues to avoid a clear analytical gaze, though its pungent scent can be smelled in the bush.

But how did I come to be fascinated by subjects like natural history when I was a child? And how and why did I then become a physicist?

A Curious Child

I owe most of my early education to my brother Ben, who is nine years older. It was he who taught me to read when I was three (from a Sunshine cracker box) and who introduced me to bird and mammal watching, botanizing, and insect collecting. We lived in New York City, principally in Manhattan, but nature study was possible even there. I thought of New York as a hemlock forest that had been logged too heavily, and we spent much of our time in the small portion that still remained, just north of the Bronx Zoo. Fragments of other habitats survived in places such as Van Cortlandt Park, with its freshwater marsh; the New Dorp area of Staten Island, with its beach and salt marsh; and even, right in our neighborhood, Central Park, which possessed some interesting bird life, especially during the spring and fall migrations.

I became aware of the diversity of nature and the striking way in which that diversity is organized. If you walk along the edge of a swamp and see a northern yellow-throat or hear one singing "Wichita, Wichita, Wichita," you know that you are likely to find another one further on. If you dig up a fossil, you are likely to run across another fossil of the same type near by. After becoming a physicist, I puzzled for some time over how the fundamental laws of physics lay the groundwork for such phenomena. It turns out that the answer is connected with the way history is treated in quantum mechanics, and that the ultimate explanation lies in the condition of the early universe. But apart from such deep physical questions, the less abstruse issue of speciation as a phenomenon in biology is well worth pondering.

It is not at all a trivial matter that there are such things as species; and they are not just artifacts of the biologist's mind, as has sometimes been claimed. Ernst Mayr, the great ornithologist and biogeographer, likes to recount how, as a young researcher in New Guinea, he counted a hundred and twenty-seven species of birds nesting in the valley where he was working. The members of the local tribe counted a hundred and twenty-six; the only difference between their list and his was that they lumped together two very similar species of gerygone that Ernst, with his scientific training, was able to distinguish from each other. Even more important than the agreement among different sorts of people is the fact that the birds themselves can tell whether or not they belong to the same species. Animals of different species are not usually in the habit of mating with one another, and in the rare cases where they do, the hybrids they produce are likely to be sterile. In fact, one of the most successful definitions of what constitutes a species is the statement that there is no effective exchange of genes by ordinary means between members of different species.

On my early nature walks, I was impressed by the fact that the butterflies, birds, and mammals we saw really did fall neatly into species. If you go for a stroll, you may see song sparrows, swamp sparrows, field sparrows, and white-throated sparrows, but you are not likely to see any sparrows that fall in between those categories. Disputes about whether or not two populations belong to the same species arise mostly when the populations are found in different places or else when they belong to different time periods, with at least one being in the fossil record. Ben and I loved to talk about how species are all related by evolution, like the leaves on an evolutionary "tree," with groupings such as genera, families, and orders representing attempts to specify further the structure of that tree. How distant the relationship is between two different species depends on how far down the tree it is necessary to go in order to find a common ancestor.

Ben and I did not spend all our time together in the out-of-doors. We also visited art museums, including those rich in archaeological material (such as the Metropolitan Museum of Art) and those containing objects from medieval Europe (like the Cloisters). We read history books. We learned to read some inscriptions in Egyptian hieroglyphics. We studied Latin and French and Spanish grammars, just for fun, and we noticed how French and Spanish words (and many "loan" words in English) were derived from Latin. We read about the Indo-European language family and learned how many Latin, Greek, and native English words had a common origin, with fairly regular transformation laws. For instance, English "salt" corresponds to Latin *"sal"* and ancient Greek *"hals,"* while English "six" corresponds to Latin *"sex"* and ancient Greek *"hex"*; the initial s in English and Latin goes with rough breathing in ancient Greek, which we indicate by "h." Here was another evolutionary tree, this one for languages.

Historical processes, evolutionary trees, organized diversity, and individual variation were all around. In addition to exploring diversity, I also learned that in many cases it was in danger. Ben and I were early conservationists. We saw how the few areas around New York that were more or less natural were becoming still fewer as swamps, for example, were drained and paved over.

Back in the 1930s, we were already acutely aware of the finiteness of the Earth, of the encroachment of human activities on plant and animal communities, and of the importance of population limitation, as well as soil conservation, forest protection, and the like. Naturally, I didn't yet connect the need for all these reforms in attitude and practice with the evolution of human society on a planetary scale toward greater sustainability, although that is the way I look at it today. But even then I did have some thoughts about the future of the human race, especially in connection with the textbooks and the scientific romances of H. G. Wells.

I loved to read his novels as well. I also devoured books of short stories, and Ben and I read English poetry from anthologies. We went to concerts occasionally, and even the Metropolitan Opera, but we were very poor and had to be content most of the time with activities that were free of charge. We made crude attempts to play the piano and to sing operatic arias as well as songs from Gilbert and Sullivan. We listened to the radio and tried to hear distant stations, both long- and shortwave, and when we succeeded we wrote to them for "verification cards." I remember vividly the ones we received from Australia, with pictures of the kookaburra bird.

Ben and I wanted to understand the world and enjoy it, not to slice it up in some arbitrary way. We didn't differentiate sharply among such categories as the natural sciences, the social and behavioral sciences, the humanities, and the arts. In fact, I have never believed in the primacy of such distinctions. What has always impressed me is the unity of human culture, with science being an important part. Even the distinction between nature and human culture is not a sharp one; we human beings need to remember that we are a part of nature.

Specialization, although a necessary feature of our civilization, needs to be supplemented by integration of thinking across disciplines.

One obstacle to integration that keeps obtruding itself is the line separating those who are comfortable with the use of mathematics from those who are not. I was fortunate enough to be exposed to quantitative thinking from an early age.

Although Ben was interested in physical science and mathematics, it was principally my father who encouraged me to study those subjects. An immigrant from Austria-Hungary in the early years of the century, he had interrupted his studies at the University of Vienna to come to the United States and help out his parents. They had immigrated a few years earlier and were living in New York but having trouble making ends meet. My father's first job was at an orphanage in Philadelphia, where he picked up English and baseball from the orphans. Though already a young adult when he started to learn English, his grammar and pronunciation became perfect. When I knew him, the only way one might have guessed he was foreign-born was by noticing that he never made any mistakes.

After exploring a number of career opportunities, he finally settled, in the 1920s, on the Arthur Gell-Mann School of Languages, where he attempted to teach other immigrants to speak flawless English. He also taught German and hired teachers for French, Spanish, Italian, and Portuguese. The school was a modest success, but things changed in 1929, the year I was born. Not only did the stock market crash, but also a law went into effect severely limiting immigration to the United States. From then on, potential pupils for my father's school were reduced in number by the new quotas and impoverished by the Depression. By the time I was three years old, the school had failed and my father had to find a low-paying routine job in a bank to keep us alive. I was brought up to think of the days before I was born as the good old days.

Father was intrigued by mathematics, physics, and astronomy, and he would spend hours each day locked in his study poring over books on special and general relativity and on the expanding universe. He encouraged my interest in mathematics, which I came to love, studying it on my own and admiring its coherence and rigor.

During my senior year in high school, filling out the application form for admission to Yale, I had to name my probable major subject. When I discussed the choice of field with my father, he scorned my plans to study archaeology or linguistics, saying I would starve. Instead, he suggested engineering. I replied that I would rather starve, and also that whatever I designed would probably fall apart. (Later on, I was told, after an aptitude test, "Anything but engineering!") My father then proposed that we compromise on physics.

I explained that I had taken a course in physics in high school, that it was the dullest course in the curriculum, and that it was the only subject in which I had done badly. We had had to memorize such things as the seven kinds of simple machine: the lever, the screw, the inclined plane, and so on. Also, we had studied mechanics, heat, sound, light, electricity, and magnetism, but with no hint of any connections among those topics.

My father now switched from economic arguments to promoting physics on the basis of its intellectual and aesthetic appeal. He promised me that advanced physics would be more exciting and satisfying than my high school course, and that I would love special and general relativity and quantum mechanics. I decided to humor the old man, knowing that I could always change my major subject if and when I arrived in New Haven. Once I got there, however, I was too lazy to do so right away. Then, before very long, I was hooked. I began to enjoy theoretical physics. My father had been right about relativity and quantum mechanics. I began to understand, as I studied them, that the beauty of nature is manifested just as much in the elegance of these fundamental principles as in the cry of a loon or in trails of bioluminescence made by porpoises at night.

Complex Adaptive Systems

A wonderful example of the simple underlying principles of nature is the law of gravity, specifically Einstein's general-relativistic theory of gravitation (even though most people regard that theory as anything but simple). The phenomenon of gravitation gave rise, in the course of the physical evolution of the universe, to the clumping of matter into galaxies and then into stars and planets, including our Earth. From the time of their formation, such bodies were already manifesting complexity, diversity, and individuality. But those properties took on new meanings with the emergence of complex adaptive systems. Here on Earth that development was associated with the origin of terrestrial life and with the process of biological evolution, which has produced such a striking diversity of species. Our own species, in at least some respects the most complex that has so far evolved on this planet, has succeeded in discovering a great deal of the underlying simplicity, including the theory of gravitation itself.

Research on the sciences of simplicity and complexity, as carried out at the Santa Fe Institute and elsewhere around the world, naturally includes teasing out the meaning of the simple and the complex, but also the similarities and differences among complex adaptive systems, functioning in such diverse processes as the origin of life on Earth, biological evolution, the behavior of organisms in ecological systems, the operation of the mammalian immune system, learning and thinking in animals (including human beings), the evolution of human societies, the behavior of investors in financial markets, and the use of computer software and/or hardware designed to evolve strategies or to make predictions based on past observations.

The common feature of all these processes is that in each one a complex adaptive system acquires information about its environment and its own interaction with that environment, identifying regularities in that information, condensing those regularities into a kind of "schema" or model, and acting in the real world on the basis of that schema. In each case, there are various competing schemata, and the results of the action in the real world feed back to influence the competition among those schemata.

Each of us humans functions in many different ways as a complex adaptive system. (In fact the term "schema" has long been used in psychology to mean a conceptual framework such as a human being always uses to grasp data, to give them meaning.)

Imagine you are in a strange city during the evening rush hour, trying to flag down a taxi on a busy avenue leading outward from the center. Taxis rush by you, but they don't stop. Most of them already have passengers, and you notice that those cabs have their roof lights turned off. Aha! You must look for taxis with roof lights on. Then you discover some in that condition and indeed they lack passengers, but they don't stop either. You need a modified schema. Soon you realize that the roof lights have an inner and an outer part, with the the latter marked "Out of Service." What you need is a taxi that has only the inner part of the roof light illuminated. Your new idea receives confirmation when two taxis discharge their passengers a block ahead and then their drivers turn on just the inner roof lights. Unfortunately, those taxis are immediately grabbed by other pedestrians. A few more cabs finish their trips nearby, but they too are snapped up. You are impelled to cast your net wider in your search for a successful schema. Finally, you observe, on the other side of the avenue, going in the opposite direction, many taxis cruising with just their inner roof lights on. You cross the avenue, hail one, and climb in.

As a further illustration, imagine that you are a subject in a psychology experiment in which you are shown a long sequence of pictures of familiar objects. The pictures represent various things, and each one may be shown many times. You are asked from time to time to predict what the next few images will be, and you keep trying to construct mental schemata for the sequence, inventing theories about how the sequence is structured, based on what you have seen. Any such schema, supplemented by the memory of the last few pictures shown, permits you to make a prediction about the next ones. Typically, those predictions will be wrong the first few times, but if the sequence has an easily grasped structure, the discrepancy between prediction and observation will cause you to reject unsuccessful schemata in favor of ones that make good predictions. Soon you may be foreseeing accurately what will be shown next.

Now imagine a similar experiment run by a sadistic psychologist who exhibits a sequence with no real structure at all. You are likely to go on making up schemata, but this time they keep failing to make good predictions, except occasionally by chance. In this case the results in the real world afford no guidance in choosing a schema, other than the one that says, "This sequence seems to have no rhyme or reason." But human subjects find it hard to accept such a conclusion.

Whether putting together a business plan for a new venture, refining a recipe, or learning a language, you are behaving as a complex adaptive system. If you are training a dog, you are watching a complex adaptive system in operation and you are functioning as one as well (if it is mainly the latter that is happening, then the dog may be training you, as is often the case). When you are investing in a financial market, you and all the other investors are individual complex adaptive systems participating in a collective entity that is evolving through the efforts of all the component parts to improve their positions or at least survive economically. Such collective entities can be complex adaptive systems themselves. So can organized collective entities such as business firms or tribes. Humanity as a whole is not yet very well organized, but it already functions to a considerable extent as a complex adaptive system.

It is not only learning in the usual sense that provides examples of the operation of complex adaptive systems. Biological evolution provides many others. While human beings acquire knowledge mainly by individual or collective use of their brains, the other animals have acquired a much larger fraction of the information they need to survive by direct genetic inheritance; that information, evolved over millions of years, underlies what is sometimes rather vaguely called "instinct." Monarch butterflies hatched in parts of the United States "know" how to migrate, in enormous numbers, to the pine-clad slopes of a particular mountain in Mexico to spend the winter. Isaac Asimov, the late biochemist, popularizer of science, and science fiction author, told me that he once had an argument with a theoretical physicist who denied that a dog could know Newton's laws of motion. Isaac asked indignantly, "You say that, even after watching a dog catch a Frisbee with its mouth?" Obviously, the physicist and he were using "knowing" to mean different things: in the case of the physicist, mostly the result of learning in the cultural context of the human scientific enterprise; in Isaac's case, information stored in the genes, supplemented by some learning from the experience of the individual.

That capacity to learn from experience, whether in paramecia or dogs or people, is itself a product of biological evolution. Furthermore, evolution has given rise not only to learning but to other new types of complex adaptive systems as well, such as the immune system in mammals. The immune system undergoes a process very similar to biological evolution itself, but on a time scale of hours or days instead of millions of years, enabling the body to identify in a timely fashion an invading organism or an alien protein and produce an immune response.

Complex adaptive systems, it turns out, have a general tendency to generate other such systems. For example, biological evolution may lead to an "instinctive" solution to a problem faced by an organism, but it may also produce enough intelligence for an organism to solve a similar problem by learning. The diagram on the following page illustrates how various complex adaptive systems on Earth are related to one another. Certain chemical reactions involving reproduction and some transmit-

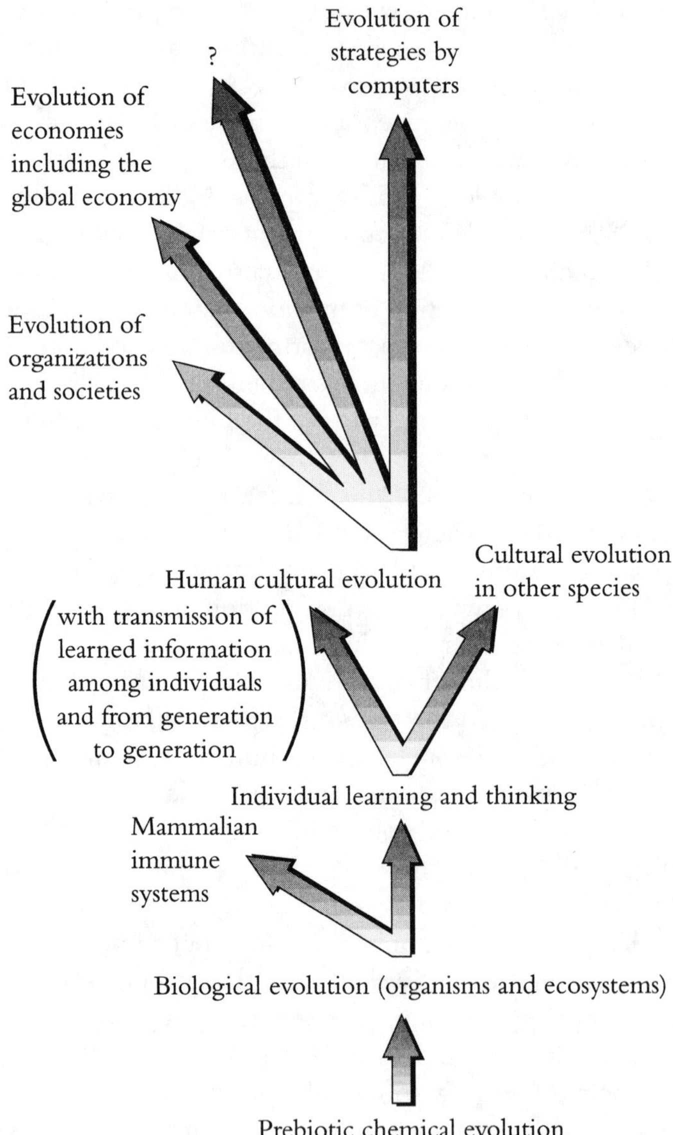

Some complex adaptive systems on Earth.

ted variation led, around four billion years ago, to the appearance of the first life forms and then to diverse organisms constituting ecological communities. Life then gave rise to further complex adaptive systems such as the immune system and the learning process. In human beings, the development of the capacity for symbolic language expanded learning into an elaborate cultural activity, and new complex adaptive systems arose within human culture: societies, organizations, economies, and the scientific enterprise, to name but a few. Now that rapid and powerful computers have emerged from human culture, we can make it possible for them to act as complex adaptive systems as well.

In the future, human beings may create new kinds of complex adaptive systems. One example, which has appeared in science fiction, was first brought to my attention as a result of a conversation that took place in the early 1950s. The late, great Hungarian-American physicist Leo Szilard invited a colleague and me to attend an international meeting on arms control. My colleague, "Murph" Goldberger (later president of Caltech and then director of the Institute for Advanced Study in Princeton), replied that he could attend only the second half of the meeting. Leo turned to me, and I said that I could attend only the first half. Murph and I then asked if we could share an invitation. Leo thought for a moment and then told us, "No, it is no good; your neurons are not interconnected."

Some day, for better or for worse, such interconnections might be possible. A human being could be wired directly to an advanced computer, (not through spoken language or an interface like a console), and by means of that computer to one or more other human beings. Thoughts and feelings would be completely shared, with none of the selectivity or deception that language permits. (Voltaire is supposed to have remarked that "Men . . . employ speech only to conceal their thoughts.") My friend Shirley Hufstedler says that being wired up together is not something she would recommend to a couple about to be married. I am not sure that I would recommend such a procedure at all (although if everything went well it might alleviate some of our most intractable human problems). But it would certainly create a new form of complex adaptive system, a true composite of many human beings.

Gradually, students of complex adaptive systems are becoming familiar with their general properties as well as with the distinctions among them. Although they differ widely in their physical attributes, they resemble one another in the way they handle information. That common feature is perhaps the best starting point for exploring how they operate.

CHAPTER

3

INFORMATION AND CRUDE COMPLEXITY

In studying any complex adaptive system, we follow what happens to the information. We examine how it reaches the system in the form of a stream of data. (For example, if a subject in a psychological experiment is shown a sequence of images, they constitute the data stream.) We notice how the complex adaptive system perceives regularities in the data stream, sorting them out from features treated as incidental or arbitrary and condensing them into a schema, which is subject to variation. (In the example, the subject makes up and continually modifies conjectured rules that are supposed to describe the regularities governing the sequence of images.) We observe how each of the resulting schemata is then combined with additional information, of the same kind as the incidental information that was put aside in abstracting regularities from the data stream, to generate a result with applications to the real world: a description of an observed system, a prediction of events, or a prescription for behavior of the complex adaptive system itself. (In the psychological experiment, the subject may combine a tentative schema based on the past succession of images with the information provided by the next few and thus make a prediction about what images will be shown later. In this case, as often happens, the additional special information comes from a later portion of the same data stream as the one from which the schema was abstracted.) Finally, we see how the description, prediction, or behavior has consequences in the real world that feed back to exert "selection pressures" on the competition among the various schemata; some are demoted in a hierarchy or eliminated altogether, while one or more manage to survive and may be promoted. (In the example, a schema that makes predictions contradicted by the succeeding images is presumably discarded by the subject, while one that gives correct predictions is retained and assigned a high value. Here the testing of the schemata is carried out using later portions of the very same data stream that gave rise to the schemata in the first place and that supplied the additional special information used in making predictions.) The operation of a complex adaptive system can be represented by a diagram such as the one on page 25, in which the flow of information is emphasized.

Like everything else, complex adaptive systems are subject to the laws of nature, which themselves rest on the fundamental physical laws of matter and the universe. Moreover, of all the physical situations permitted by those laws, only specific conditions permit complex adaptive systems to exist.

When studying the universe and the structure of matter we can follow the same practice that we follow when studying complex adaptive systems: concentrate on the information. What are the regularities and where do accidents and the arbitrary enter in?

Indeterminacy from Quantum Mechanics and from Chaos

According to classical physics of a century ago, exact knowledge of the laws of motion and of the configuration of the universe at any one moment in time would, in principle, permit the complete history of the universe to be predicted. We now know that to be quite false. The universe is quantum-mechanical, which implies that even if its initial state and the fundamental laws of matter are known, only a set of probabilities for different possible histories of the universe can be calculated. Moreover, the degree of this quantum-mechanical "indeterminacy" goes far beyond what is usually discussed. Many people are familiar with the Heisenberg uncertainty principle, which prohibits, for

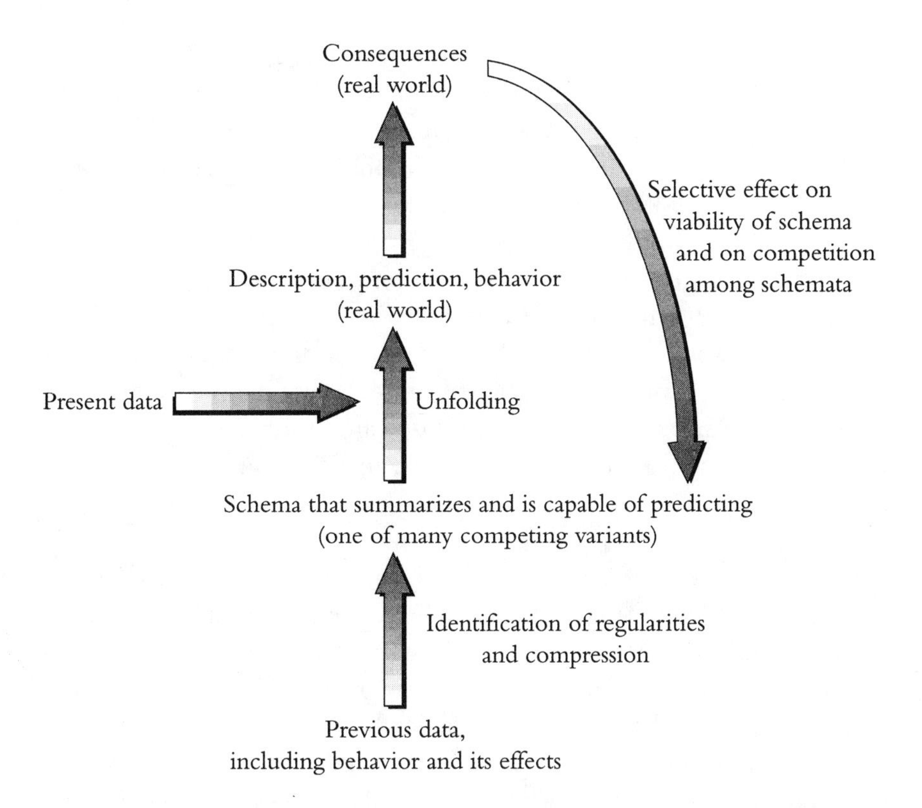

How a complex adaptive system works.

example, exact specification of both the position and momentum of a particle at the same time. While that principle has received wide publicity (sometimes in quite misleading terms) over many decades, the additional indeterminacy that quantum mechanics requires is rarely mentioned. We shall take it up in some detail further on.

Even when the classical approximation is justified and quantum-mechanical indeterminacy is correspondingly ignored, there remains the widespread phenomenon of chaos, in which the outcome of a nonlinear dynamical process is so sensitive to initial conditions that a minuscule change in the situation at the beginning of the process results in a large difference at the end.

Some contemporary statements about classical determinism and classical chaos were anticipated in this passage written in 1903 by the French mathematician Henri Poincaré in his book *Science and Method* (as cited by Ivars Peterson in *Newton's Clock*):

> If we knew exactly the laws of nature and the situation of the universe at the initial moment, we could predict exactly the situation of that same universe at a succeeding moment. But even if it were the case that the natural laws had no longer any secret for us, we could still only know the initial situation *approximately*. If that enabled us to predict the succeeding situation with the same approximation, that is all we require, and we should say that the phenomenon had been predicted, that it is governed by laws. But it is not always so; it may happen that small differences in the initial conditions produce very great ones in the final phenomena. A small error in the former will produce an enormous error in the latter. Prediction becomes impossible, and we have the fortuitous phenomenon.

One of the papers that called attention to chaos in the 1960s was written by a meteorologist, Edward N. Lorenz. In fact, meteorology often supplies examples of chaos that strike close to home. Although satellite photography and calculations using powerful computers have made weather prediction fairly reliable for many purposes, meteorological reports still cannot always tell us correctly what many of us most want to know—whether or not it will rain *here tomorrow*. Exactly where a given storm system will pass and when it will drop rain may be arbitrarily sensitive to the details of winds and of the position and physical state of clouds a few days or even a few hours earlier. The slightest imprecision in the meteorologist's knowledge of those data can render a prediction of tomorrow's weather useless for planning a company picnic.

Since nothing can ever be measured with perfect accuracy, chaos gives rise to effective indeterminacy at the classical level over and above the indeterminacy in principle of quantum mechanics. The interaction between these two kinds of unpredictability is a fascinating and still rather poorly studied aspect of contemporary physics. The challenge of understanding the relationship between quantum unpredictability and the classical chaotic kind even impressed the editorial staff of the *Los Angeles Times* so much that in 1987 they printed an editorial on the subject! The writer pointed to the apparent paradox that some theorists studying the quantum mechanics of systems that exhibit chaos in the classical limit have been unable to find the chaotic kind of indeterminacy superposed on the quantum-mechanical kind.

Fortunately, the issue is now being clarified through the work of various theoretical physicists, including a student of mine named Todd Brun. His results seem to confirm that for many purposes it is useful to regard chaos as a mechanism that can amplify to macroscopic levels the indeterminacy inherent in quantum mechanics.

Recently, there has been a great deal of careless writing about chaos. From the name of a technical phenomenon in nonlinear dynamics, the word has been turned into a kind of catchall expression for any sort of real or apparent complexity or uncertainty. When I give a public lecture on complex adaptive systems, for example, and mention the phenomenon perhaps once, or maybe not at all, I am bound to be congratulated at the end on having given an interesting talk about chaos.

It seems to be characteristic of the impact of scientific discovery on the literary world and on popular culture that certain items of vocabulary, interpreted vaguely or incorrectly, are often the principal survivors of the journey from the technical publication to the popular magazine or paperback. The important qualifications and distinctions, and sometimes the actual ideas themselves, tend to get lost along the way. Witness the popular uses of "ecology" and "quantum jump," to say nothing of the New Age expression "energy field." Of course, one can argue that words like "chaos" and "energy" antedate their use as technical terms, but it is the technical meanings that are being distorted in the process of vulgarization, not the original senses of the words.

In the face of what appear to be increasingly efficient literary mechanisms for turning certain useful concepts into meaningless clichés, an effort should be made to prevent the same fate from befalling the various notions of complexity. We will have to tease some of them apart and try to see where each one applies.

Meanwhile, what about the word "complex" in the term "complex adaptive system" as it is used here? Really, "complex" need have no precise significance in this phrase, which is merely a conventional one. Still, the presence of the word implies the belief that any such system possesses at least a certain minimum level of complexity, suitably defined.

Simplicity refers to the absence (or near-absence) of complexity. Whereas the former word is derived from an expression meaning "once folded," the latter comes from an expression meaning "braided together." (Note that both "plic-" for fold and "plex-" for braid come from the same Indo-European root *"plek."*)

Different Kinds of Complexity

What is really meant by the opposing terms simplicity and complexity? In what sense is Einsteinian gravitation simple while a goldfish is complex? These are not easy questions—it is not simple to define "simple." Probably no single concept of complexity can adequately capture our intuitive notions of what the word ought to mean. Several different kinds of complexity may have to be defined, some of which may not yet have been conceived.

What are some cases where the question of defining complexity comes up? One is the computer scientist's concern about the time a computer requires to solve a certain kind of problem. In order to keep that time from depending on the cleverness of the programmer, scientists focus on the shortest possible solution time, which is often called the "computational complexity" of the problem.

Even the minimum time still depends on the choice of computer, however. Such "context dependence" keeps cropping up in attempts to define different kinds of complexity. But the computer scientist is particularly interested in what happens to a set of problems that are similar except for size; furthermore, his or her main concern is with what happens to computational complexity as the size of the problem gets larger and larger without limit. How does the minimum solution time depend on size as the size tends to infinity? The answer to such a question can be independent of the details of the computer.

Computational complexity has proved to be quite a useful notion, but it does not correspond very closely to what we usually mean when we employ the word complex, as in a highly complex story plot or organizational structure. In those contexts, we may be more concerned with how long a message would be required to describe certain properties of the system in question than with how long it would take to solve some problem on a computer.

For example, a debate has been going on for decades in the science of ecology over whether "complex" ecosystems like tropical forests are more or less resilient than comparatively "simple" ones such as the forest of oaks and conifers found high in the San Gabriel Mountains behind Pasadena. Here resilience refers to the likelihood of surviving (or even deriving benefit from) major disturbances from climate change, fire, or some other environmental alteration, whether human in origin or not. Currently those ecological scientists seem to be winning the argument who claim that, up to a point, the more complex ecosystem is more resilient. But what do they mean by simple and complex? The answer is certainly related in some way to the length of a description of each forest.

To arrive at a very elementary notion of complexity for forests, ecologists might count the number of species of trees in each type (less than a dozen in a typical high-mountain forest of the temperate zone versus hundreds in a lowland tropical forest). They might also count the number of species of birds or mammals; again, the comparison would greatly favor the tropical lowlands. With insects the results would be even more striking—imagine how many insect species there must be in an equatorial rain forest. (The number has always been thought to be very large, and recently the estimates have been greatly increased. Starting with the work of Terry Erwin of the Smithsonian Institution, experiments have been carried out in which all the insects in a single rain forest tree are killed and collected. The number of species was found to be on the order of ten times that previously estimated, and many of the species were new to science.)

One need not count only species. Ecologists would also include interactions among organisms in the forest, such as those between predator and prey, parasite and host, pollinator and pollinated, and so on.

Coarse Graining

But down to what level of detail would they count? Would they look at microorganisms, even viruses? Would they look at very subtle interactions as well as obvious ones? Evidently, they have to stop somewhere.

Hence, when defining complexity it is always necessary to specify a level of detail up to which the system is described, with finer details being ignored. Physicists call that "coarse graining." The image that inspires the name is probably that of the grainy photograph. When a detail in a photograph is so small that it needs to be greatly enlarged to be identified, the enlargement may show the individual photographic grains. Instead of a clear picture of the detail, there will then be only a few dots to convey a rough image of it. In the Antonioni film *Blow-Up*, the title refers to just such an enlargement. The graininess of a photograph supplies a bound on the amount of information it can yield. If the film is very grainy, the best the whole picture can do is to give a rough impression of what was photographed; it is exhibiting very coarse graining. If a spy satellite takes a picture of a previously unknown weapons "complex," the measure of complexity that can be assigned to it will depend on the graininess of the photograph.

Having established the importance of coarse graining, we are still faced with the question of how to define the complexity of the system being investigated. For instance, what characterizes a simple or a complex pattern of communication among a certain number of people (say, N people)? Such a question might arise for a psychologist or a student of organizations who is trying to compare how well or how rapidly some problem is solved by the N people under different conditions of communication. At one extreme (call it case A), each person works alone and there is no communication at all. At the other extreme (call it case F), each person is free to communicate with every other person. Case A is obviously simple. Is case F much more complex or is it about as simple as case A?

As to level of detail (coarse graining), suppose all the people are treated alike, having no individual characteristics, and are represented in a diagram simply as dots, in such a way that the positions of the dots do not matter and all the dots are interchangeable. Communication between any two people is either allowed or not allowed, with no gradations in between, and each two-way communication link is represented as a line (with no directionality) connecting two dots. The resulting diagram is what mathematicians call an "undirected graph."

The Length of the Description

With the level of detail specified in this way, it is possible to explore what is meant by the complexity of a pattern of connection. First take the case of a small number of dots, say eight ($N = 8$). Here it is easy to draw some of the patterns, including some trivial ones. The diagrams on page 31 show some of the possible communication patterns among eight individuals. In A, none of the dots is connected to any other. In B, some of the dots, but not all, have connections. In C, all the dots are

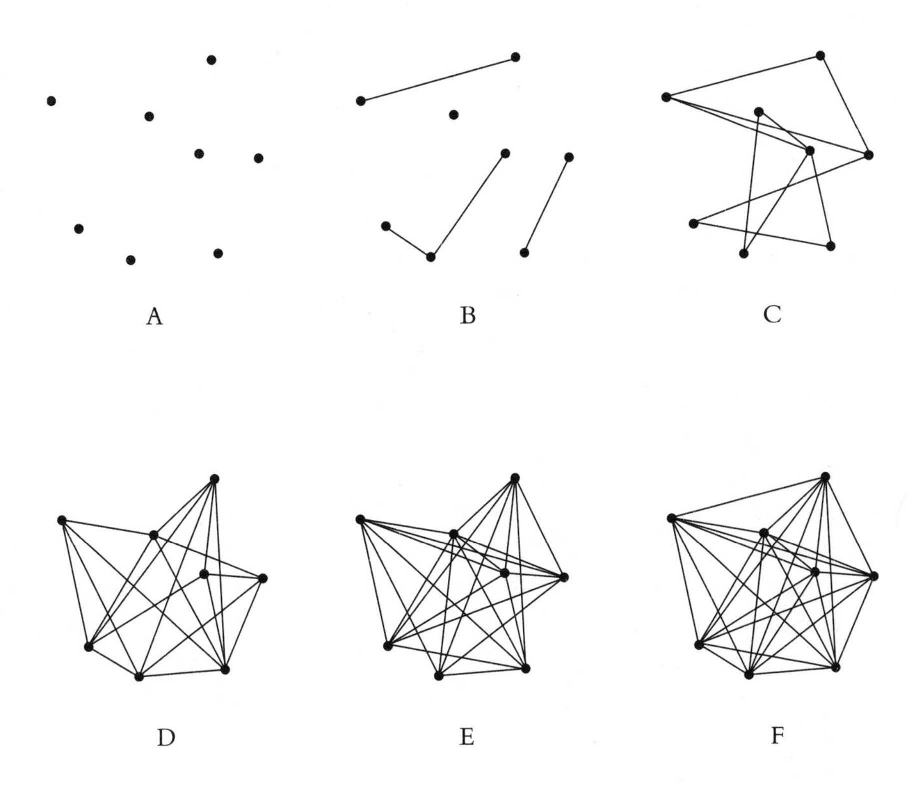

Some patterns of connection of eight dots.

connected, but not in all possible ways. In D, the connections that are present in C are absent and those that are absent in C are present; D is what we might call the complement of C and vice versa. Similarly, E and B are complements of each other. So are F and A: Pattern A has no connections, while F has all possible connections. Which patterns are to be assigned higher complexity than which others?

Everyone will agree that A, with no connections, is simple, and that B, with some connections, is more complex or less simple than A. But what about the others? One particularly interesting case is that of F. An initial reaction to F might be that it is the most complex of all, since it has the most connections. But is that sensible? Isn't the property of having all dots connected just as simple as that of having no dots connected? Maybe F belongs at the bottom of the complexity scale, along with A.

Such reasoning brings us back to the suggestion that at least one way of defining the complexity of a system is to make use of the length of its description. Pattern F would then really be about as simple as its complement, pattern A, since the phrase "all dots connected" is of about the same length as the phrase "no dots connected." Moreover, the complexity of E is not all that different from that of its complement, B, since adding the word "complement" doesn't make the description significantly longer. The same is true of D and C. In general, complementary patterns will have about the same complexity.

Patterns B and E are evidently more complex than A and F, and so are C and D. The comparison of B and E with C and D is trickier. It may seem that C and D are more complex, using the simple criterion of description length, but whether they really are depends to some extent on the vocabulary available for the description.

Before going further with the notion that complexity is related to the length of a description, it is worth noting that the same diagrams that we have applied to patterns of communication among people can also be applied to another situation, one that is of great significance in science, technology, and business today. These days, computer scientists are making rapid progress in the construction and utilization of "parallel processing" computers, which are much more effective in solving certain kinds of problems than conventional machines. Instead of a single giant computer that works away steadily at a problem until it is finished, parallel processing uses an array of many smaller computing units, all operating simultaneously, with some pattern of communication links joining certain pairs of units. Here again one can ask, What does it mean for one pattern of communication hookups to be more complex than another? In fact, it was a physicist designing a parallel processing computer who asked me that very question years ago and renewed my interest in the problem of defining complexity.

We previously considered the possibility of counting the number of species, interactions, and so forth in order to define simple and complex ecological communities. If all the kinds of trees occurring in the community were listed, for example, the length of that part of the description would be roughly proportional to the number of tree species. Hence, in that case too the length of the description was effectively being used as a measure.

Context Dependence

If complexity is defined in terms of the length of a description, then it is not an intrinsic property of the thing described. Obviously, the length of a description may depend on who or what is doing the describing. (That reminds me of James Thurber's story *The Glass in the Field*, in which a goldfinch gives a concise account to other birds of a collision with a pane of glass: "I was flying across a meadow when all of a sudden the air crystallized on me.") Any definition of complexity is necessarily context-dependent, even subjective. Of course, the level of detail at which the system is being described is already somewhat subjective—it too depends on the observer or the observing equipment. In actuality, then, we are discussing one or more definitions of complexity that depend on a description of one system by another system, presumably a complex adaptive system, which could be a human observer. Suppose, for present purposes, that the describing system is, in fact, a human observer.

To refine the notion of the length of a description, we should avoid describing something by pointing to it; clearly it is just as easy to point to a complex system as to a simple one. Therefore, we are concerned with a description that is being communicated to someone at a distance. Also, it is easy to give a name like "Sam" or "Judy" to something extremely complicated, making its description trivially short. The descriptive language must be previously agreed upon and not include special terms made up for the purpose.

Of course, many kinds of arbitrariness and subjectivity still remain. The length of the description will vary with the language used, and also with the knowledge and understanding of the world that the correspondents share. If, for example, a rhinoceros is to be described, the message can be shortened if both parties already know what a mammal is. If the orbit of an asteroid is to be described, it makes a great deal of difference whether both know Newton's law of gravitation and his second law of motion—it may also matter to the length of the description whether the orbits of Mars, Jupiter, and the Earth are already known to both parties.

Conciseness and Crude Complexity

But what if a description is unnecessarily long because words are being wasted? I recall the story of the grade school teacher who assigned a 300-word composition to her class as homework. A pupil who had spent the weekend playing managed to scribble on Monday morning the following essay: "Yesterday the neighbors had a fire in their kitchen and I stuck my head out of the window and yelled, 'Fire! Fire! Fire!...'" The child repeated the word "fire" until the essay was 300 words long. However, if it hadn't been for that requirement, the child could have written instead, "... yelled 'Fire!' 280 times" and conveyed the same meaning. For our definition of complexity we are therefore concerned with the length of the shortest possible message describing a system.

These points can be integrated into a definition of what may be called "crude complexity": the length of the shortest message that will describe a system, at a given level of coarse graining, to someone at a distance, employing language, knowledge, and understanding that both parties share (and know they share) beforehand.

Certain familiar ways of describing a system do not yield anything like the shortest message. For example, if we describe the parts of a system separately (say the pieces of a car or the cells in a human body) and also tell how the whole is composed of the parts, we have ignored many opportunities to compress the message. Those opportunities would make use of similarities among the parts. For example, most of the cells in a human body share the same genes and have many other features in common, while the cells in a given tissue are even more similar. The shortest description would take that into account.

Algorithmic Information Content

Certain experts in information theory utilize a quantity that is much like crude complexity, although their definition is more technical and naturally involves computers. They envisage a description to a given level of coarse graining that is expressed in a given language and then encoded by some standard coding procedure into a string of 1s and 0s. Each choice of a 1 or 0 is known as a "bit." (Originally, that was a contraction of "binary digit." It is binary because there are only two possible choices, whereas with the usual digits of the decimal system there are ten: 0, 1, 2, 3, 4, 5, 6, 7, 8, 9.) It is that string of bits, or "message string," with which they are concerned.

The quantity they define is called "algorithmic complexity," "algorithmic information content," or "algorithmic randomness." Nowadays the word "algorithm" refers to a rule for calculating something and, by extension, to a program for computing something. Algorithmic information content refers, as we shall see, to the length of a computer program.

Originally, algorithm meant something different. The word sounds as if it is derived from the Greek, like "arithmetic," but, in fact, that is only the result of a disguise. The "th" was introduced by analogy with the "th" in "arithmetic," although it doesn't really belong there. A spelling that better reflects the etymology would be "algorism." It comes from the name of the man whose book introduced the idea of zero into Western culture. He was the ninth century Arab mathematician Muhammad ibn Musa al-Khwarizmi. The surname indicates that his family came from the province of Khorezm, south of the Aral Sea, now part of the newly independent republic of Uzbekistan. He wrote a mathematical treatise the title of which contains the Arabic phrase "*al jabr,*" meaning "the transposition," from which we get the word "algebra." Originally, the word "algorism" referred to the decimal system of notation, which is thought to have passed from India to Europe largely through the translation into Latin of al-Khwarizmi's "Algebra."

Algorithmic information content (AIC) was introduced in the 1960s by three authors working independently. One was the great Russian mathematician Andrei N. Kolmogorov. Another was an American, Gregory Chaitin, who was only fifteen years old at the time. The third was another American, Ray Solomonoff. Each assumes an idealized all-purpose computer, treated as essentially infinite in storage capacity (or else finite but able to acquire additional capacity as needed). The computer is equipped with specified hardware and software. They then consider a particular message string and ask what programs will cause the computer to print out that string and then stop computing. The length of the shortest such program is the AIC of the string.

We have seen that subjectivity or arbitrariness is inherent in the definition of crude complexity, arising from such sources as coarse graining and the language used to describe the system. In AIC, additional sources of arbitrariness have been introduced, namely the particular coding procedure that turns the description of the system into a bit string, as well as the particular hardware and software associated with the computer.

None of this arbitrariness bothers the mathematical information theorists very much, because they are usually concerned with limits in which finite arbitrariness becomes comparatively insignificant. They like to consider sequences of similar bit strings of increasing length, studying how the AIC behaves as the length approaches infinity. (That is reminiscent of how computer scientists like to treat the computational complexity of a sequence of similar problems as the problem size approaches infinity.)

Let us return to the idealized parallel processing computer made up of units, represented by dots, connected by communication links represented by lines. Here Kolmogorov, Chaitin, and Solomonoff would not be very interested in the AIC of various possible patterns of connections among a mere eight points. Instead, they would ask questions about the connections among N points as N tends to infinity. Under those conditions, certain differences in the behavior of the AIC (for example, between the simplest pattern of connections and the most complex) dwarf any differences that result from the use of one computer instead of another, one coding procedure instead of another, or even one language instead of another. An information theorist cares whether a certain AIC keeps growing as N approaches infinity and, if so, how fast. He or she is not much concerned with the comparatively negligible differences between one AIC and another that are introduced by various kinds of arbitrariness in the descriptive equipment.

We can learn an interesting lesson from those theorists. Even if we don't confine ourselves to systems that become infinitely large, it is important to understand that discussions of simplicity and complexity tend to become more and more meaningful as the bit strings become longer and longer. At the other extreme, say for a string of one bit, it is evidently meaningless to differentiate between simplicity and complexity.

Information Defined

It is high time to make clear the distinction between algorithmic information content and information, as discussed, for example, by Claude Shannon, the founder of modern information theory. Basically, information is concerned with a selection from alternatives, and it is most simply expressed if those alternatives can be reduced to a sequence of binary choices, each of which is between two equally probable alternatives. For example, if you learn that a coin toss resulted in tails instead of heads, you will have learned one bit of information. If you learn that three successive coin tosses resulted in heads, then tails, then heads again, you will have acquired three bits of information.

The game Twenty Questions provides a beautiful opportunity to express the most varied sorts of information in the form of successive binary choices between equally probable alternatives, or as close to equally probable as the questioner can get. It is played by two people, the first of whom dreams up something that the second player has to guess in twenty questions or less, after having been told whether it is animal, vegetable, or mineral. The questions have to be answered either "yes" or "no"; each is a binary choice. For the second player, it is advantageous to make the questions as close as possible to a choice between equally probable alternatives. Knowing that the thing is mineral, for example, the questioner would be ill-advised to ask right away whether it is the Hope diamond. Instead, he or she might ask, "Is it natural [as opposed to being manufactured or modified by humans]?" Here, the probabilities of affirmative and negative responses are about equal. If the answer is "No," the next question might be, "Is it a specific object as opposed to a class of objects?" When the probabilities of a yes and no answer are equal, each question will elicit one bit of information (the most that such a question can extract). Twenty bits of information correspond to a choice from among 1,048,576 equally probable alternatives, the product of multiplying together 20 factors of 2. That product is the number of different bit strings of length 20.)

Note that bit strings are employed differently depending on whether AIC or information is being discussed. In the case of algorithmic information content, a single bit string (preferably a long one) is considered, and its internal regularities are measured by the length (in bits) of the shortest program that will cause a standard computer to print out the bit string and then stop. By contrast, in the case of information, one may consider a choice among all the different bit strings of a given length. If they are all equally probable, their length is the number of bits of information.

One can also deal with a set of bit strings, for example equally probable ones, each with a particular value of AIC. In that case it is often useful to define an amount of information, determined by the number of strings, as well as a value of AIC averaged over the set.

Compression and Random Strings

Algorithmic information content has a very curious property. To discuss it, we have to look first at the relative "compressibility" of different message strings. For a bit string of a given length (say a very long one), we can ask when the algorithmic complexity is low and when it is high. If a long string has the form 110110110110110110 . . . 110110, it can be produced by a very short program that says to print 110 a particular number of times. Such a bit string has a very low AIC, even though it is long. This means it is highly compressible.

By contrast, it can be shown mathematically that most bit strings of a given length are incompressible. In other words, the shortest program that will produce one of those strings (and then have the computer stop) is one that says PRINT followed by the string itself. Such a string has a maximum AIC for its length. There is no rule, no algorithm, no theorem that will simplify the description of that bit string and allow it to be described by a shorter message. It is called a "random" string precisely because it contains no regularity that will permit it to be compressed. The fact that algorithmic information content is maximal for random strings explains the alternative name algorithmic randomness.

The Uncomputability of AIC

The curious property is that AIC is not computable. Even though most bit strings are random, there is no way of knowing exactly which ones are. In fact, we cannot, in general, be sure that the AIC of a given string isn't lower than we think it is. This is because there may always be a theorem we will never find, an algorithm we will never discover, that would permit the string to be further compressed. More precisely, there is no procedure for finding all the theorems that would permit further compression. That was proved some years ago by Greg Chaitin, in work that is reminiscent of part of a famous result of Kurt Gödel.

Gödel was a mathematical logician who stunned the world of mathematics in the early 1930s with his discoveries about the limitations of systems of axioms in mathematics. Until his time, mathematicians had hoped it might be possible to formulate a system of axioms for mathematics that could be proved consistent and used in principle to derive the truth or falsity of all mathematical propositions. Gödel showed that neither of those goals is attainable.

Negative results like that often represent monumental advances in mathematics or in science. We might compare Albert Einstein's discovery that there can be no absolute definition of time or space, but only a combined space-time. In fact, Gödel and Einstein were good friends. At the Institute for Advanced Study in Princeton, New Jersey, in the early 1950s, I used to see them walk to work together, and they made a strange-looking couple, like Mutt and Jeff. Gödel was so tiny that he made Einstein look quite tall. Did they discuss deep mathematical or physical questions? (Gödel worked from time to time on problems related to general relativity.) Or was their conversation mainly about the weather and their health problems?

The part of Gödel's conclusion that is relevant to our discussion is the one about undecidability: given any system of axioms for mathematics, there will always be propositions that are undecidable on the basis of those axioms. In other words, there are propositions that cannot, in principle, be shown to be either true or false.

The most celebrated kind of undecidable proposition is a statement that is independent of the axioms. One can use such a proposition to enlarge the set of axioms by introducing either the proposition or its contrary as a new axiom.

But there are other undecidable propositions that have a different character. Suppose, for example, that an undecidable proposition relating to positive whole numbers is of the form, "Every even number greater than 2 has the following property. . . ." If there were any exception to such a proposition, in principle we could find it, given enough time, by trying out every even number in succession (4, 6, 8, 10, ...) until we hit a number not possessing the property in question. That would immediately disprove the proposition, but it would also contradict its undecidability, since undecidability means precisely that the proposition cannot be proved or disproved. Thus, *there is no exception* to the proposition. In the ordinary sense of the word "true," the proposition is true.

We can make this more concrete by considering a proposition that has never been proved, after centuries of effort, although no exception to it has ever been found. The proposition is Goldbach's conjecture, which states that every even number larger than 2 is the sum of two prime numbers. A prime number is a number greater than 1 that is not divisible by any number except itself and 1. The first few prime numbers, therefore, are 2, 3, 5, 7, 11, 13, 17, 19, 23, 29, 31, and 37. It is easy to see from this list how every even number between 4 and 62 can be expressed in at least one way as the sum of two primes. Computer calculations have verified that every even number up to some unbelievably large value has the same property. However, no such computation can *prove* the conjecture, which could always fail for some still larger even number. Only a rigorous mathematical demonstration can turn the conjecture into a proved theorem.

There is no reason to believe that Goldbach's conjecture is undecidable, but suppose it is. It would then be true, even though unprovable, because there could be no exception to it. The existence of any even number greater than 2 that is not the sum of two primes would disprove the conjecture and therefore contradict its undecidability.

The fact that such true but unprovable theorems are always lurking in the background means, as Chaitin has shown, that there may be one that will permit a long message string to be compressed when we think it is incompressible, or to be further compressed when we think we have found the shortest program that will cause the computer to print it out and then stop. Thus, in general, one cannot be sure of the value of algorithmic information content; one can only place an upper bound on it, a value that it cannot exceed. Since the value may be below that bound, AIC is uncomputable.

The property of uncomputability may be awkward, but a different property is what prevents us from using algorithmic information content to define complexity. Although AIC is useful for introducing useful notions like coarse graining, compressibility of message strings, and the length of a description generated by an observing system, it has one really bad flaw: the alternative name algorithmic randomness gives it away. Algorithmic information content is largest for random strings. It is a measure of randomness, and randomness is not what is usually meant by complexity, either in ordinary discourse or in most scientific usage. Thus AIC is not true or effective complexity.

It turns out that care is required when discussing randomness, however, because the word does not always mean exactly the same thing. I first became aware of that pitfall a long time ago, in my contacts with the RAND Corporation.

CHAPTER

4

RANDOMNESS

When I first came to work at Caltech in the 1950s, I needed a consulting job to pay some bills. Caltech professors are allowed to consult once a week, and I inquired among my colleagues to find out what the possibilities were. One or two suggested the RAND Corporation, located in Santa Monica near the famous pier and Muscle Beach.

The RAND Corporation had started out, shortly after the Second World War, as Air Force Project RAND (said to be an acronym for research and no development). It was to advise the U.S. Air Force on matters such as matching strategy to mission (that is, to the tasks assigned to the service), and devising rational methods of procurement. After a while, its role was broadened to include advice to government on a variety of matters, many of them connected with defense strategy. Project RAND continued to be important, but it provided only part of the financial support for the organization, which became a not-for-profit corporation and branched out into civilian work of many kinds. RAND employs specialists in a great many fields, including political science, economics, physics, mathematics, and operations research.

The physics department, consisting mostly of theoreticians, hired me as a consultant, and I started to earn money doing unclassified research. Three of us from Caltech formed a car pool and spent every Wednesday at RAND.

The Meanings of "Random"

One of the things I remember best from my early visits to RAND was being handed a small pile of recently produced reports, so that I could become familiar with some of the work that was going on. One of the reports in the stack was the "RAND Table of Random Numbers," which was undoubtedly useful though not very exciting to read (I am told, however, that the subtitle, "And 100,000 Normal Deviates," led some librarians to shelve it under abnormal psychology).

What I found interesting about the report was a small piece of paper (a "blow-in") that fluttered out of it and fell to the floor. I picked it up and found it was an errata sheet. The RAND mathematicians were supplying corrections to some of the random numbers! Were they catching random errors in the random numbers? For a long while, I regarded this incident as just one more scene in the human comedy, but as I speculated about it later it focused my attention on an important fact: even to mathematicians and scientists the word "random" means several different things.

As we have been using the word, applied for instance to a single string of a thousand bits, random means that the string is incompressible. In other words, it is so irregular that no way can be found to express it in shorter form. A second meaning, however, is that it has been generated by a random process, that is, by a chance process such as a coin toss, where each head gives 1 and each tail 0. Now those two meanings are not exactly the same. A sequence of a thousand coin tosses *could* produce a string of a thousand heads, represented as a bit string of a thousand 1s, which is as far from being a random bit string as it is possible to get. Of course, a sequence of all heads is not at all probable. In fact, its chance of turning up is only one in a very large number of about three hundred digits. Since most long strings of bits are incompressible (random) or nearly so, a set of a thousand tosses will *often* lead to a random bit string, *but not always.* One way to avoid confusion would be to refer to chance processes as "stochastic" rather than random, reserving the latter term mainly for incompressible strings.

But what does random signify in the RAND table of random numbers? How could the table be equipped with an errata sheet? And of what use is a table of random numbers in the first place?

One of the activities of the RAND physics department in 1956 and 1957 was an unclassified project, with applications to astrophysics, that required a calculation in rather basic physics. I undertook to carry it out, receiving some help from another consultant, an old friend named Keith Brueckner. Part of the calculation involved doing a couple of difficult sums approximately, and one of the most interesting RAND physicists, Jess Marcum, offered to do them by what is known as the Monte Carlo method, utilizing the table of random numbers.

Random Numbers and the Monte Carlo Method

The method was very suitable for Jess because he was a gambler as well as a physicist. In his early years, he had won a good deal of money at blackjack in casinos. He used the "student method," betting lightly on most games, when the odds were slightly against him, and then heavily when the odds were in his favor, for example when all the ten-counting cards (tens and picture cards) were in one part of the deck. That method of play was possible only as long as a single deck was being used. After a while, all the casinos adjusted their procedure (adapting to the "students") and began to use many decks at a time. Jess moved on to other pursuits.

At one point, he took a leave of several months from RAND to play the horses. His method was to handicap the handicappers. He didn't claim to be an expert on the horses themselves, but merely studied the racing forms to see how well the odds given by each handicapper corresponded with the actual results. He then followed the advice of the successful handicappers. However, he added another wrinkle. Just before each race, he checked the tote board to see whether the quoted odds (reflecting the bets received up to that time) corresponded with those of the good handicappers. If they didn't, it meant that the crowd was following other advice, probably that of bad handicappers. Jess rushed into the gap between the quoted odds and the odds given by the best predictors, betting heavily. In this way, he made steady money at the racetrack. But after a while he concluded that his RAND salary paid at least as much with less risk, and so he went back to work. That is how Jess happened to be available to help me.

The Monte Carlo method of doing sums is applied when there is a really huge set of quantities to be added; a rule (an algorithm!) is given for computing the first quantity from the number 1, the second quantity from the number 2, the third quantity from the number 3, and so forth; the rule is such that the quantity varies rather smoothly from one number to the next; and the computation of each quantity from the corresponding number is long and tedious, so that one does not want to do any more of those computations than necessary. (These days, with enormously rapid and powerful computers easily available, many such sums are computed directly, although the computers of thirty-five years ago required tricks like the Monte Carlo method.)

Suppose we have to add 100 million quantities after calculating each of them from the corresponding number, which runs, of course, from 1 to 100 million. To employ the Monte Carlo approximation, we use a table of random numbers to obtain, say, 5,000 numbers between 1 and 100 million, chosen by chance. In each of the 5,000 cases, every number between 1 and 100 million has an equal probability of turning up. We then calculate the quantities corresponding to the 5,000 numbers and add them up, taking them to be a representative sample of the whole 100 million quantities to be added. Finally, we multiply the result by 100 million divided by 5,000 (that is, 20,000). In this way, we have approximated our lengthy calculation by a much shorter one.

Random or Pseudorandom?

The table of random numbers is supposed to be a set of whole numbers between one and some fixed large value, with each number chosen by a chance process in which every number in the range has an equal probability of occurring. In fact such a table is not usually generated that way, but is instead a table of pseudorandom numbers! These are reeled off by a computer using some definite mathematical rule, but one so messy that it is supposed to simulate a chance process (for example, a rule might be used that is chaotic in the technical sense). The resulting list of numbers may then be tested to ascertain whether it meets some of the statistical criteria that a list obtained by a true chance process would be expected to meet in most cases. In the case of the RAND table, were the numbers really pseudorandom? Did a last minute check reveal that one such criterion was not quite satisfied? Was that why an errata sheet had to be "blown in"? It turns out that the answers to these questions are in the negative. After all, a table of random numbers *can* be generated by a truly stochastic process, for example, one that makes use of quantum-mechanical phenomena. In fact, the RAND table was prepared in a stochastic way, using noise from a vacuum tube. Moreover, the errata sheet referred to the 100,000 normal deviates, and not to the table of random numbers itself! The mystery that was so instructive was really no mystery at all. Stochastic methods require a great deal of work, however, and it is more convenient to let a computer reel off a sequence using a deterministic rule and then to ensure that the resulting unwanted regularities in the sequence are comparatively harmless in the situations where the numbers are to be used. Still, experience has shown that using such pseudorandom sequences as if they were random can be dangerous.

I read recently about a set of pseudorandom numbers used in numerous laboratories that turned out to be seriously nonrandom. As a result, certain kinds of calculations performed with those numbers came out badly in error. This incident can serve to remind us that sequences of numbers arising from deterministic chaotic or near-chaotic processes can possess a considerable amount of regularity.

Regularities in Price Fluctuations

Sometimes sequences thought to be stochastic turn out to be partially pseudorandom instead. For example, many neoclassical economists have preached for years that the price fluctuations in financial markets around the values dictated by market fundamentals constitute a "random walk," a stochastic process. At the same time, advice on market investments has been available from "chartists" who pore over squiggles in graphs of prices versus time and claim to derive from those squiggles better-than-chance predictions of how prices will behave in the near future. I once read an article by an economist who expressed his fury at the very idea of someone pretending to utilize such evidence in defiance of economists' insistence that the fluctuations amount to nothing but a chance process.

But it has now been shown convincingly that the chance process idea is wrong. These fluctuations are in part pseudorandom, as in deterministic chaos; in principle, they contain enough regularities for one to make money off them. That does not mean that every financial nostrum peddled by chartists will make you a fortune; much of their advice is probably worthless. But the idea that price fluctuations amount to more than a chance process is not in itself a crazy one, as that angry economist believed. (Doyne Farmer and Norman Packard, two physicists belonging to the Santa Fe Institute family, have actually quit their jobs in scientific research to start an investment firm. They used to work on the theory of deterministic chaos and of nearly chaotic processes. They moved on to study computer-based adaptive systems such as the neural nets and genetic algorithms described in Chapter 20. Now they use systems like those to find regularities in price fluctuations (especially changes in volatility), and they invest accordingly. They started off by practicing for a few months with play money and then went on to invest real funds provided by a large bank. So far, they are doing quite well.)

We have encountered three different technical uses of the word random:

1. A random bit string is one so irregular that there is no rule for compressing its description.
2. A random process is a chance or stochastic process. In generating long bit strings of a given length, it will often produce random, completely incompressible strings; sometimes strings containing a few regularities so that they are somewhat compressible, and very occasionally strings that are exceedingly regular, highly compressible, and not at all random.
3. A table of random numbers is usually generated by a pseudorandom process—a deterministic computational process that does not really use chance at all, but is so messy (chaotic, for example) that it simulates a stochastic process fairly well for many purposes, and satisfies some of the statistical criteria that a stochastic process would usually satisfy. When such pseudorandom processes are used to generate bit strings, the strings resemble to a considerable extent the results of a chance process of generation.

Shakespeare and the Proverbial Monkeys

Now we are equipped to discuss why algorithmic randomness or algorithmic information content does not fully match our intuitive idea of what complexity is. Consider the famous monkeys at the typewriters, who it is assumed would hit the various keys in a stochastic manner, with an equal chance of typing any symbol or a space with each stroke. I doubt if real monkeys would behave that way, but for our purposes it doesn't matter. The question is, what are the chances that the monkeys would, in a certain period of time, type the works of Shakespeare (or else all the books in the British Museum—the part now called the British Library). Obviously, there is a non-zero chance that if a certain number of monkeys were each to type sufficiently many pages, the total text would include a connected passage comprising the works of Shakespeare (say the Folio Edition). However, that chance is inconceivably small. If all the monkeys in the world were typing eight hours a day for ten thousand years, the chance that the resulting text would include a connected part that was the Folio Edition of Shakespeare is utterly negligible.

In a story by Russell Maloney called *Inflexible Logic*, which appeared in *The New Yorker* magazine some years ago, six chimpanzees began systematically typing the books in the British Museum, one after another, with no hesitation and no mistakes. However, those apes came to a bad end: a scientist killed them in order to preserve his conception of the laws of probability. The last chimp, in his death agonies, "was slumped before his typewriter. Painfully, with his left hand, he took from the machine the completed last page of Florio's Montaigne. Groping for a fresh sheet, he inserted it, and typed with one finger, 'UNCLE TOM'S CABIN, by Harriet Beecher Stowe. Chapte . . . ' Then he, too, was dead."

Consider a non–*New-Yorker* monkey of the proverbial kind typing material equal in length to the Folio Edition, and compare a typical product of that monkey with the work of Shakespeare. Which has the greater algorithmic information content? Obviously, the work of the monkey. By means of a chance process (the second of our meanings of random), the monkey is extremely likely to produce a random or near-random sequence of symbols (in the first sense of random). If the work of the monkey is encoded in some standard manner as a bit string, the chances are excellent that the bit string will have maximal or nearly maximal algorithmic randomness for a string of its length. The works of Shakespeare are obviously less random. The rules of English grammar, spelling conventions (despite Shakespeare's sloppy use of an already sloppy system), the need to make sense, and many other factors all contribute to nonrandomness in Shakespeare's text, thus giving it a much lower algorithmic information content (or algorithmic randomness) than any probable, equally long passage typed by the monkey. And all that is true of any author in English; we have not yet taken into account the uniqueness of Shakespeare!

Effective Complexity

Evidently, AIC or algorithmic randomness, even though it is sometimes called algorithmic complexity, does not correspond to what is meant by complexity in most situations. To define effective complexity, one needs something quite different from a quantity that achieves its maximum in random strings. In fact, it is just the nonrandom aspects of a system or a string that contribute to its effective complexity, which can be roughly characterized as the length of a concise description of the regularities of that system or string. Crude complexity and AIC fail to correspond to what we usually understand by complexity because they refer to the length of a concise description of the whole system or string—including all its random features—not of the regularities alone.

In order to discuss more fully the concept of effective complexity, it is essential to examine in detail the nature of complex adaptive systems. We shall see that their learning or evolution requires, among other things, the ability to distinguish, to some extent, the random from the regular. Effective complexity is then related to the description of the regularities of a system by a complex adaptive system that is observing it.

CHAPTER

5

A CHILD LEARNING A LANGUAGE

When my daughter was learning to speak, one of her first sentences was "Daddy go car-car," which she would recite every morning when I left for work. I was flattered that the sentence was about me and delighted that she was actually talking, even if her English still needed some work. It has struck me only recently how certain features of English grammar were already present in that utterance. Take word order, for example. In English, the subject comes before the verb (while in some other languages, such as Welsh, Hawaiian, and Malagasy, it does not). The order of subject and verb was already correct, as was the position of the phrase "car-car." In the grammatical English sentence "[Daddy] [is going away] [in his car]," the order of the three elements is exactly the same as in the baby's approximation.

As my daughter grew older, her grammar naturally kept improving and in a few years, like other children, she was speaking correctly. Any normal young child with a caregiver, such as a parent, who speaks a particular language and uses it regularly to address the child, will learn, over a period of years, how to speak that language grammatically. (Of course, some Americans think that this statement fails to apply to many U.S. high school students.) In fact, most children are capable of learning two or even three languages with native fluency, especially if each of two or three caregivers uses just one of the languages correctly and habitually with the child. This is true even if a child's only exposure to a language is through a single speaker of it. But how does the child come to know, for a given language, which ways of constructing a sentence are grammatical and which are not?

Imagine that there are only fifty thousand possible sentences, and that a mother and child systematically try out fifty new ones every day for a thousand days, the mother patiently indicating "OK" or "bad sentence" for each one. If we assume this absurd scenario, plus a perfect memory on the part of the child, then after three years the youngster would know exactly which of the fifty thousand sentences are grammatical.

A computer scientist might say that this fictitious child had constructed in his or her mind a "look-up table" in which each candidate sentence was listed, along with the label "grammatical" or "ungrammatical." Clearly a real child does not prepare such a table. For one thing, fifty thousand sentences are far too few.

In any human language, there are an unlimited number of possible sentences, which can contain arbitrarily many clauses, each loaded with modifying words and phrases. Sentence length is limited only by the time available and by the patience and memory capacity of the speaker and listener. Moreover, there is typically a vocabulary of many thousands of words to work with. There is no chance that a child will hear or try to speak every possible sentence and enter it in a look-up table. Yet at the end of the real learning process, a child can tell whether a *previously unheard* sentence is grammatical.

Children must make up, without being fully aware of it, provisional sets of rules for what is grammatical and what is not. Then, as they continue to hear grammatically correct sentences and (occasionally) try out a sentence and have it corrected, they keep altering the set of rules, again without necessarily being fully aware of it. For example, in English it is easy for a child to acquire the regular or "weak" construction of the simple past tense by adding -ed or -d to a verb. Then, after running across "sing" and "sang" (the present and past of a "strong" verb) the child tries a modified set of rules that includes this exception. That new set, however, may lead the child to say "bring" and "brang," which will eventually be corrected to "bring" and "brought." And so on. Gradually, the internal set of rules is improved. The child is constructing a kind of grammar in its mind.

A child learning a language does indeed make use of grammatical information, acquired over the years from examples of grammatical and ungrammatical sentences. But instead of constructing a look-up table, a child somehow compresses this experience into a set of rules, an internal grammar, which works even for new sentences that have never been encountered before.

But is the information obtained from the outside world, for example from a parent who speaks the language in question, sufficient to construct such an internal grammar? That question has been answered in the negative by Noam Chomsky and his followers, who conclude that the child must come already equipped at birth with a great deal of information applicable to the grammar of any natural human language. The only plausible source of such information is a biologically evolved innate proclivity to speak languages with certain general grammatical features, shared by all natural human languages. The grammar of each individual language also contains additional features, not biologically programmed. Many of those vary from language to language, although some are probably universal like the innate ones. The additional features are what the child has to learn.

Grammar as a Partial Schema

Of course, whether a declarative sentence is grammatical is largely independent of whether it is factual. Speakers of English know that it is grammatically correct to say "the sky is green, with purple and yellow stripes," even though it is unlikely to be true, at least on Earth. But there are many circumstances other than mere veracity that influence the choice of which grammatical sentence one utters on a particular occasion.

In constructing an internal grammar, a child effectively separates grammatical features from all the other factors, some of them stochastic, that have led to the particular sentences he or she hears. Only in that way is compression into a manageable set of grammatical rules possible.

The child who does this has exhibited the first characteristic of a complex adaptive system. He or she has compressed certain regularities identified in a body of experience into a schema, which includes rules that govern that experience but omits the special circumstances in which the rules have to be applied.

Grammar, however, does not encompass all the regularities encountered in a language. There are also the rules of sounds (constituting what linguists call the "phonology" of a language), the rules of semantics (relating to what makes sense and what does not), and others. The grammatical schema is therefore not the entire set of rules for speaking a language, and grammar is not all that is left when arbitrary features of the linguistic data stream have been put aside. Nevertheless, a child's acquisition of grammar is an excellent example of the construction of a schema—a partial schema.

The process of learning grammar also demonstrates the other features of a complex adaptive system in operation. A schema is subject to variation, and the different variants are tried out in the real world. In order to try them out, it is necessary to fill in details, such as the ones that were thrown away in creating the schema. That makes sense, since in the real world the same kind of data stream is encountered again as that from which the schema was abstracted in the first place. Finally, what happens in the real world influences which variants of the schema survive.

In the acquisition of English grammar the schema is varied, for example, when the rule for constructing the simple past tense of a verb with -ed or -d is modified by exceptions such as those for sing-sang and bring-brang. In order to try out these variants, the child must use the schema in an actual sentence, thus restoring special circumstances of the sort pared away to make the schema possible. For example, the child may say, "We sang a hymn yesterday morning." That sentence passes muster. If, however, the child says, "I brang home something to show you," the parent may reply, "It's very nice of you to show me that cockroach you found at Aunt Bessie's, but you ought to say 'I *brought* home something . . . '." That experience would probably result in the child's trying out a new schema, one that allows for both sing-sang and bring-brought. (In very many cases, of course, a child tries out a schema simply by waiting for someone else to speak.)

Complex Adaptive Systems and Effective Complexity

The operation of a complex adaptive system was shown in the diagram on page 25. Since a complex adaptive system separates regularities from randomness, it affords the possibility of defining complexity in terms of the length of the schema used by a complex adaptive system to describe and predict the properties of an incoming data stream. Those data typically relate, of course, to the functioning of some other system that the complex adaptive system is observing.

Utilizing the length of a schema does not signify a return to the concept of crude complexity, because the schema is not a complete description of the data stream or of the observed system, but only of the identified regularities abstracted from the available data. In some cases, such as grammar, only regularities of a certain type are included while the rest are put aside, so the result is a partial schema.

One may think of grammatical complexity in terms of a textbook of grammar. Roughly speaking, the longer the textbook, the more complex the grammar. This agrees very well with the notion of complexity as the length of a schema. Every nasty little exception adds to the length of the book and the grammatical complexity of the language.

As usual, there are sources of arbitrariness such as coarse graining and shared initial knowledge or understanding. In the case of a textbook of grammar, the coarse graining corresponds to the level of detail achieved by the text. Is it a very elementary grammar that leaves out many obscure rules and lists of exceptions, covering only the main points needed by a traveler who doesn't mind making a mistake now and then? Or is it a weighty academic tome? If so, is it one of the old, familiar kind or a currently fashionable generative grammar? Obviously the length of the book will depend on such distinctions. As to the level of initial knowledge, consider an old-fashioned grammar of a foreign language written in English for English speakers. It will not have to introduce so many new grammatical ideas to the reader if it is a grammar of Dutch (fairly similar to English and closely related) rather than of Navajo, which is very different from English in structure. The grammar of Navajo should be longer. Similarly, a hypothetical grammar of Dutch written for speakers of Navajo would presumably have to be longer than a grammar of Dutch written for English speakers.

Even taking these factors into account, it is still reasonable to relate the grammatical complexity of a language to the length of a textbook describing its grammar. However, it would be more interesting if it were possible instead to look inside the brain of a native speaker (as advancing technology may some day make possible) and see how the grammar is encoded there. The length of the schema represented by that internal grammar would provide a somewhat less arbitrary measure of grammatical complexity. (Naturally, the definition of length in this case may be a subtle one, depending on how the bits of grammatical information are actually encoded. Are they inscribed locally in neurons and synapses or distributed somehow over a whole network?)

We define the effective complexity of an entity, relative to a complex adaptive system that is observing it and constructing a schema, as the length of a concise description of the entity's regularities identified in the schema. We can use the term "internal effective complexity" when the schema somehow governs the system under discussion (as grammar stored in the brain regulates speech), rather than merely being used by an external observer, such as the author of a grammatical text.

Separating Regularity from Randomness

The usefulness of the concept of effective complexity, especially when it is not internal, depends on whether the observing complex adaptive system does a good job of identifying and compressing regularities and discarding what is incidental. If not, the effective complexity of the observed system has more to do with the particular observer's shortcomings than with the properties of the system observed. It turns out that the observer is often fairly efficient, but the concept of efficiency raises deep issues. We already know that the notion of optimal compression may run into the obstacle of uncomputability. But what about the actual identification of regularities, apart from compression? Is it really a well-defined problem to identify the regularities in a stream of data?

The task would be easier if the data stream were in some sense indefinitely long, as in the case of speech or text so extended that it comprises a representative sample of the possible sentences (up to a given length) that can be uttered in a given language. Here, even a rare grammatical regularity would show itself over and over under similar conditions and thus tend to be distinguishable from a false rule arising from a mere chance fluctuation. (For instance, in a short English text the past perfect tense might not occur, suggesting wrongly that it does not exist in English. In a very long text, that would not be likely to happen.)

Identifying Certain Classes of Regularities

A number of theoretical physicists, such as Jim Crutchfield of the University of California at Berkeley and the Santa Fe Institute, have made considerable progress in understanding how to distinguish regularity from randomness for an indefinitely long bit string. They define particular broad classes of regularities and show how a computer could be used, in principle, to identify any regularities belonging to those categories. Even their methods, however, do not provide an algorithm that would pick out every kind of regularity. *There is no such algorithm.* They show, however, how a computer, having found in a bit string regularities belonging to certain classes, can deduce that new regularities belonging to a broader class are present and can be identified. That is called "hierarchical learning."

Typically, a class of regularities corresponds to a set of mathematical models of how a data stream might be generated. Suppose the data stream is a string of bits known to be generated by a process that is, at least in part, stochastic—say coin tosses are involved. A very simple example of a set of models would then be a sequence of biased coin tosses, in which the probability of heads (yielding 1s in the bit string) has some fixed value between zero and one for each model, while the probability of tails (yielding 0s in the bit string) is one minus the probability of heads.

If the probability of heads is one-half, then any apparent regularity in such a sequence would be the effect of chance alone. As the data stream gets longer and longer, the probability of being fooled by such chance regularities gets smaller and smaller, and the likelihood increases of recognizing that the sequence stems from the equivalent of unbiased coin tosses. At the opposite extreme, take a two-bit string. The chance of both bits being 1s (a case of perfect regularity) is one out of four for unbiased tosses. But such a sequence could just as well come from tossing a two-headed coin. Thus, a short bit string arising from a sequence of unbiased coin tosses can often be mistaken for a heavily biased sequence. In general, the advantage of an indefinitely long data stream is that it greatly increases the chances of discriminating among models, where each model corresponds to a particular class of regularities.

Another example of models, slightly more complicated than sequences of biased coin tosses, might have the additional provision that all sequences in which two heads occur in succession would be thrown away. The resulting regularity, that the bit string never has two ones in a row, would be quite easy to recognize in a long string. A still more complicated model might consist of sequences of biased coin tosses in which any sequence containing an even number of heads in succession would be discarded.

When a complex adaptive system receives an arbitrarily long data stream, say in the form of a bit string, it can search systematically for regularities of given classes (corresponding to models of given classes), but there is no such procedure for searching for every type of regularity. Whatever regularities are identified can then be incorporated into a schema describing the data stream (or a system giving rise to that stream).

Dividing the Data Stream into Parts—Mutual Information

When identifying regularities in an incoming data stream, a complex adaptive system typically divides that stream into many parts that are in some way comparable to one another and searches for their common features. Information common to many parts, called "mutual information," is diagnostic of regularities. In the case of a stream of text in a given language, sentences could serve as the parts to be compared. Mutual grammatical information among the sentences would point to grammatical regularities.

However, mutual information is used only to identify regularities, and the quantity of it is not a direct measure of effective complexity. Instead, once regularities have been identified and a concise description of them given, the length of the description measures the effective complexity.

Large Effective Complexity and Intermediate AIC

Suppose there are no regularities in the system being described, as will often (but not always!) be the case in a passage typed by the proverbial

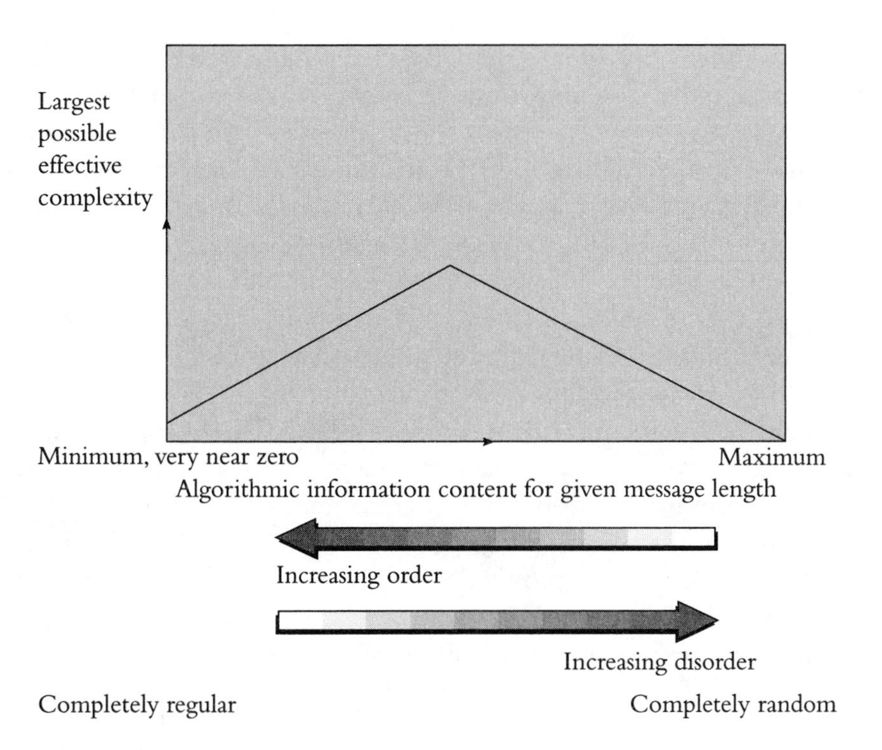

A sketch showing roughly how the largest possible effective complexity varies with AIC.

monkeys. A properly operating complex adaptive system would then be unable to find any schema, since a schema summarizes regularities and there aren't any. To put it differently, the only schema will have length zero, and the complex adaptive system will assign zero effective complexity to the random junk it is studying. That is entirely appropriate; a grammar of pure gibberish ought to have zero length. Although the algorithmic information content of a random bit string is maximal for its length, the effective complexity is zero.

At the other end of the scale of AIC, when it is near zero, the bit string is entirely regular, consisting, for example, only of 1s. The effective complexity—the length a concise description of the regularities of such a bit string—should be very close to zero, since the message "all 1s" is so short.

For effective complexity to be sizable, then, the AIC must be neither too low nor too high; in other words, the system must be neither too orderly nor too disorderly.

The sketch on this page illustrates crudely how the largest possible effective complexity of a system (relative to a properly functioning complex adaptive system as observer) varies with AIC, attaining high values only in the intermediate region between excessive order and excessive disorder. Many important quantities that occur in discussions of simplicity, complexity, and complex adaptive systems share the property that they can be large only in that intermediate region.

When a complex adaptive system observes another system and identifies some of its regularities, the algorithmic information content of the data stream coming from the observed system is expressed as the sum of two terms: the apparently regular information content and the apparently stochastic information content. The effective complexity of the observed system is essentially the same as the apparently regular portion of the information content. For a random data stream, recognized as such, that effective complexity is zero and the whole of the AIC is recognized as the product of chance. For a perfectly regular data stream, recognized as such (a long bit string consisting entirely of 1s, for example), the entire AIC is regular (with no stochastic portion) but extremely small. The interesting situations lie between these extremes, where the AIC is sizable but not maximal (for the length of the data stream), and is the sum of two appreciable terms, the apparently regular portion (the effective complexity) and the apparently stochastic portion.

Learning with the Genes or the Brain

Although our examination of complex adaptive systems began with the example of learning in a human child, it is not necessary to invoke anything so sophisticated to illustrate the concept. Our fellow primates—the ones caricatured in the typewriter story—could be used just as well. So could a dog. In fact, one way we encounter learning in other mammals is through training our pets.

Teaching a dog to stay involves applying an abstraction to a great many different situations: staying in a sitting position on the ground, staying in a car when the door is opened, staying nearby instead of pursuing a tempting squirrel. The dog learns, by means of rewards and/or punishments, the schema for the command to stay. Alternative schemata, for example, one that makes an exception for chasing cats, are (at least in theory) rejected as training proceeds. But even if the dog adopts a schema with the exception, a complex adaptive system is still at work. A schema other than the one the trainer intended has survived, as a result of competing pressures from training and the instinct to chase cats.

When given the command to stay, the trained dog fills in the details appropriate to the particular situation and carries the schema into the real world of behavior, where the reward or punishment occurs and helps to determine whether the schema survives. However, the tendency to chase cats or squirrels, which also influences the competition among schemata, has not been learned by the individual dog. Instead, it has been genetically programmed as a result of biological evolution.

All organisms have such programs. Consider an ant wandering around in search of a morsel of food. It is following a built-in procedure evolved over millions of years. Herbert Simon, the distinguished expert on psychology, economics, and computer science at Carnegie-Mellon University, long ago used the ant's movements to illustrate the meaning of what I call effective complexity. The path followed by the ant may appear complex, but the rules of the search process are simple. The intricate path of the ant manifests a great deal of algorithmic complexity (AIC), of which only a little comes from the rules, which correspond more or less to the regularities of the search. That small part, though, constitutes (at least approximately) the entire effective complexity. The remainder of the AIC, the bulk of the apparent complexity, comes from incidental, largely random features of the terrain that the ant is exploring. (Recently I discussed the ant story with Herb, who exclaimed with a grin, "I got a lot of mileage out of that ant!")

In a sequence of less and less sophisticated organisms, say a dog, a goldfish, a worm, and an amoeba, individual learning plays a smaller and smaller role compared to that played by the instincts stored up through biological evolution. But biological evolution itself can also be described as a complex adaptive system, even in the humblest organisms.

CHAPTER

6

BACTERIA DEVELOPING DRUG RESISTANCE

When I was young, I had the habit of browsing through encyclopedias (a habit that persists to this day, provoking merriment in my family). At one point I ran into an article on the bronze disease, which started me thinking for the first time about some of the issues central to this book.

The bronze disease is a set of chemical reactions that can corrode bronze surfaces, creating greenish-blue spots that grow and spread. Under moist conditions, the reactions can actually spread the disease through the air from one surface to another and ruin a whole collection of bronze objects. Since Chinese bronze vessels of the Shang dynasty, for instance, may be worth a million dollars apiece, protection against the bronze disease is not an unimportant matter. However, when I first read about it, as a poor boy, I was obviously not thinking from the collector's point of view.

Instead I was wondering, "How is the bronze disease different from a plague caused by a living organism? Is it that the bronze disease is merely obeying the laws of physics and chemistry?" But even as a child I rejected, as serious scientists have done for generations, the idea that life is characterized by any special "vital forces" outside of physics and chemistry. No, a bacterium too obeys the laws of physics and chemistry. But then what is the difference? It occurred to me that bacteria (like all other living things) exhibit variation that is heritable and subject to natural selection, whereas for the bronze disease there is no evidence of any such thing. Indeed, that distinction is critical.

To explore the same distinction further, consider the example of a turbulent flow of fluid down a pipe. For more than a century it has been known that energy is dissipated from large turbulent eddies into smaller and smaller ones. In describing those eddies, physicists have often quoted Jonathan Swift:

> So, Nat'ralists observe, a Flea
> Hath smaller fleas that on him prey,
> And these have smaller fleas to bite 'em,
> And so proceed ad infinitum.

Moreover, the physicist and polymath L. F. Richardson created his own bit of doggerel specifically applicable to eddies:

> Big whorls have little whorls,
> Which feed on their velocity;
> And little whorls have lesser whorls,
> And so on to viscosity.

In a sense, the larger eddies give birth to smaller ones. If the pipe has bends and constrictions, there may be some large eddies that come to grief without having offspring, while others survive to spawn many smaller ones, which generate still smaller ones, and so on. The eddies thus seem to exhibit a kind of variation and selection. Yet no one has ever suggested that they resemble life. What important feature do the turbulent eddies lack that living organisms possess? What really distinguishes turbulent flow from biological evolution?

The difference lies in how information is handled in the two cases. There is no indication that in turbulent flow any significant information processing goes on, any compression of regularities. In biological evolution, however, the experience represented by past variation and natural selection is handed down to future generations in a highly compressed package of information, the "genome" (or set of genes) of an organism. Each gene can have different alternative forms, which in some connections are called "alleles." The set of particular alleles of all the genes in a given organism is known as the "genotype."

Biologists emphasize the distinction between the genotype, which describes the inherited information contained in the genes of an individual organism, and the phenotype, which describes the appearance and behavior of the organism during the course of its life history. Changes in the genotype, such as an alteration in a given gene from one allele to another, can, of course, affect the phenotype through the gene's influence on chemical processes in the organism. But the phenotype is also influenced during the development of the organism by a huge variety of other circumstances, many of them random. Think of all the accidental circumstances that affect the development of a human being, from the single-cell and fetal stages through infancy and childhood, before procreation becomes possible in adulthood. The genotype of an individual human being is like a basic recipe, allowing for wide variations in the actual dish prepared by the cook. A single genotype permits one of many different possible alternative adults to emerge from the development process. In the case of identical twins, who always share the same genotype, two of the different alternative adults co-exist. When raised separately, they can supply precious information about the roles of "nature" and "nurture" in the formation of the adult phenotype.

In the course of biological evolution, random changes take place in the genotype from generation to generation. They contribute, along with the accidents of development that occur in a given generation, to phenotypic changes that help to determine whether an organism is viable and able to reach maturity, to reproduce, and to pass on its genotype, in whole or in part, to its descendants. The distribution of genotypes in the population is thus the result of chance combined with natural selection.

The Evolution of Drug Resistance in Bacteria

One case of biological evolution of great significance for contemporary humanity is the development of resistance to antibiotics in bacteria. For example, after a number of decades of widespread use of penicillin to control certain species of pathogenic (disease-producing) bacteria, strains of those organisms have appeared that are not particularly sensitive to the drug. In order to cope with the diseases caused by these altered germs, new types of antibiotics are required, and much human suffering and even death may occur while new drugs are being perfected. Similarly, the bacillus that causes tuberculosis yielded for decades to certain antibiotics, but in recent years has developed resistant strains. Tuberculosis is again a major health menace, even in places where it was formerly controlled.

In the acquisition of drug resistance by bacteria, an important role is often played by the exchange of genetic material between two individual bacteria as they come together, merge, and then separate again. This process, which is as close to sexual activity as such primitive organisms get, was first observed by Joshua Lederberg when he was a graduate student at Yale. I was an undergraduate there at the time, and I remember how much public attention was drawn to the discovery of sex in the realm of germs; there was even an item in *Time* magazine. Josh was launched on his career, which led eventually to the presidency of Rockefeller University. In discussing bacterial drug resistance, I shall, for the sake of simplicity, ignore sex (with apologies to Josh).

For the same reason, I propose to ignore another very important mechanism for transfer of genetic material between cells, in which the carrier is a virus—a bacteriophage (or "phage")—that infects bacteria. Experiments on this process, called transduction, were precursors of the first work in genetic engineering.

Careful research on bacteria has centered on the species *Escherichia coli* (or *E. coli*), common, harmless, and even useful in the human intestines, but often pathogenic when infecting other parts of the body (and also, in certain mutant forms, dangerous even in the digestive tract). Each *E. coli* organism is a single cell with genetic material consisting of a few thousand genes. A typical gene is a sequence of something like a thousand "nucleotides" (known collectively as DNA). Those DNA constituents, which make up all genes in all organisms, come in four kinds, called A, C, G, and T for the initial letters of their chemical names. Any gene is part of a longer strand composed of nucleotides and paired with another strand in the famous double helix. The double helix structure was worked out in 1953 by Francis Crick and James Watson, utilizing the work of Rosalind Franklin and Maurice Wilkins. In *E. coli,* there are two helical strands of around five million nucleotides each.

The nucleotides strung out along one strand are complementary to those on the other strand in the sense that A and T always occur opposite each other, as do G and C. Since either helix is determined by the other, we need look at only one of them in order to read the complete message.

Suppose the number of nucleotides in the strand is really five million. A can be encoded as 00, C as 01, G as 10, and T as 11, so that the five million nucleotides are represented by a string of ten million 0s and 1s, in other words, by a bit string with ten million bits. That string stands for the information that each *E. coli* bacterium transmits to its progeny, which come into being by means of the splitting of the cell into two cells, the double helix giving rise to two new double helices, one for each new cell.

Each of the several thousand genes in the bacterium can exist in many forms. The mathematical possibilities are extremely numerous, of course. For a string of one thousand nucleotides, say, the number of different conceivable sequences equals $4 \times 4 \times 4 \times \ldots \times 4 \times 4$, with a thousand factors of 4. This number, when written in the usual decimal system, has around six hundred digits! Only a tiny fraction of those theoretically possible sequences can be found in nature (the existence of all of them would require far more atoms than the universe contains). In practice, at any given time, each gene may have some hundreds of different alleles that actually occur with significant probability in the bacterial population and are distinguished by their different chemical and biological effects.

Any gene can undergo mutation from one form to another as a result of various kinds of accidents, for example, the random passage of a cosmic ray or the presence of some strong chemical in the environment. Even a single mutation can have a significant effect on cell behavior. For example, the mutation of a certain gene in an *E. coli* cell to a certain new allele could, in principle, lead to the resistance of that cell to a drug such as penicillin. That resistance would then be passed on to the cell's progeny, as they multiply through repeated cell divisions.

Mutations are typically chance processes. Suppose a single bacterium in a host tissue produces a colony of descendants, all of the same genotype. Mutations can then occur in that colony, with the mutant forms giving rise to colonies of their own. In that way the population of bacteria in the tissue comes to contain various genotypes. If penicillin is introduced in sufficient quantity, only the colonies resistant to it will continue to grow. The important point is that the resistant mutant bacteria are often already there by chance, usually through a mutation in an ancestor, when the drug starts to exert selection pressure in their favor. Even if they are not already present, they exist somewhere else, or at the very least they have come into existence from time to time through chance processes and then disappeared. The mutations are not induced by the penicillin, as Lederberg showed long ago.

The mutation of a gene to an allele corresponding to drug resistance presumably has some unfavorable effects on the operation of an *E. coli* cell. Otherwise that allele would almost certainly have been present already in a large number of *E. coli* bacteria, and penicillin would not have worked in the first place. However, as penicillin continues to be widely used, the survival of the penicillin-resistant strain is favored, while the selective disadvantage, whatever it is, is outweighed by the advantage of drug resistance. (A different antibiotic, not occurring so widely in nature as penicillin, might serve even better as an example, since the species of bacterium would have had fewer contacts with it before it was introduced into medicine.)

The development of drug resistance takes place, then, through a change in the genotype, the string of some ten million bits that is transmitted by the cell to its descendants. It is through the genes that the bacterium "learns" to cope with this menace to its survival. But the genotype contains a huge amount of other information that permits the bacterium to function. The genes contain the lessons learned over billions of years of biological evolution about how to survive as a bacterium.

The experience of the species *E. coli* and its ancestral life forms was not merely recorded for reference in a look-up table; regularities in that experience were identified and compressed into the string represented by the genotype. Some of them are regularities experienced only recently, like the prevalence of antibiotics. Most of them are quite ancient. The genotype varies to some extent from individual to individual (or from one colony of genetically identical individuals to another), and mutations can occur by accident at any time and be transmitted to progeny.

This kind of learning differs in an interesting way from the kind that takes place through the use of a brain. We have emphasized that mutant forms of a bacterium exhibiting resistance to an antibiotic may easily be present by chance when the drug is introduced and that in any case those forms have existed from time to time in the past. Ideas, however, more often arise in response to a challenge rather than being already available when the challenge is presented. (There is some slight evidence for genetic mutations in biology occasionally arising in response to need, but if the phenomenon really exists it is comparatively insignificant compared to chance mutation.)

Evolution as a Complex Adaptive System

To what extent can the evolutionary process be described as the operation of a complex adaptive system? The genotype satisfies the criteria for a schema, encapsulating in highly compressed form the experience of the past and being subject to variation through mutation. The genotype itself does not usually get tested directly by experience. It controls, to a great extent, the chemistry of the organism, but the ultimate fate of each individual depends also on environmental conditions that are not at all under the control of the genes. The phenotype, in other words, is co-determined by the genotype and by all those external conditions, many of them random. Such an unfolding of schemata, with input from new data, to produce effects in the real world is characteristic of a complex adaptive system.

Finally, the survival of a particular genotype for a single-celled organism is related to whether cells with that genotype survive until they divide, whether their offspring survive until they divide, and so forth. That fulfills the requirement of a feedback loop involving selection pressures. The bacterial population is certainly a complex adaptive system.

The effective complexity of the bacterium, in our sense of the length of a schema, is evidently related to the length of the genome. (If parts of the double DNA helix are just fillers and contribute no genetic information, as seems to be the case in higher organisms, the length of those parts would not be counted.) The length of the relevant part of the genome provides a crude internal measure of effective complexity. It is internal because it relates to the schema that the organism uses to describe its own heritage to its descendants, rather than a schema devised by some outside observer. (This measure resembles the length of the internal grammar in the brain of a child learning its native language, as opposed to the length of a book describing the grammar of that language.) It is only a crude measure because biological evolution, like other complex adaptive systems, performs the task of compression of regularities with varying efficiency in different cases. Sometimes such variations can invalidate the measure, as in certain organisms that are obviously rather simple but have anomalously long genomes, which do not give a concise description of the relevant regularities.

But a comparison of the genomes of different organisms reveals deficiencies in the whole notion of using effective complexity, based on the length of a schema, as the only measure of the complexity of a species. For instance, in considering subtle but important differences such as those that distinguish our species from closely related great apes, we have to include more sophisticated ideas.

Those comparatively few genetic changes that permit an apelike creature to develop language, advanced thinking, and elaborate culture, all manifesting great effective complexity, have greater significance than most comparable sets of alterations in the genetic material. The effective complexity of the new (human) genome, as measured crudely by its length, is not by itself a satisfactory measure of the complexity of the corresponding organisms (people), since the slightly altered genome can give rise to so much effective complexity of a novel kind (cultural complexity).

We will therefore find it necessary to supplement effective complexity with the concept of *potential complexity*. When a modest change in a schema permits a complex adaptive system to create a great deal of new effective complexity over a certain period of time, the modified schema can be said to have a greatly increased value of potential complexity with respect to that time interval. We shall pursue this subject later on, but for now let us return to the idea of adaptation to drug resistance as a complex adaptive system and compare that picture with an incorrect theory of how such resistance comes about.

Direct Adaptation

Today it seems obvious that drug resistance develops largely through a genetic mechanism such as we have been considering. However, that was not always the case. In the 1940s, when penicillin was just coming into use and the sulfa drugs were still the glamorous weapons in the battle against bacterial infection, drug resistance was already a problem, and some scientists offered very different models for its development. One of those was the distinguished English chemist Cyril (later Sir Cyril) Hinshelwood. I remember seeing his book on the subject when I was a student, and being quite skeptical, even then, of his ideas on this particular subject.

Hinshelwood's erroneous theory of drug resistance was, naturally, a chemical theory. His book was full of equations describing the rates of chemical reactions. The general idea was that the presence of the drug caused changes in the chemical balance of the bacterial cell that were detrimental to the cell's reproduction. However, prolonged exposure of the bacteria to high doses produced, by direct chemical means, adjustments in the cell's metabolism that limited the detrimental effect of the drug and permitted cells to survive and divide. In cell division, the theory went, this simple form of drug resistance was passed on mechanically to the daughter cells through the chemical composition of the ordinary cell material. The proposed mechanism was straight negative feedback in a set of chemical reactions. (If your car starts to go off the road and you supply a corrective turn of the steering wheel, that is another example of negative feedback.)

In Hinshelwood's theory, the bacterial genes were not involved. There was no complex adaptive system underlying the development of bacterial drug resistance: no compression of information, no schema, no chance variation, and no selection. In fact, a chapter of the book is devoted to refuting the idea of selection of spontaneous variants.

We can describe Hinshelwood's theory as involving "direct adaptation." Such processes are very common. Consider the operation of a thermostat set for a particular temperature; the device causes a heating system to go on when the temperature falls below the set point and off when it reaches it again. In place of a set of competing and evolving schemata, the thermostat has a single fixed program, and a very simple one at that. The device just keeps mumbling to itself, "It's too cold. It's too cold. It's a little too hot. It's too cold . . . ," and acting accordingly.

It is useful to contrast direct adaptation with the operation of complex adaptive systems, but I do not mean to suggest that direct adaptation is uninteresting. Indeed, most of the excitement about cybernetics in the aftermath of the Second World War was related to processes of direct adaptation, especially the stabilization of systems by means of negative feedback. The basic principle is the same as that of a thermostat, but the problems it presents can be much more challenging.

Direct Adaptation, Expert Systems, and Complex Adaptive Systems

The word "cybernetics" was introduced by a great but eccentric mathematics professor at the Massachusetts Institute of Technology, Norbert Wiener, who as a child had been considered a prodigy and never got over the need to show off in bizarre ways. As a graduate student at MIT, I would occasionally find him asleep on the stairs, creating a real obstacle to traffic with his portly figure. Once he stuck his head in the door of my dissertation adviser, Viki Weisskopf, and uttered some words that Viki found completely incomprehensible. "Oh, I thought all European intellectuals knew Chinese," said Wiener, and hurried off down the hall.

The word is derived from the ancient Greek *kubernetes* meaning helmsman. That word begins with the Greek letter "*kappa,*" and it is responsible for the "kappa" in "Phi Beta Kappa," the academic honor society, the motto of which means "Philosophy, life's helmsman." After borrowing from Greek into Latin and later from French into English, the same word gave rise to the verb "govern," and indeed cybernetics relates to both steering and governing, as in controlling a robot. But in the days of cybernetics robots were not usually able to create an evolving schema out of their sense impressions. Only now are we entering the age of robots that are really complex adaptive systems.

Take a mobile robot, for instance. In the early cybernetic era, it might have been equipped with sensors to indicate the presence of a nearby wall and to activate a device for avoiding it. Other sensors might have detected bumps just ahead and caused the form of locomotion to change in some predetermined way so the robot could get over them. The point of the design was to provide a direct response to environmental signals.

The next era was that of the "expert system," in which information supplied by human experts in a field was fed into a computer in the form of an "internal model" that could be used to interpret incoming data. The advances in robot design achieved by such methods were not dramatic, but an example from a different field can be used to illustrate the approach. Medical diagnosis can be automated to some extent by obtaining the expert advice of physicians and constructing a "decision tree" for the computer, with a definite rule for decision-making at each branch based on particular data concerning the patient. Such an internal model is fixed, unlike the schemata of complex adaptive systems. The computer can diagnose illnesses, but it does not learn more and more about diagnosis from its experience with successive patients. It continues to use the same internal model developed by consulting the experts.

The experts can be consulted again, of course, and the internal model redesigned to take account of the successes and failures of the computer diagnoses. In that case, the extended system consisting of the computer, the model designers, and the experts can be regarded as a complex adaptive system, an artificial one with "humans in the loop."

Today we are entering the era of computers and robots functioning as complex adaptive systems without humans in the loop. Many future robots will have elaborate schemata subject to variation and selection. Consider a six-legged mobile robot having in each leg a set of sensors that detect obstacles and an information processor that responds in some prearranged manner to the signals from those sensors to control the motions of that leg, moving it up or down and forward or backward. Such legs resemble a set of old-fashioned cybernetic devices.

Nowadays robot design might include a form of communication among the legs, but not through a governing central processing unit. Instead, each leg would have the capacity to influence the behavior of the others by means of communication links. The pattern of strengths of influence of the legs on one another would be a schema, subject to variations produced, for example, by input from a generator of pseudorandom numbers. The selection pressures influencing the adoption and rejection of candidate patterns might originate from additional sensors that measure what is happening not just to an individual leg, but also to the robot as a whole, such as whether it is moving forward or backward and whether its belly is far enough off the ground. In this way the robot would tend to develop a schema that yielded a gait suited to the terrain on which it was traveling and that was subject to alteration when the character of that terrain changed. Such a robot may be regarded as at least a primitive form of complex adaptive system.

I am told that a six-legged robot something like this has been built at MIT and that it has discovered, among other gaits, one that is commonly used by insects: the front and back legs on one side move together with the middle leg on the other side. When the robot uses that gait depends on the terrain.

Now consider, in contrast to a robot that learns a few useful properties of the terrain it needs to traverse, a complex adaptive system exploring the general properties, as well as a host of detailed features, of a much grander terrain, namely the whole universe.

CHAPTER

7

THE SCIENTIFIC ENTERPRISE

The concept of the complex adaptive system is beautifully illustrated by the human scientific enterprise. The schemata are theories, and what takes place in the real world is the confrontation between theory and observation. New theories have to compete with existing ones, partly on the basis of coherence and generality, but ultimately according to whether they explain existing observations and correctly predict new ones. Each theory is a highly condensed description of a whole class of situations and, as such, needs to be supplemented with the details of one or more situations in order to make specific predictions.

The role of theory in science should be fairly obvious, and yet in my own case it took me a long time to get a real feeling for it, even though I was to devote my whole career to theoretical science. It was only when I entered graduate school at MIT that it finally dawned on me how theoretical physics works.

As an undergraduate at Yale, I had managed to get high grades in science and math courses without always understanding the point of what I was learning. Sometimes, it seemed, I was able to get by merely by regurgitating on examinations what I had been fed in class. My views changed when I attended one of the sessions of the Harvard–

MIT theoretical seminar. I had thought of the seminar as some sort of glorified class. In fact, it was not a class at all, but a serious discussion group on subjects in theoretical physics, particularly the physics of atomic nuclei and elementary particles. Professors, post-docs, and graduate students from both institutions attended; one theorist would lecture and then there would be a general discussion of the topic he had presented. I was unable to appreciate such scientific activity properly because my way of thinking was still circumscribed by notions of classes and grades and pleasing the teacher.

The speaker on this occasion was a Harvard graduate student who had just completed his Ph.D. dissertation on the character of the lowest energy state of a nucleus called boron ten (B^{10}), composed of five protons and five neutrons. By an approximation method that seemed promising but was not guaranteed to work, he had found that the lowest state should have a "spin" angular momentum of one quantum unit, as was generally believed to be the case. When he finished his talk, I wondered what kind of impression his approximate derivation of the expected result had made on the distinguished theoreticians in the front row. The first person to comment was not a theoretician at all, however, but a little man with a three days' growth of beard who looked as if he had just crawled out of the basement of MIT. He said, "Hey, da spin ain't one. It's t'ree. Dey measured it!" Suddenly, I understood the main function of the theoretician: not to impress the professors in the front row but to agree with observation. (Of course, experimentalists can make mistakes; in this case, however, the observation to which the scruffy man referred turned out to be correct.)

I was ashamed of not having been fully aware earlier that the scientific enterprise worked that way. The process by which theories are selected according to their agreement with observation (as well as their coherence and generality) is not so different from biological evolution, where genetic patterns are selected according to whether they tend to lead to organisms that have progeny. But I was not to appreciate fully the parallel between the two processes until many years later, when I had learned more about simplicity and complexity and about complex adaptive systems.

Today most physicists are either theoreticians or experimentalists. Sometimes the theoreticians are ahead, having formulated a highly successful body of theory capable of making predictions that are repeatedly confirmed by observation. At other times the experimentalists find an unexpected result, and the theoreticians have to go back to the drawing board. But the existence of these two distinct classes of researchers should not be taken for granted. It was not always the case in physics, and in many fields—including cultural anthropology, archaeology, and most parts of biology—there are still only a few professional theorists and they are not necessarily treated with great respect. In molecular biology, a very prestigious subject today, most theoretical puzzles that have turned up have yielded rather easily to the ingenuity of the experimenters. As a result, many prominent molecular biologists are not impressed with the need for theorists in biology.

By contrast, the field of population biology has a long and honorable tradition of mathematical theory, personified in such distinguished figures as Sir Ronald Fisher, J. B. S. Haldane, and Sewall Wright. Through their work and that of many other theorists, numerous detailed predictions have been made and confirmed by observation in population genetics, and even the mathematical literature has been enriched.

Theory tends to emerge as a profession as a science matures and as the depth and power of theoretical methods increase. But the roles of theory and observation should be regarded as distinct whether or not there are separate classes of practitioners for the two activities. Let us see how the interaction between the two fits in with the notion of a complex adaptive system.

Theories typically arise as a result of a multitude of observations, in the course of which a deliberate effort is made to sort out the wheat from the chaff, to separate out rules from special or accidental circumstances. A theory is formulated as a simple principle or set of principles, expressed in a comparatively short message. As Stephen Wolfram has emphasized, it is a compressed package of information, applicable to many cases. There are, in general, competing theories, each of which shares these characteristics. To make predictions about a particular case, each theory must be unfolded or re-expanded; that is, the compressed general statement that constitutes the theory must be supplemented with detailed information about the special case. The theories may then be tested by further observations, often made in the course of experiments. How well each theory does, in competition with the others, at predicting the results of those observations helps to determine whether it survives. Theories in serious disagreement with the outcome of careful and well-designed experiments (especially experiments that have been repeated with consistent results) tend to be displaced by better ones, while theories that successfully predict and explain observations tend to be accepted and used as a basis for further theorizing (that is, as long as they are not themselves challenged by later observations).

Falsifiability and Suspense

It has often been emphasized, particularly by the philosopher Karl Popper, that the essential feature of science is that its theories are falsifiable. They make predictions, and further observations can verify those predictions. When a theory is contradicted by observations that have been repeated until they are worthy of acceptance, that theory must be considered wrong. The possibility of failure of an idea is always present, lending an air of suspense to all scientific activity.

Sometimes the delay in confirming or disproving a theory is so long that its proponent dies before the fate of his or her idea is known. Those of us working in fundamental physics during the last few decades have been fortunate in seeing our theoretical ideas tested during our lifetimes. The thrill of knowing that one's prediction has actually been verified and that the underlying new scheme is basically correct may be difficult to convey but it is overwhelming.

It has often been said that theories, even if contradicted by new evidence, die only when their proponents die. Although that remark is usually directed at the physical sciences, my impression is that, if it applies at all, it is more to the difficult and complex life sciences and behavioral sciences. My first wife Margaret, a student of classical archaeology, found it applicable to her field in the 1950s. She was astonished to discover that many physicists actually change their minds when confronted with evidence contradicting their favorite ideas.

When suspense seems to be lacking in a particular field, controversy may erupt about whether it is really scientific. Psychoanalysis is frequently criticized as not falsifiable, and I tend to agree. Here I mean psychoanalysis as a theory describing how human behavior is influenced by mental processes outside of conscious awareness and how those mental processes themselves are initiated by experiences, especially early ones. (I will not discuss treatment, which is an almost entirely separate issue. Treatment could be helpful because of a constructive relationship between analyst and analysand without confirming the ideas of psychoanalysis. Likewise treatment might be ineffective even if many of those ideas are correct.)

I believe that there is probably a considerable amount of truth in the body of lore developed by psychoanalysis, but that it does not constitute a science at the present time precisely because it is not falsifiable. Is there any statement that might be made by a patient, or any behavior that might be demonstrated by a patient, that could not somehow be reconciled with the underlying ideas of psychoanalysis? If not, those ideas cannot really amount to a scientific theory.

Back in the 1960s, I toyed with the idea of switching from theoretical physics to observational psychology or psychiatry. I wanted to isolate a subset of the ideas of psychoanalysis that were falsifiable and could therefore amount to a theory, and then try to find ways of testing such a theory. (The set of ideas might not correspond exactly to those of a particular psychoanalytic school, but at least they would be closely related to those of psychoanalysis in general. They would be concerned with the role of mental processes outside of awareness in the everyday life of reasonably normal people as well as in the patterns of repeated, apparently maladaptive behavior exhibited by people labeled neurotic.)

For some months I visited various distinguished psychoanalysts and academic psychologists (at that time still strongly influenced by behaviorism—cognitive psychology was only in its infancy). Both discouraged me, although for opposite reasons. Many of the psychologists tended to feel that unconscious mental processes were unimportant or too difficult to study or both, and that psychoanalysis was so silly that it was not worth any serious attention. The analysts tended to feel that their discipline was so well established that it did not require any elaborate effort to incorporate some of its ideas into science, and that any research needed to refine its precepts could best be carried out by psychoanalysts themselves in the course of their work with patients. I finally gave up and continued to work in physics, but many years later I had the opportunity to make an indirect contribution to a renewed effort aimed at incorporating into science certain ideas about conscious and unconscious mental processes and their effects on patterns of behavior. That effort is yielding some encouraging results.

Selection Pressures on the Scientific Enterprise

In practice, the scientific enterprise does not precisely conform to any clearcut model of how it ought to work. Ideally, scientists perform experiments either in an exploratory mode or in order to test serious theoretical proposals. They are supposed to judge a theory by how accurate, general, and coherent a description it gives of the data. They should not exhibit such traits as selfishness, dishonesty, or prejudice.

But the practitioners of science are, after all, human beings. They are not immune to the normal influences of egotism, economic self-interest, fashion, wishful thinking, and laziness. A scientist may try to steal credit, knowingly initiate a worthless project for gain, or take a conventional idea for granted instead of looking for a better explanation. From time to time scientists even fudge their results, breaking one of the most serious taboos of their profession.

Nevertheless, the occasional philosopher, historian, or sociologist of science who seizes upon these lapses from scientific rectitude or ideal scientific practice in order to condemn the whole enterprise as corrupt has failed to understand the point of science. The scientific enterprise is, by its nature, self-correcting and tends to rise above whatever abuses occur. Extravagant and baseless claims like those made for polywater or cold fusion are soon discounted. Hoaxes like the Piltdown man are eventually exposed. Prejudices such as those against the theory of relativity are overcome.

A student of complex adaptive systems would say that in the scientific enterprise the selection pressures that *characterize* science are accompanied by the familiar selection pressures that generally occur in human affairs. But the characteristically scientific selection pressures play the crucial role in advancing the understanding of nature. Repeated observations and calculations (and comparisons between them) tend to weed out, especially in the long run, imperfections (that is, features that are imperfect from the scientific standpoint) introduced by the other pressures.

While the historical details of any scientific discovery are usually somewhat messy, the net result can sometimes be a brilliant and general clarification, as in the formulation and verification of a unifying theory.

Theories That Unify and Synthesize

Sometimes a theory accomplishes a remarkable synthesis, compressing into a brief and elegant statement the regularities found in a whole range of phenomena previously described separately and inadequately. A splendid example from basic physics is the work done by James Clerk Maxwell in the 1850s and 1860s on the theory of electromagnetism.

Since ancient times some people had been familiar with certain simple phenomena of static electricity, for instance that amber (elektron in Greek) has the power, when rubbed on a cat's fur, of attracting fragments of feathers. Likewise they had known about certain properties of magnetism, including the fact that the mineral magnetite (an iron oxide named after Magnesia in Asia Minor, where it is common) is capable of attracting bits of iron and also of magnetizing them so that they can do the same to other bits of iron. The early modern scientist William Gilbert included some important observations on electricity in his famous treatise on magnetism in 1600. But electricity and magnetism were still regarded as separate classes of phenomena; they were not understood to be closely related until the nineteenth century.

The invention by Alessandro Volta of the first battery (the voltaic pile) around 1800 resulted in experiments on electric currents that opened the way for the discovery of the interaction between electricity and magnetism. In about 1820, the subject of electromagnetism was born when Hans Christian Ørsted discovered that an electric current in a wire produces a magnetic field that curls around the wire. In 1831, Michael Faraday found that a changing magnetic field can induce an electric current in a loop of wire; this effect was later interpreted to mean that a magnetic field changing with time gives rise to an electric field.

By the 1850s, when Maxwell began his work on a comprehensive mathematical description of electromagnetic effects, most of the individual pieces of the electromagnetic puzzle had been formulated as scientific laws. What Maxwell did was to write down a set of equations that reproduced those laws, as shown here on the facing page. In textbooks for undergraduates today, they are usually written as four equations. The first one restates Coulomb's law describing how electric charges produce an electric field. The second equation embodies Ampère's conjecture that there are no true magnetic charges (and hence all magnetism can be attributed to electric currents). The third equation reformulates Faraday's law describing the generation of an electric field by a changing magnetic field. The fourth equation, as Maxwell first wrote it, merely reproduced Ampère's law describing how an electric current gives rise to a magnetic field. Looking at his four equations, Maxwell saw that there was something wrong with them and he corrected the flaw by altering the last equation. The reasoning that he actually used at the time looks fairly obscure to us today, but there is a modified version of his argument that appeals to the contemporary mind and makes clear what kind of alteration was needed.

The conservation of total electric charge (its constancy in time) is a beautiful and simple law, well established by observation, that was already an important principle in Maxwell's time. However, his original equations did not conform to that principle. What kind of change in the equations would make them obey it? The third equation has a term describing the generation of an electric field by a changing magnetic field. Why couldn't a corresponding term be inserted into the fourth equation that would describe the generation of a magnetic field by a changing electric field? Sure enough, for a particular value of the coefficient (multiplier) of the new term, the equation became consistent with the conservation of electric charge. Moreover, that value was small enough that Maxwell could safely put in the term without contradicting the results of any known experiment. With the new "displacement current" term, Maxwell's equations were complete. The subjects of electricity and magnetism were fully unified by means of an elegant and consistent description of electromagnetic phenomena.

The consequences of the new description could now be explored. It soon turned out that the equations, with the new term included, had "wave solutions"—electromagnetic waves of all frequencies were generated in a calculable way by accelerating electric charges. In a vacuum, the waves would all travel at the same speed. Computing that speed, Maxwell found it to be identical, within the margin of error that prevailed at the time, to the famous speed of light, about 186,000 miles

In notation somewhat like that used in undergraduate textbooks of today:

$$\nabla \cdot \mathbf{E} = 4\pi\rho \qquad (1)$$

$$\nabla \cdot \mathbf{B} = 0 \qquad (2)$$

$$\nabla \times \mathbf{E} + \frac{1}{c}\dot{\mathbf{B}} = 0 \qquad (3)$$

$$\nabla \times \mathbf{B} - \frac{1}{c}\dot{\mathbf{E}} = \frac{4\pi}{c}\mathbf{j} \qquad (4)$$

In less compressed notation like that used when Maxwell began his work:

$$\frac{\partial E_x}{\partial x} + \frac{\partial E_y}{\partial y} + \frac{\partial E_z}{\partial z} = 4\pi\rho \qquad (1)$$

$$\frac{\partial B_x}{\partial x} + \frac{\partial B_y}{\partial y} + \frac{\partial B_z}{\partial z} = 0 \qquad (2)$$

$$\left.\begin{aligned} \frac{\partial E_y}{\partial x} - \frac{\partial E_x}{\partial y} + \frac{1}{c}\dot{B}_z = 0 \\ \frac{\partial E_z}{\partial y} - \frac{\partial E_y}{\partial z} + \frac{1}{c}\dot{B}_x = 0 \\ \frac{\partial E_x}{\partial z} - \frac{\partial E_z}{\partial x} + \frac{1}{c}\dot{B}_y = 0 \end{aligned}\right\} \qquad (3)$$

$$\left.\begin{aligned} \frac{\partial B_y}{\partial x} - \frac{\partial B_x}{\partial y} - \frac{1}{c}\dot{E}_z = \frac{4\pi}{c}j_z \\ \frac{\partial B_z}{\partial y} - \frac{\partial B_y}{\partial z} - \frac{1}{c}\dot{E}_x = \frac{4\pi}{c}j_x \\ \frac{\partial B_x}{\partial z} - \frac{\partial B_z}{\partial x} - \frac{1}{c}\dot{E}_y = \frac{4\pi}{c}j_y \end{aligned}\right\} \qquad (4)$$

In more compressed relativistic notation:

$$\partial_\nu F^{\mu\nu} = \frac{4\pi}{c} j^\mu \qquad \text{(1 and 4)}$$

$$\varepsilon^{\mu\nu\kappa\lambda}\,\partial_\nu F_{\kappa\lambda} = 0 \qquad \text{(2 and 3)}$$

Three ways of writing Maxwell's equations.

per second. Could light consist of electromagnetic waves in a certain band of frequencies? That conjecture had been made before, in a vaguer form, by Faraday, but it gained enormously in clarity and plausibility from Maxwell's work. Although it took years to prove by experiment, the idea was absolutely correct. Maxwell's equations also required the existence of waves of higher frequencies than those of visible light (what we now call ultra-violet rays, X-rays, etc.) and of lower frequencies than the visible (what we now call infrared rays, microwaves, radio waves, etc.). Eventually all those forms of electromagnetic radiation were discovered experimentally, not only confirming the theory but also leading to the extraordinary achievements in technology with which we are all familiar.

The Simplicity of Great Unifying Theories

Maxwell's equations describe the behavior of electromagnetism throughout the universe in a few lines. (The exact number depends on the compactness of the notation, as in the figure.) Given the boundary conditions and the charges and currents, the electric and magnetic fields can be calculated. The universal aspects of electromagnetism are captured by the equations—only the special details need to be supplied. The equations identify the regularities with precision and compress them into a tiny mathematical package of immense power. What could be a more elegant example of a schema?

As the length of the schema is practically zero, so too is the effective complexity, as we have defined it. In other words, the laws of electromagnetism are extremely simple.

A critic might complain that while Maxwell's equations are indeed short, it nevertheless takes some education to understand the notation in which they are expressed. When Maxwell first published the equations, he used a less compressed way of writing them than is used in teaching undergraduates today and the set of equations was somewhat longer. Correspondingly, we can now use a relativistic notation that makes them shorter. (Both the longer and the shorter versions are illustrated.) The critic might demand that in each case we include in the schema not only the equations but also an explanation of the notation.

This is not an unreasonable demand. We have already said, in connection with crude complexity, that it would be misleading to employ special language just to reduce the length of a description. In fact, the underlying mathematics of Maxwell's equations is not particularly difficult to explain, but even if that were not the case, it would still require only a finite amount of interpretation. This pales into insignificance when we consider that the equations apply to all electric and magnetic fields, everywhere in the universe. The compression achieved is still enormous.

Universal Gravitation—Newton and Einstein

Another remarkable universal law is that of gravitation. Isaac Newton developed the first version, which was followed two and a half centuries later by the more accurate general-relativistic gravitational theory of Albert Einstein.

Newton gained his brilliant insight into the universality of gravitation when he was a young man of 23. In 1665 Cambridge University was forced to close because of the plague and Newton, a fresh B.A., returned to his family home in Woolsthorpe, Lincolnshire. There, in 1665 and 1666, he began to develop the integral and differential calculus, as well as the law of gravitation and his three laws of motion. In addition, he carried out the famous experiment with a prism, showing that white light is made up of the colors of the rainbow. Each one of those pieces of work was a major innovation, and though historians of science like to emphasize nowadays that Newton didn't complete them all in one *annus mirabilis,* or marvelous year, they still admit that he made a good start on each of them during that time. As my wife, the poet Marcia Southwick, likes to say, he could have written an impressive essay entitled "What I Did During My Vacation."

Legend associates Newton's conception of a universal law of gravitation with the fall of an apple. Was there really such an incident? Historians of science are not sure but they do not rule it out, because there are four independent sources that refer to it. One of them, Conduitt, wrote:

> In the year 1666 he retired again from Cambridge . . . to his mother in Lincolnshire & whilst he was musing in a garden it came into his thought that the power of gravity (w[ch] brought an apple from the tree to the ground) was not limited to a certain distance from the earth but that this power must extend much farther than was usually thought. Why not as high as the moon said he to himself & if so that must influence her motion & perhaps retain her in orbit, whereupon he fell a calculating what would be the effect of that supposition but being absent from books & tak-

> ing the common estimate in use among Geographers & our seamen before Norwood had measured the earth, that 60 English miles were contained in one degree of latitude on the surface of the Earth, his computation did not agree with his theory & inclined him then to entertain the notion that together with the force of gravity there might be a mixture of that force wch the moon would have if it was carried along in a vortex . . .

In this story we see a number of processes at work that may occur from time to time in the life of a theoretical scientist. An idea strikes at an odd moment. It makes possible a connection between two sets of phenomena previously thought to be distinct. A theory is then formulated. Some of its consequences can be predicted; in physics the theorist "falls a calculating" in order to make those predictions. The predictions may fail to agree with observation even though the theory is correct, either because of an error in the reported observations (as in Newton's case) or because the theorist has made a conceptual or mathematical mistake in applying the theory. The theorist may then modify the correct theory (with its simplicity and elegance) and cobble together a more complicated one to accommodate the error. Witness the bit at the end of the Conduitt quotation about the messy "vortex" force that Newton thought of adjoining to the force of gravity!

After a long while, the discrepancy between theory and observation was straightened out and Newton's theory of universal gravitation was accepted, until it was replaced in about 1915 by Einstein's general-relativistic theory, which agrees exactly with Newton's in the limit in which all bodies are moving very slowly compared with the speed of light. In the solar system, planets and satellites move at speeds on the order of ten miles per second, while the speed of light is about 186,000 miles per second. The Einsteinian corrections to Newton's theory are therefore very tiny and have so far been detectable only in a very few observations. In all the tests that have been conducted, Einstein's theory checks out.

The replacement of an excellent theory by an even better one is described in a particular way by Thomas Kuhn in his book *The Structure of Scientific Revolutions*, and his point of view has become extremely influential. He pays special attention to "paradigm shifts," using "paradigm" in a rather special sense (some might say he is misusing the word!). His approach emphasizes changes in matters of principle as the improved theory takes over.

In the case of gravitation, Kuhn might point to the fact that Newton's theory employs "action at distance," that is, a gravitational force that acts instantaneously, whereas in Einstein's theory gravitational influences propagate, like electromagnetic ones, at the speed of light. In Newton's nonrelativistic theory, space and time are treated as separate and absolute, and gravitation is not connected with geometry; in Einstein's theory space and time are mixed together (as they always are in relativistic physics), and Einsteinian gravitation can be regarded as intimately connected with the geometry of space-time. Also, general relativity, unlike Newtonian gravitation, is based on the principle of equivalence, which states that it is impossible to distinguish locally between a gravitational field and an accelerated frame of reference (as in an elevator); the only thing one can feel or measure locally is the difference between one's acceleration and the local acceleration of gravity.

The paradigm shift approach is concerned with such profound differences in philosophy and language between an old theory and a new one. Kuhn does not like to stress the fact (although he mentions it, of course) that the old theory may still provide a sufficiently good approximation for calculations and predictions in the domain for which it was developed (in this case the limit of weak fields and very low relative velocities). I should like to call special attention to that feature, however, in order to point out that in the competition of schemata in the scientific enterprise, the triumph of one schema over another does not necessarily mean that the loser is abandoned and forgotten. In fact, it may be utilized far more often than its more accurate and sophisticated successor. That is certainly true for Newtonian versus Einsteinian mechanics of the solar system. Victory in the competition between scientific theories may be more a matter of demotion of the old theory and promotion of the new one over its head than a matter of the death of the vanquished. (Of course, it does often happen instead that the old theory no longer has any value at all, and then it is mainly historians of science who bother to discuss it.)

Einstein's general-relativistic equation

$$G_{\mu\nu} = 8\pi\kappa T_{\mu\nu}$$

does for gravitation what Maxwell's equations do for electromagnetism. The left-hand side of the equation relates to the curvature of space-time (and thus to the gravitational field) and the right-hand side to the energy density, etc., of matter other than the gravitational field. It expresses in a single short formula the universal features of gravitational fields throughout the cosmos. Given the masses, positions, and velocities of matter, one can calculate the gravitational field (and thus the effect of gravitation on the motion of a test body) at any place and time. We are dealing with a remarkably powerful schema, which has compressed into a brief message the general properties of gravitation everywhere.

Again, a critic might demand that we include as part of the schema not only the formula but also an explanation of the symbols. My father, who struggled as an educated layman to understand Einstein's theory, used to say, "Look how simple and beautiful this theory is, but what are $T_{\mu\nu}$ and $G_{\mu\nu}$?" As in the case of electromagnetism, even if a course of study is included in the schema, Einstein's equation is still a bargain in terms of compression, since it describes the behavior of all gravitational fields everywhere. The schema is still remarkably short, and its complexity low. Hence, Einstein's general-relativistic theory of gravitation is simple.

CHAPTER

8

THE POWER OF THEORY

The mind-set of the theoretical scientist is useful not only for probing the ultimate secrets of the universe but for many other tasks as well. All around us are facts that are related to one another. Of course, they can be regarded as separate entities and learned that way. But what a difference it makes when we see them as part of a pattern! Many facts then become more than just items to be memorized—their relationships permit us to use a compressed description, a kind of theory, a schema, to apprehend and remember them. They begin to make some sense. The world becomes a more comprehensible place.

Pattern recognition comes naturally to us humans; we are, after all, complex adaptive systems ourselves. It is in our nature, by biological inheritance and also through the transmission of culture, to see patterns, to identify regularities, to construct schemata in our minds. However, those schemata are often promoted or demoted, accepted or rejected, in response to selection pressures that are far different from those operating in the sciences, where agreement with observation is so critical.

Unscientific approaches to the construction of models of the world around us have characterized much human thinking since time immemorial, and they are still widespread. Take, for example, the version of sympathetic magic based on the idea that similar things must be connected. It seems natural to many people around the world that, when in need of rain, they should perform a ceremony in which water is procured from a special place and poured on the ground. The similarity between the action and the desired phenomenon suggests that there should be a causal connection. The selection pressures that help to maintain such a belief do not include objective success, the criterion that is applied in science (at least when science is working properly). Instead, other kinds of selection are operating. For instance, there may be powerful individuals who carry out the ceremony and encourage the belief in order to perpetuate their authority.

The same society may also be familiar with sympathetic magic that works through an effect on people, say eating a lion's heart to increase the bravery of a warrior. Now that can achieve some objective success, simply by psychological means: if a man believes he has eaten something that will increase courage, that may give him the confidence to act bravely. Similarly, destructive witchcraft (whether based on sympathetic magic or not) can be objectively successful if the victim believes in it and knows that it is being practiced. Say that you want me to suffer pain and you make a wax effigy of me, with some hair and nail parings of mine embedded in it, and you then stick pins into the effigy. If I believe, even a little, in the efficacy of such magic and I know you are engaged in it, I may feel pain in the appropriate places and become ill (and, in an extreme case, perhaps even die) through psychosomatic effects. The occasional (or frequent!) success of sympathetic magic in such cases may then encourage the belief that such magic works even when, as in the ceremony for the production of rain, it cannot achieve objective success except by chance.

We shall return to the subject of unscientific models, as well as the many reasons for their appeal, in the chapter on superstition and skepticism, but for now we are concerned with the value of theorizing *in the scientific manner* about the world around us, seeing how connections and working relationships fall into place.

"Merely Theoretical"

Many people seem to have trouble with the idea of theory because they have trouble with the word itself, which is commonly used in two quite distinct ways. On the one hand, it can mean a coherent system of rules and principles, a more or less verified or established explanation accounting for known facts or phenomena. On the other hand, it can refer to speculation, a guess or conjecture, or an untested hypothesis, idea, or opinion. Here the word is used with the first set of meanings, but many people think of the second when they hear "theory" or "theoretical." One of my colleagues on the board of directors of the John D. and Catherine T. MacArthur Foundation is likely to comment, when some fairly bold research project is proposed for funding, "I think we should take a chance and support that, but let's be careful not to spend money on anything *theoretical.*" To a professional theorist those ought to be fighting words, but I understand that he and I use theoretical in different senses.

Theorizing about Place Names

It can be useful to theorize on almost any aspect of the world around us. Take place names. In California, people familiar with the Spanish place names along the coast are not surprised that many of them relate to the Roman Catholic religion, of which the Spanish explorers and settlers were usually devout communicants. However, I believe few people ask why each place has been given its particular name. Yet it is not unreasonable to inquire whether there might be some system behind the giving of saints' names like San Diego, Santa Catalina, and Santa Bárbara, as well as other religious names such as Concepción (Conception) or Santa Cruz (Holy Cross), to islands, bays, and points of land along the coast. A hint is available when we notice on the map Point Año Nuevo (New Year). Could other names also refer to days of the year? Of course! In the Roman Catholic calendar we find, in addition to the New Year on January 1, the following:

San Diego	(Saint Didacus of Alcalá de Henares) November 12
Santa Catalina	(Saint Catherine of Alexandria) November 25
San Pedro	(Saint Peter of Alexandria) November 26
Santa Bárbara	(Saint Barbara, Virgin and Martyr) December 4

San Nicolás	(Saint Nicholas of Myra) December 6
La Purísima Concepción	(The Immaculate Conception) December 8

Perhaps on a voyage of discovery the geographical features were named in order from southeast to northwest, according to the days on which they were sighted. Sure enough, scholars have verified from the historical record that in 1602 the explorer Sebastián Viscaíno named San Diego Bay on November 12, Santa Catalina Island on November 25, San Pedro Bay on November 26, the bay at Santa Barbara on December 4, San Nicolas Island on December 6, and Point Concepcion on December 8. Point Año Nuevo was apparently the first point sighted in the new year 1603, although on January 3 rather than New Year's Day. On January 6, the day of the Three Kings, Viscaíno named Point Reyes (Kings).

The theory works, but is it general? What about Santa Cruz? The day of the Holy Cross is September 14, which doesn't fit in with the sequence. Was it named on a different voyage of discovery? The schema is beginning to acquire a little complexity. In fact, many of the Spanish religious names along the coast correspond to dates on just a few voyages of discovery, so the effective complexity is not that great.

In this kind of theorizing, the construction of rough schemata to describe the results of human activity, arbitrary exceptions may be encountered, which fortunately do not plague schemata such as Maxwell's equations for electromagnetism. San Quentin, for example, located north of San Francisco and well known for its state prison, sounds as if it might have been named by some early Spanish explorer on Saint Quentin's Day. However, research by place-name experts reveals that the "San" was added in error to the earlier name Quentin, for the Spanish Quintín, the name of an Indian chief who was captured there in 1840.

Empirical Theory—Zipf's Law

In the place-name example, theorizing has led not only to the identification of regularities but also to a plausible explanation of them and a confirmation of that explanation. That is the ideal situation. Often, however, we encounter less than ideal cases. We may find regularities, predict that similar regularities will occur elsewhere, discover that the prediction is confirmed, and thus identify a robust pattern: however, it may be a pattern for which the explanation continues to elude us. In such a case we speak of an "empirical" or "phenomenological" theory, using fancy words to mean basically that we see what is going on but do not yet understand it. There are many such empirical theories that connect together facts encountered in everyday life.

Suppose we pick up a book of statistical facts, like the *World Almanac*. Looking inside, we find a list of U.S. metropolitan areas in order of decreasing population, together with the population figures. There may also be corresponding lists for the cities in individual states and in other countries. In each list every city can be assigned a rank, equal to 1 for the most populous city, 2 for the next most populous, and so on. Is there a general rule for all these lists that describes how the population decreases as the rank increases? Roughly speaking, yes. With fair accuracy, the population is inversely proportional to the rank; in other words, the successive populations are roughly proportional to 1, 1/2, 1/3, 1/4, 1/5, 1/6, 1/7, 1/8, 1/9, 1/10, 1/11, and so on.

Now let us look at the list of the largest business firms in decreasing order of volume of business (say the monetary value of sales during a given year). Is there an approximate rule that describes how the sales figures of the firms vary with their ranks? Yes, and it is the same rule as for populations. The volume of business is approximately in inverse proportion to the rank of the firm.

How about the exports from a given country in a given year in decreasing order of monetary value? Again, we find the same rule is a fair approximation.

An interesting consequence of that rule is easily verified by perusing any of the lists mentioned, for example a list of cities with their populations. First let us look at, say, the third digit of each population figure. As expected, the third digit is randomly distributed; the numbers of 0s, 1s, 2s, 3s, etc. in the third place are all roughly equal. A totally different situation obtains for the distribution of first digits, however. There is an overwhelming preponderance of 1s, followed by 2s, and so forth. The percentage of population figures with initial 9s is extremely small. That behavior of the first digit is predicted by the rule, which , if

Rank n	City	Population (1990)	Unmodified Zipf's law 10,000,000 divided by n	Modified Zipf's law 5,000,000 divided by $(n-2/5)^{3/4}$
1	New York	7,322,564	10,000,000	7,334,265
7	Detroit	1,027,974	1,428,571	1,214,261
13	Baltimore	736,014	769,231	747,639
19	Washington, D.C.	606,900	526,316	558,258
25	New Orleans	496,938	400,000	452,656
31	Kansas City, Mo.	434,829	322,581	384,308
37	Virginia Beach, Va.	393,089	270,270	336,015
49	Toledo	332,943	204,082	271,639
61	Arlington, Texas	261,721	163,934	230,205
73	Baton Rouge, La.	219,531	136,986	201,033
85	Hialeah, Fla.	188,008	117,647	179,243
97	Bakersfield, Calif.	174,820	103,093	162,270

Populations of U.S. cities from the *1994 World Almanac* compared with Zipf's original law and a modified version of it.

exactly obeyed, would give a proportion of initial 1s to initial 9s of 45 to 1.

What if we put down the World Almanac and pick up a book on secret codes, containing a list of the most common words in a certain kind of English text arranged in decreasing order of frequency of occurrence? What is the approximate rule for the frequency of occurrence of each word as a function of its rank? Again, we encounter the same rule, which works for other languages as well.

Many of these relationships were noticed in the early 1930s by a certain George Kingsley Zipf, who taught German at Harvard, and they are all aspects of what is now called Zipf's law. Today, we would say that Zipf's law is one of many examples of so-called scaling laws or power laws, encountered in many places in the physical, biological, and behavioral sciences. But in the 1930s such laws were still something of a novelty.

In Zipf's law the quantity under study is inversely proportional to the rank, that is, proportional to 1, 1/2, 1/3, 1/4, etc. Benoit Mandelbrot has shown that a more general power law (nearly the most general) is obtained by subjecting this sequence successively to two kinds of modification. The first alteration is to add a constant to the rank, giving 1/(1 + constant), 1/(2 + constant), 1/(3 + constant), 1/(4 + constant), etc. The further change allows, instead of these fractions, their squares or their cubes or their square roots or any other powers of them. The choice of the squares, for instance, would yield the sequence $1/(1 + \text{constant})^2$, $1/(2 + \text{constant})^2$, $1(3 + \text{constant})^2$, $1(4 + \text{constant})^2$, etc. The power in the more general power law is 1 for Zipf's law, 2 for the squares, 3 for the cubes, 1/2 for the square roots, and so on. Mathematics gives a meaning to intermediate values of the power as well, such as 3/4 or 1.0237. In general, we can think of the power as 1 plus a second constant. Just as the first constant was added to the rank, so the second one is added to the power. Zipf's law is then the special case in which those two constants are zero.

Mandelbrot's generalization of Zipf's law is still very simple: the additional complexity lies only in the introduction of the two new adjustable constants, a number added to the rank and a number added to the power 1. (An adjustable constant, by the way, is called a "parameter," a word that has been widely misused lately, perhaps under the influence of the somewhat similar word "perimeter." The modified power law has two additional parameters.) In any given case, instead of comparing data with Zipf's original law, one can introduce those two constants and adjust them for an optimal fit to the data. We can see in the chart on page 94 how a slightly modified version of Zipf's law fits some population data significantly better than Zipf's original rule (with both constants set equal to zero), which already works fairly well. "Slightly modified" means that the new constants have rather small values in the altered power law used for the comparison. (The constants in the chart were chosen by mere inspection of the data. An optimal fit would have yielded even better agreement with the actual populations.)

When Zipf first described his law, at a time when very few other scaling laws were known, he tried to make an important issue of how his principle distinguished the behavioral from the physical sciences, where such laws were supposedly absent. Today, after so many power laws have been discovered in physics, those remarks tend to detract from

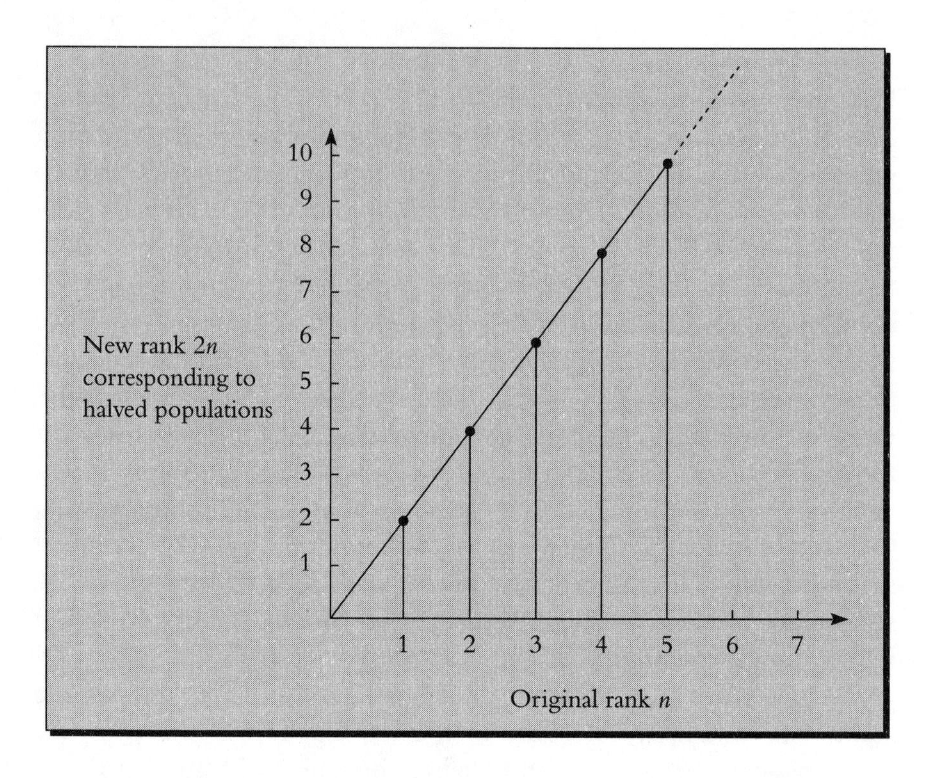

How a power law (in this case Zipf's original law) exhibits scaling.

Zipf's reputation rather than enhance it. Another circumstance is said to have worked against his reputation as well, namely that he indicated a certain sympathy with Hitler's territorial rearrangements of Europe, perhaps justifying his attitude by arguing that those conquests tended to make the populations of European countries conform more closely to Zipf's law.

Whether the story is true or not, it teaches an important lesson about the applications of behavioral science to policy: just because certain relationships tend to occur, that doesn't mean they should necessarily be regarded as always desirable. I encountered this issue at a recent Aspen Institute seminar, where I mentioned the tendency of distributions of wealth or income to follow scaling laws under certain conditions. I was immediately asked whether such a situation should be regarded as a good thing. As I remember, I shrugged my shoulders. After all, the steepness of the distribution, which determines the sharpness of the inequalities in wealth or income, depends on what power occurs in the law.

Zipf's law remains essentially unexplained, and the same is true of a great many other power laws. Benoit Mandelbrot, who has made really important contributions to the study of such laws (especially their connection to fractals), admits quite frankly that early in his career he was successful in part because he placed more emphasis on finding and describing the power laws than on trying to explain them. (In his book *The Fractal Geometry of Nature* he refers to his "bent for stressing consequences over causes.") He is quick to point out, however, that in some fields, especially in the physical sciences, quite convincing explanations have developed. For instance, the phenomenon of chaos in nonlinear dynamics is closely associated with fractals and power laws, in ways that are quite well understood. Also, Benoit has from time to time constructed models in which power laws occur. For instance, he calculated the word frequencies in texts typed by the proverbial monkeys. They obey a modified version of Zipf's law, with the power approaching 1 (Zipf's original value) as the number of symbols gets larger and larger. (He also noticed, by the way, that when word frequencies in actual texts written in natural languages are fitted by a modified Zipf's law, the power can differ significantly from 1, with the deviation depending on the richness of the vocabulary in the text in question.)

Scale Independence

In the last few years, increasing progress has been made toward an explanation of certain power laws. One such effort involves what is called "self-organized criticality," a concept proposed by the Danish theoretical physicist Per Bak, in collaboration with Chao Tang and Kurt Wiesenfeld. Their initial application of the idea was to sand piles, such as those we might see on a desert or a beach. The piles are roughly conical, and each has a fairly well-defined slope. If we examine those slopes, we observe that they are mostly the same. How does that come about? Suppose winds keep depositing additional sand grains on the piles (or a physicist in the laboratory keeps dribbling sand from a container onto experimental piles). As each pile builds up, its sides can become steeper, but only until the slope attains a critical value. Once that critical slope is reached, the addition of more sand starts to produce avalanches that reduce the height of the pile.

If the slope is greater than the critical value, an unstable situation results, in which avalanches of sand occur very readily and reduce the slope until it reverts to the critical value. Thus the sand piles are naturally "attracted" to the critical value of the slope, without any special external adjustment being necessary (hence "self-organized" criticality).

The size of an avalanche is usually measured by the number of sand grains that participate. Observation reveals that when the slope of the pile is near its critical value, the sizes of the avalanches obey a power law to a good approximation.

In this case the constant added to the Zipf's law power is very large. In other words, if the avalanches are assigned numerical ranks in order of size, then the number of participating sand grains decreases very rapidly with the rank. The distribution of avalanches in sand piles is an example of a power law that has been successfully studied by theoretical as well as experimental means. A numerical simulation of the avalanche process by Bak and his colleagues reproduced both the law and an approximate value of the large power.

Despite the sharp decline in size as rank increases, nearly all scales of avalanche size are present to some degree. In general, a power law distribution is one that is "scale-independent." That is why power laws are also called "scaling laws." But what exactly does it mean for a distribution law to be scale-independent?

The scale independence of power laws is well illustrated by Zipf's original rule, according to which populations of cities, for instance, are proportional to 1/1: 1/2: 1/3: 1/4: 1/5 . . . For the sake of simplicity, take them to be one million, one half million, one-third million, and so on. Let us multiply those populations by a fixed fraction, say 1/2; the new populations, in millions, become 1/2, 1/4, 1/6, 1/8, 1/10 . . . They are just the original populations previously assigned ranks 2, 4, 6, 8, 10 . . . So a reduction by a factor of two in all the populations is equivalent to doubling the ranks of the cities from the sequence 1, 2, 3, 4, . . . to the sequence 2, 4, 6, 8, . . . If the new ranks are plotted on a graph against the old ones, a straight line results, as in the diagram shown on page 96.

That straight line relationship can serve as the definition of a scaling law for any type of quantity: reduction of all the quantities by any constant factor (1/2 in the example) is equivalent to choosing new ranks in the original set of quantities, such that the new ranks graphed against the old ones yield a straight line plot. (The new ranks will not always come out whole numbers, but in every case the formula for size against rank will give a simple smooth curve that can be used to interpolate between whole numbers.)

In the case of sand pile avalanches, since their sizes are distributed according to a power law, a reduction of all the sizes by any common factor is equivalent to a simple reassignment of the ranks in the original sequence of avalanches. It is evident that in such a law no particular scale is being picked out, except at the two ends of the size spectrum, where the obvious limitations are encountered. No avalanche can have less than one grain participating; evidently, the power law must break down at the scale of a single grain. At the other end of the spectrum, no avalanche can be larger than the sand pile in question. But the largest avalanche is picked out anyway by having rank one.

Thinking about the largest avalanche recalls a frequent feature of power law distributions for sizes of events in nature. The largest or most catastrophic ones, with very low numerical ranks, even though they fall more or less on the curve dictated by the power law, can be regarded as individual historical events with a great many noteworthy consequences, while small events of very high numerical rank are usually considered merely from the statistical point of view. Huge earthquakes, registering around 8.5 on the Richter scale, are recorded in screaming newspaper headlines and in history books (especially if large cities are involved). The records of the multitudes of earthquakes of size 1.5 on the Richter scale languish in the data banks of seismologists, destined mostly for statistical studies. Yet the energy releases in earthquakes do obey a power law, discovered long ago by Charles Richter himself and his mentor Beno Gutenberg, both deceased Caltech colleagues. (Gutenberg was the professor who was so deep in conversation with Einstein about seismology one day in 1933 that neither of them noticed the Long Beach earthquake shaking the Caltech campus.) Similarly, the very small meteorites that are always striking the Earth are noted mainly in statistical surveys by specialists, while the huge collision that helped to cause the Cretaceous extinction sixty-five million years ago is considered a major individual event in the history of the biosphere.

Since power laws have been shown to operate in cases of self-organized criticality, the already popular expression "self-organized" has gained even greater currency, often coupled with the word "emergent." Scientists, including many members of the Santa Fe Institute family, are trying hard to understand the ways in which structures arise without the imposition of special requirements from the outside. In an astonishing variety of contexts, apparently complex structures or behaviors emerge from systems characterized by very simple rules. These systems are said to be self-organized and their properties are said to be emergent. The grandest example is the universe itself, the full complexity of which emerges from simple rules plus the operation of chance.

In many cases the study of emergent structures has been greatly facilitated by modern computers. The emergence of new features is often easier to follow by means of such machines than by writing equations on a piece of paper. The results are often especially striking in cases where a great deal of real time elapses during the course of emergence, because the computer can effectively speed up the process by a gigantic factor. Still, the computation may require a great many steps, and that raises a whole new issue.

Depth and Crypticity

In our discussions of complexity so far, we have considered compressed descriptions of a system or its regularities (or short computer programs for generating encoded descriptions), and we have related various kinds of complexity to the lengths of those descriptions or programs. However, we have paid little attention to the time, labor, or ingenuity necessary to accomplish the compression or to identify the regularities. Since the work of a theoretical scientist consists precisely of recognizing regularities and compressing their description into theories, we have effectively been setting at nought the value of a theorist's labor, clearly a monstrous crime. Something must be done to rectify that error.

We have seen that several different concepts are needed to capture adequately our intuitive notions of what complexity is. Now it is time to supplement our definition of effective complexity with definitions of other quantities, which will characterize how long it takes a computer to get from a short program to a description of a system and vice versa. (To some extent these quantities must resemble the computational complexity of a problem, which we defined earlier as the shortest time in which a computer can produce a solution.)

Such additional concepts have been studied by a number of people, but they are treated in an especially elegant way by Charles Bennett, a brilliant thinker at IBM, who is given time to generate ideas, publish them, and give talks about them here and there. I like to compare his peregrinations with those of a twelfth-century troubadour traveling from court to court in what is now the south of France. Instead of courtly love, Charlie "sings" of complexity and entropy, of quantum computers and quantum encipherment. I have had the pleasure of working with him in Santa Fe and also in Pasadena, where he spent a term visiting our group at Caltech.

Two particularly interesting quantities, labeled "depth" and "crypticity" by Charlie, are both related to computational complexity and have a reciprocal relation to each other. The study of each quantity throws light on the case of an apparently complex system that nevertheless has low algorithmic information content and low effective complexity because its description can be generated by a brief program. The catch lies in the answers to the following questions: 1) How laborious is it to go from the brief program or a highly compressed schema to a full-blown description of the system itself or of its regularities? 2) How laborious is it, starting with the system, to compress its description (or a description of its regularities) into a program or schema?

Very roughly, depth is a measure of the first kind of difficulty and crypticity is a measure of the second. Evidently, it is crypticity that is related to the value assignable to the work of a theorist (although a subtler description of theorizing might include an attempt to distinguish between the ingenious and the merely laborious).

A Hypothetical Example

To illustrate how a good deal of simplicity may be associated with great depth, let us return to Goldbach's conjecture that every even number greater than 2 is the sum of two prime numbers. As mentioned earlier, the conjecture has never been proved or disproved, but it has been verified for all even numbers up to some huge bound, set by the capabilities of the computer used and the patience of the investigators.

Previously we allowed ourselves to imagine that Goldbach's conjecture was technically undecidable (on the basis of the axioms of number theory) and therefore really true. This time let us imagine instead that Goldbach's conjecture is false. In that case, some gigantic even number g is the smallest even number greater than 2 that is *not* the sum of two prime numbers. That hypothetical number g has a very simple description, which we have just given. Likewise, there are very short programs for calculating it. For example, one can simply search methodically for larger and larger prime numbers and test Goldbach's conjecture on all even numbers up to 3 plus the largest prime found. In that way, the smallest even number *g* violating Goldbach's conjecture will eventually be discovered.

If Goldbach's conjecture really is false, it is likely that the running time for any such short program to find *g* is very long indeed. In this hypothetical case, then, the number *g* has a fairly low algorithmic information content and effective complexity but very considerable depth.

A Deeper Look at Depth

Charlie's technical definition of depth involves a computer, the same kind that we introduced in connection with algorithmic information content: an ideal all-purpose computer that can increase its memory capacity as much as it needs to at any time (or that has infinite memory to begin with). He starts with a message string of bits describing the system being studied. Then he considers not just the shortest program that will cause the computer to print out that string and then stop (as in the definition of algorithmic information content), but a whole range of fairly short programs that have the same effect. For each of these programs, he asks how much computing time is required to go from the program to the message string. Finally, he averages that time over the programs, using an averaging procedure that emphasizes the shorter programs.

Bennett has also recast the definition in a slightly different form, using Greg Chaitin's metaphor. Imagine that the proverbial monkeys are set to work typing not prose but computer programs. Let us concentrate our attention just on those rare programs that will cause the computer to print out our particular message string and then halt. Out of all such programs, what is the probability that the computer time required will be less than some particular time *T*? Call that probability *p*. The depth *d* will then be a particular kind of average of *T*, an average that depends on the curve of *p* versus *T*.

The illustration on page 103 gives a rough picture of how the probability *p* varies with the maximum permitted computer time *T*. When *T* is very small, it is very unusual for the monkeys to produce a program that will compute the desired result within such a short time,

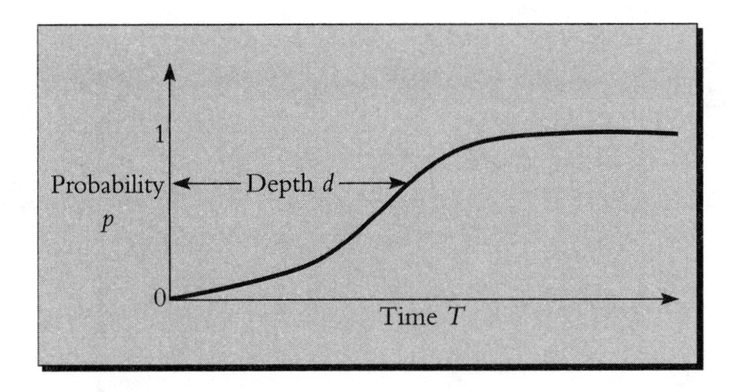

Depth as a rise time.

so p is near 0. When T is extremely long, the probability naturally approaches 1. The depth d is, roughly speaking, the rise time of the curve of T against p. It tells us what maximum permitted running time T is required in order to pick up a good fraction of the programs that will cause the computer to print out the message string and then halt. Depth is thus a crude measure of how long it takes to generate the string.

When a system occurring in nature has a great deal of depth, that is an indication that it took a long time to evolve or that it stems from something that took a long time to evolve. People who show an interest in the conservation of nature or in historic preservation are trying to protect both depth and effective complexity, as manifested in natural communities or in human culture.

But depth, as Charlie has shown, tends to communicate itself to *by-products* of long processes of evolution. We can find evidence of depth not only in today's life forms, including humans; in glorious works of art produced by human hands; and in fossilized remains of dinosaurs or Ice Age mammals; but also in a beer can pull tab on a beach or in the graffiti spray-painted on a canyon wall. Preservationists need not defend all manifestations of depth.

Depth and AIC

Although depth is an average of running time over program lengths, with the average weighted so as to emphasize the shorter programs, we

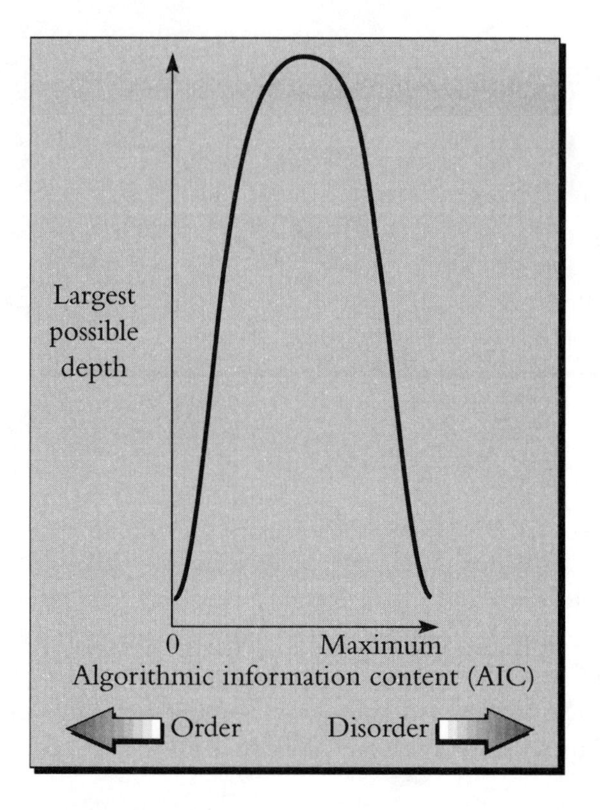

Largest possible depth crudely sketched as a function of algorithmic information content.

can often get a good idea of depth by looking at the running time for the shortest program. Suppose, for example, that the message string is completely regular, with algorithmic information content near zero. Then the running time of the shortest program is not very long—the computer doesn't have to do much "thinking" to execute a program such as "PRINT twelve trillion zeroes" (of course, the printing may take some time if the printer is slow). If the algorithmic information content is very low, the depth is low.

What about a random string, which has maximal algorithmic information content for a given message length? To go from the shortest program—PRINT followed by the string—to actually printing out the string will again require no "thinking" on the part of the computer. So the depth is low when the algorithmic information content is maximal, as well as when it is very small. The situation bears some resemblance to the way the maximum effective complexity varies with algorithmic information content, as in the sketch on page 59. Here, we can see very roughly how the maximum depth varies with algorithmic information content. It is low right near both ends but can be sizable anywhere in between, in the intermediate region between order and disorder. Of course, in that in-between region, the depth doesn't *have* to be large.

Note that this figure has a different form from the one on page 59. Even though both are just crude sketches, they show that depth can be great even for values of AIC fairly close to complete order or complete disorder, where effective complexity is still small.

Crypticity and Theorizing

The definition of crypticity refers to an operation that is the reverse of the one figuring in the definition of depth. The crypticity of a message string is the minimum time required for the standard computer, starting from the string, to find one of the shorter programs that will cause the machine to print out the string and then stop.

Suppose the message string results from the encoding of a data stream studied by a theorist. The crypticity of the string is then a rough measure of the difficulty of the theorist's task, which is not so very different from that of the computer in the definition. The theorist identifies as much regularity as possible, in the form of mutual information relating different parts of the stream, and then constructs hypotheses, as simple and coherent as possible, to explain the observed regularities.

Regularities are the compressible features of the data stream. They stem partly from the fundamental laws of nature and partly from particular outcomes of chance events that could have turned out otherwise. But the data stream also has random features, which arise from chance events that have *not* resulted in regularities. Those features are incompressible. Thus the theorist, in compressing the regularities of the data stream as much as is practicable, is at the same time discovering a concise description of the entire stream, a description made up of compressed regularities and incompressible random supplementary information. Similarly, a brief program that causes the computer to print out the message string (and then stop) may be arranged to consist of a basic program, which describes the regularities of the string, supplemented by inputs of information describing specific accidental circumstances.

Although our discussions of theory have only scratched the surface of the subject, we have already mentioned theorizing about place names, about empirical formulae for statistics, about the heights of sand piles, and about classical electromagnetism and gravitation. While there is a great deal of formal similarity among these various kinds of theorizing, they involve discoveries at many different levels, among which it is useful to distinguish. Are the basic laws of physics being studied? Or approximate laws that apply to messy physical objects such as sand piles? Or rough but general empirical laws about human institutions like cities and business firms? Or specific rules, with many exceptions, about place names used by people in a particular geographical area? Clearly there are important differences in accuracy and in generality among the various kinds of theoretical principles. Frequently those differences are discussed in terms of which are more fundamental than others. But what does that mean?

CHAPTER 9

WHAT IS FUNDAMENTAL?

The quark and the jaguar find themselves almost at opposite ends of the scale of what is fundamental. Elementary particle physics and cosmology are the most basic scientific disciplines, whereas the study of very complex living things is much less basic, although obviously of the greatest importance. In order to discuss this hierarchy of the sciences, it is necessary to disentangle at least two different threads, one having to do with mere convention and the other with real relationships among the different subjects.

I have been told that the faculty of sciences of a French university used to discuss the business related to the various subjects in a fixed order: first mathematics, then physics, then chemistry, then physiology, and so on. It would seem that the concerns of the biologists must often have been somewhat neglected under that arrangement.

Similarly, in the will of the Swedish dynamite magnate Alfred Nobel, who established the Nobel prizes, the science prizes are listed with physics first, chemistry second, and physiology and medicine third. As a result, the physics prize is always awarded at the beginning of the ceremony in Stockholm. If there is just one physics prize winner and that winner is a married man, it is his wife who comes into dinner on the arm of the King of Sweden. (When my friend Abdus Salam, a citizen of Pakistan and a Muslim, received a share of the physics prize in 1979, he turned up in Sweden with his two wives, no doubt causing some problems of protocol to arise.) The winner or winners in chemistry rank second in protocol, and those in physiology and medicine third. Mathematics is omitted from Nobel's will for reasons that are not really understood. There is a persistent rumor that Nobel was angry with a Swedish mathematician, Mittag-Leffler, for stealing the affections of a woman, but, as far as I know, it is only a rumor.

This hierarchy of subjects can in part be traced to the nineteenth-century French philosopher Auguste Comte, who argued that astronomy was the most fundamental scientific subject, physics the second, and so forth. (He regarded mathematics as a logical tool rather than a science.) Was he right? And if so, in what sense? Here it is necessary to put aside questions of prestige and try to understand what such a hierarchy really means in scientific terms.

The Special Character of Mathematics

First of all, it is true that mathematics is not really a science at all, if a science is understood to be a discipline devoted to the description of nature and its laws. Mathematics is more concerned with proving the logical consequences of certain sets of assumptions. For this reason, it can be omitted altogether from the list of sciences (as it was from Nobel's will) and treated as an interesting subject in its own right (pure mathematics) as well as an extremely useful tool for science (applied mathematics).

Another way to look at mathematics is to regard applied mathematics as concerning itself with the structures that occur in scientific theory, while pure mathematics covers not only those structures but also all the ones that might have occurred (or might yet turn out to occur) in science. Mathematics is then the rigorous study of hypothetical worlds. From that point of view, it *is* a kind of science—the science of what might have been or might be, as well as what is.

Treated in that way, is mathematics then the most fundamental science? And what about the remaining subjects? What is meant by the statement that physics is more fundamental than chemistry or chemistry more fundamental than biology? What about the different parts of physics: aren't some more fundamental than others? In general, what makes one science more fundamental than another?

I suggest that science A is more fundamental than science B when

1. The laws of science A encompass in principle the phenomena and the laws of science B.
2. The laws of science A are more general than those of science B (that is, those of science B are valid under more special conditions than are those of science A).

If mathematics is considered to be a science, it is then, according to these criteria, more fundamental than any other. All conceivable mathematical structures lie within its province, while the ones that are useful in describing natural phenomena are only a tiny subset of those that are or may be studied by mathematicians. Through that subset, the laws of mathematics do cover all the theories used in the other sciences. But what about those other sciences? What are the relations among them?

Chemistry and the Physics of the Electron

When the remarkable English theoretical physicist Paul Adrien Maurice Dirac published his relativistic quantum-mechanical equation for the electron in 1928, he is said to have remarked that his formula explained most of physics and the whole of chemistry. Of course he was exaggerating. Still, we can understand what he meant, particularly regarding chemistry, which is mostly concerned with the behavior of objects such as atoms and molecules, themselves composed of heavy nuclei with light electrons moving around them. A great many of the phenomena of chemistry are governed largely by the behavior of the electrons as they interact with the nuclei and with one another through electromagnetic effects.

Dirac's equation, describing the electron in interaction with the electromagnetic field, gave rise within a very few years to a full-blown relativistic quantum-mechanical theory of the electron and electromagnetism. The theory is quantum electrodynamics, or QED, and it has been verified by observation to a huge number of decimal places in many experiments (and so fully deserves its abbreviation, which reminds some of us of our school days, when we used "QED" at the end of a mathematical proof to mean *quod erat demonstrandum*, Latin for "which was to be demonstrated").

QED does explain, in principle, a huge amount of chemistry. It is rigorously applicable to those problems in which the heavy nuclei can be approximated as fixed point particles carrying an electric charge. Simple extensions of QED permit the treatment of nuclear motions and nonzero size as well.

In principle, a theoretical physicist using QED can calculate the behavior of any chemical system in which the detailed internal structure of atomic nuclei is not important. Wherever calculations of such chemical processes are practical using justified approximations to QED, they are successful in predicting the results of observation. In most cases in fact, one particular well-justified approximation to QED will do. It is called the Schrödinger equation with Coulomb forces, and it is applicable when the chemical system is "nonrelativistic," meaning that the electrons as well as the nuclei are moving very slowly compared to the speed of light. That approximation was discovered in the very early days of quantum mechanics, three years before the appearance of Dirac's relativistic equation.

In order to derive chemical properties from fundamental physical theory, it is necessary, so to speak, to ask chemical questions of the theory. One must put into the calculation not only the basic equations but also the conditions that characterize the chemical system or process in question. For instance, the lowest energy state of two hydrogen atoms is the hydrogen molecule H_2. An important question in chemistry is the amount of binding energy in that molecule; that is, how much lower the energy of the molecule is than the sum of the energies of the two atoms of which it is composed. The answer can be calculated from QED. But first it is necessary to "ask the equation" about the properties of the lowest energy state of that particular molecule.

The conditions of low energy under which such chemical questions arise are not universal. In the center of the sun, at a temperature of tens of millions of degrees, hydrogen atoms would be ripped apart into their constituent electrons and protons. Neither atoms nor molecules are present there with any significant probability. There is, so to speak, no chemistry in the center of the sun.

QED meets the two criteria for being considered more fundamental than chemistry. The laws of chemistry can in principle be derived from QED, provided the additional information describing suitable chemical conditions is fed into the equations; moreover, those conditions are special—they do not hold throughout the universe.

Chemistry at Its Own Level

In practice, even with the aid of the largest and fastest computers available today, only the simplest chemical problems are amenable to actual calculation from basic physical theory. The number of such amenable problems is growing, but most situations in chemistry are still described using concepts and formulae at the level of chemistry rather than that of physics.

In general, scientists are accustomed to developing theories that describe observational results in a particular field without deriving them from the theories of a more fundamental field. Such a derivation, though possible in principle when the additional special information is supplied, is at any given time difficult or impossible in practice for most cases.

For example, chemists are concerned with different kinds of chemical bonds between atoms (including the bond between the two hydrogen atoms in a hydrogen molecule). In the course of their experience, they have developed numerous practical ideas about chemical bonds that enable them to predict the behavior of chemical reactions. At the same time, theoretical chemists endeavor to derive those ideas, as much as they can, from approximations to QED. In all but the simplest cases they are only partially successful, but they don't doubt that in principle, given sufficiently powerful tools for calculation, they could succeed with high accuracy.

Staircases (or Bridges) and Reduction

We are thus led to the common metaphor of different levels of science, with the most fundamental at the bottom and the least fundamental at the top. Non-nuclear chemistry occupies a level "above" QED. In very simple cases, an approximation to QED is used to predict directly the results at the chemical level. In most cases, however, laws are developed at the upper level (chemistry) to explain and predict phenomena at that level, and attempts are then made to derive those laws, as much as possible, from the lower level (QED). Science is pursued at both levels and in addition efforts are made to construct staircases (or bridges) between them.

The discussion need not be restricted to non-nuclear phenomena. Since QED was developed around 1930, it has been greatly generalized. A whole discipline of elementary particle physics has grown up. Elementary particle theory, on which I have worked for most of my life, has as its task the description not only of the electron and electromagnetism, but of all the elementary particles (the fundamental building blocks of all matter) and all the forces of nature. Elementary particle theory describes what goes on inside the atomic nucleus as well as among the electrons. Therefore, the relationship between QED and the part of chemistry that deals with electrons can now be regarded as a special case of the relationship between elementary particle physics (as a whole) at the more fundamental level and chemistry (as a whole, including nuclear chemistry) at the less fundamental one.

The process of explaining the higher level in terms of the lower is often called "reduction." I know of no serious scientist who believes that there are special chemical forces that do not arise from underlying physical forces. Although some chemists might not like to put it this way, the upshot is that chemistry is in principle derivable from elementary particle physics. In that sense, we are all reductionists, at least as far as chemistry and physics are concerned. But the very fact that chemistry is more special than elementary particle physics, applying only under the particular conditions that allow chemical phenomena to occur, means that information about those special conditions must be fed into the equations of elementary particle physics in order for the laws of chemistry to be derived, even in principle. Without that caveat, the notion of reduction is incomplete.

One lesson to be learned from all this is that, while the various sciences do occupy different levels, they form part of a single connected structure. The unity of that structure is cemented by the relations among the parts. A science at a given level encompasses the laws of a less fundamental science at a level above. But the latter, being more special, requires further information in addition to the laws of the former. At each level there are laws to be discovered, important in their own right. The enterprise of science involves investigating those laws at all levels, while also working, from the top down and from the bottom up, to build staircases between them.

Such considerations apply within physics, too. The laws of elementary particle physics are valid for all matter, throughout the universe, under all conditions. Nuclear physics, however, was not really applicable during the earliest moments of the expansion of the universe because the density was too high to allow separate nuclei, or even neutrons and protons, to form. Still, nuclear physics is crucial for understanding what goes on in the center of the sun, where thermonuclear reactions (something like those in a hydrogen bomb) are producing the sun's energy, even though the conditions there are too harsh for chemistry.

Condensed matter physics, which is concerned with systems such as crystals, glasses, and liquids, or superconductors and semiconductors, is likewise a very special subject, applicable only under the conditions (such as low enough temperature) that permit the existence of the structures that it studies. Only when those conditions are specified is condensed matter physics derivable, even in principle, from elementary particle physics.

The Information Required for Reduction of Biology

What about the relation to physics and chemistry of another level in the hierarchy, that of biology? Are there today, as there used to be in past centuries, serious scientists who believe that there are special "vital forces" in biology that are not physico-chemical in origin? There must be very few, if any. Virtually all of us are convinced that life depends in principle on the laws of physics and chemistry, just as the laws of chemistry arise from those of physics, and in that sense we are again reductionists of a sort. Still, like chemistry, biology is very much worth studying on its own terms and at its own level, even as the work of staircase construction goes on.

Moreover, terrestrial biology is extremely special, referring as it does to living systems on this planet, which may differ widely from many of the diverse complex adaptive systems that surely exist on planets revolving around distant stars in various parts of the universe. On some of those planets, perhaps the only complex adaptive systems are ones that we would not necessarily describe as alive if we encountered them. (To take a trivial example from science fiction, imagine a society composed of very advanced robots and computers descended from ones built long ago by an extinct race of beings that we would have described as "living" while they existed.) Even if we restrict our attention to "living" beings, however, many of them presumably exhibit very different properties from those on Earth. An enormous amount of specific additional information must be supplied, over and above the laws of physics and chemistry, in order to characterize terrestrial biological phenomena.

To begin with, many features common to all life on Earth may be the results of accidents that occurred early in the history of life on the planet but could have turned out differently. (Life forms for which those accidents did turn out differently may also have existed long ago on Earth.) Even the rule that genes must be made up of the four nucleotides abbreviated A, C, G, and T, which seems to be true of all life on our planet today, may not be universal on a cosmic scale of space and time. There may be many other possible rules, followed on other planets; and beings obeying other rules may also have lived on Earth a few billion years ago until they lost out to life based on the familiar A, C, G, and T.

Biochemistry—Effective Complexity Versus Depth

It is not only the particular set of nucleotides characterizing the DNA of all terrestrial life today that may or may not be unique. The same question is debated in connection with each general property that characterizes the chemistry of all life on Earth. Some theorists claim that the chemistry of life must take numerous forms on different planets scattered through the universe. The terrestrial case would then be the result of a large number of chance events, each having contributed to the remarkable regularities of biochemistry on Earth, which would thereby have acquired a good deal of effective complexity.

On the other side of the question are those who believe that biochemistry is essentially unique, that the laws of chemistry, based on the fundamental laws of physics, leave little room for a chemistry of life other than that found on Earth. The proponents of this view are saying in effect that going from the fundamental laws to the laws of biochemistry involves almost no new information, and thus contributes very little effective complexity. However, a computer might have to do a great deal of calculating to derive the near-uniqueness of biochemistry as a theoretical proposition from the fundamental laws of physics. In that case, biochemistry would still have a good deal of depth, if not much effective complexity. Another way to phrase the question about the near-uniqueness of terrestrial biochemistry is to ask whether biochemistry depends mainly on asking the right questions about physics or also depends in an important way on history.

Life: High Effective Complexity—Between Order and Disorder

Even if the underlying chemistry of terrestrial life depends little on history, there is still an enormous amount of effective complexity in biology, far more than in such subjects as chemistry or condensed matter physics. Consider the immense number of evolutionary changes that have occurred by chance during the four billion years or so since the origin of life on Earth. Some of those accidents (probably a small fraction, but still a great many) have played major roles in the subsequent history of life on this planet and in the character of the diverse life forms that enrich the biosphere. The laws of biology do depend on the laws of physics and chemistry, but they also depend on a vast amount of additional information about how those accidents turned out. Here, much more than in the case of nuclear physics, condensed matter physics, or chemistry, one can see a huge difference between the kind of reduction to the fundamental laws of physics that is possible in principle and the trivial kind that the word "reduction" might call up in the mind of a naive reader. The science of biology is very much more complex than fundamental physics because so many of the regularities of terrestrial biology arise from chance events as well as from the fundamental laws.

But even the study of complex adaptive systems of all kinds on all planets is still rather special. The environment must exhibit sufficient regularity for the systems to exploit for learning or adapting, but at the same time not so much regularity that nothing happens. For example, if the environment in question is the center of the sun, at a temperature of tens of millions of degrees, there is almost total randomness, nearly maximal algorithmic information content, and no room for effective complexity or great depth—nothing like life can exist. Nor can there be such a thing as life if the environment is a perfect crystal at a temperature of absolute zero, with almost no algorithmic information content and again no room for much effective complexity or great depth. For a complex adaptive system to function, conditions are required that are intermediate between order and disorder.

The surface of the planet Earth supplies an environment of intermediate algorithmic information content, where depth and effective complexity can both exist, and that is part of the reason why life has been able to evolve here. Of course, only very primitive forms of life evolved at first, under the conditions that prevailed on Earth several billion years ago, but then those living things themselves altered the biosphere, particularly by adding oxygen to the atmosphere, producing a situation more like the present one and permitting higher forms of life, with more complex organization, to evolve. Conditions in between order and disorder characterize not only the environment in which life can arise, but also life itself, with its high effective complexity and great depth.

Psychology and Neurobiology—The Mind and the Brain

Complex adaptive systems on Earth give rise to several levels of science that lie "above" biology. One of the most important is the psychology of animals, and especially of the animal with the most complex psychology, the human being. Here again, it must be a rare contemporary scientist who believes that there exist special "mental forces" that are not biological, and ultimately physicochemical, in nature. Again, virtually all of us are, in this sense, reductionists. Yet in connection with such subjects as psychology (and sometimes biology), one hears the word "reductionist" hurled as an epithet, even among scientists. (For example, the California Institute of Technology, where I have been a professor for almost forty years, is often derided as reductionist; in fact I may have used the term myself in deploring what I regard as certain shortcomings of our Institute.) How can that be? What is the argument really about?

The point is that human psychology—while no doubt derivable in principle from neurophysiology, the endocrinology of neurotransmitters, and so forth—is also worth studying at its own level. Many people believe, as I do, that when staircases are constructed between psychology and biology, the best strategy is to work from the top down as well as from the bottom up. It is this proposition that does not receive universal agreement, for example at Caltech, where very little research on human psychology takes place.

Where work does proceed on both biology and psychology and on building staircases from both ends, the emphasis at the biological end is on the brain (as well as the rest of the nervous system, the endocrine system, etc.), while at the psychological end the emphasis is on the mind—that is, the phenomenological manifestations of what the brain and related organs are doing. Each staircase is a brain–mind bridge.

At Caltech, it is mostly the brain that is studied. The mind is neglected, and in some circles even the word is suspect (a friend of mine calls it the M-word). Yet very important psychological research was carried out some years ago at Caltech, particularly the celebrated work of the psychobiologist Roger Sperry and his associates on the mental correlates of the left and right hemispheres of the human brain. They used patients in whom, as a result of accident or of surgery for epilepsy, the corpus callosum connecting the left and right brains had been cut. It was known that speech tends to be associated with the left brain, along with control of the right side of the body, while control of the left side of the body is usually associated with the right brain. They found, for example, that a patient whose corpus callosum has been cut can show inability to express verbally information relating to the left side of the body, while giving indirect evidence of possessing that information.

As Sperry became less active with increasing age, the research he had begun was continued at other institutions by his former students and post-docs and by many new recruits to the field. Further evidence was found that the left brain excels not only in verbal language but also in logic and analysis, while the right brain excels in nonverbal communication, in the affective aspects of language, and in such integrative tasks as face recognition. Some researchers have linked the right brain to intuition and to seeing the big picture. Unfortunately, popularization has exaggerated and distorted many of those results, and much of the resulting discussion has ignored Sperry's cautionary remark that "the two hemispheres in the normal intact brain tend regularly to function closely together as a unit. . . . " Nevertheless, what has been discovered is quite remarkable. I am particularly intrigued with the continuing investigation of the extent to which the claim is true that amateurs typically listen to music predominantly with the right brain, while professional musicians usually listen mostly with the left brain.

Concentration on Mechanism or Explanation—"Reductionism"

Why does so little research in psychology go on at Caltech today? Granted, the school is small and can't do everything. But why so little evolutionary biology? (I sometimes say in jest that a creationist institution could scarcely do less.) Why so little ecology, linguistics, or archaeology? One is led to suspect that these subjects have something in common that puts off most of our faculty.

The scientific research agenda at Caltech tends to favor the study of mechanisms, underlying processes, and explanations. Naturally I am sympathetic to that approach, since it is the one that characterizes elementary particle physics. Indeed, the emphasis on underlying mechanisms has led to many impressive successes in a variety of fields. T. H. Morgan was brought to Caltech to found the biology division in the 1920s when he was mapping the genes of the fruit fly, thus laying the groundwork for modern genetics. Max Delbrück, who arrived in the 1940s, became one of the founders of molecular biology.

If a subject is considered too descriptive and phenomenological, not yet having reached the stage where mechanisms can be studied, our faculty regards it as insufficiently "scientific." If Caltech had existed with those same proclivities in Darwin's time, would it have invited *him* to join the faculty? After all, he formulated his theory of evolution without many clues to the underlying processes. His writings indicate that if pressed to explain the mechanism of variation, he would probably have opted for something like the incorrect Lamarckian idea (Lamarckians believed that cutting the tails off mice for a few generations would produce a strain of tailless mice, or that the long necks of giraffes are explained by generations of ancestors stretching their necks to reach higher into the yellow thorn acacias.) Yet his contribution to biology was monumental. In particular, his theory of evolution laid the groundwork for the simple unifying principle of the common descent of all existing organisms from a single ancestor. What a contrast to the complexity of the previously widespread notion of the stability of species, each specially created by supernatural means.

Even if I agreed that subjects like psychology are not yet sufficiently scientific, my preference would be to take them up in order to participate in the fun of making them more scientific. In addition to favoring, as a general rule, the bottom-up method of building staircases between disciplines—from the more fundamental and explanatory toward the less fundamental—I would, in many cases (not just that of psychology), encourage a top-down approach as well. Such an approach begins with the identification of important regularities at the less fundamental level and defers until later understanding of the underlying, more fundamental mechanisms. But the atmosphere of the Caltech campus is permeated by a strong bias in favor of the bottom-up approach, which has produced most of the spectacular achievements responsible for the reputation of the institution. That bias is what invites the charge of reductionism, with its pejorative connotation.

Subjects such as psychology, evolutionary biology, ecology, linguistics, and archaeology all bear on complex adaptive systems. They are all studied at the Santa Fe Institute, where a great deal of emphasis is laid on the similarities among those systems and on the importance of studying their properties at their own levels, not just as consequences of more fundamental scientific disciplines. In that sense, the founding of the Santa Fe Institute is part of a rebellion against the excesses of reductionism.

Simplicity and Complexity from Quark to Jaguar

While I believe Caltech is making a serious mistake by neglecting most of the "sciences of complexity," I have been pleased by the support given there to elementary particle physics and cosmology, the most fundamental sciences of all, involving the search for the basic laws of the universe.

One of the great challenges of contemporary science is to trace the mix of simplicity and complexity, regularity and randomness, order and disorder up the ladder from elementary particle physics and cosmology to the realm of complex adaptive systems. We also need to understand how the simplicity, regularity, and order of the early universe gave rise over time to the intermediate conditions between order and disorder that have prevailed in many places in later epochs, making possible, among other things, the existence of complex adaptive systems such as living organisms.

To do this, we have to examine fundamental physics from the point of view of simplicity and complexity and ask what role is played by the unified theory of the elementary particles, the initial condition of the universe, the indeterminacies of quantum mechanics, and the vagaries of classical chaos in producing the patterns of regularity and randomness in the universe within which complex adaptive systems have been able to evolve.

PART II

THE QUANTUM UNIVERSE

CHAPTER

10

SIMPLICITY AND RANDOMNESS IN THE QUANTUM UNIVERSE

How do the fundamental laws of matter and the universe stand today? How much is well established and how much is conjecture? And how do those laws look with respect to simplicity and complexity or regularity and randomness?

The fundamental laws are subject to the principles of quantum mechanics, and at every stage of our thinking we will have to refer to the quantum approach. The discovery of quantum mechanics is one of the greatest achievements of the human race, but it is also one of the most difficult for the human mind to grasp, even for those of us who have used it daily in our work for decades. It violates our intuition—or rather our intuition has been built up in a way that ignores quantum-mechanical behavior. That circumstance makes it all the more necessary to explore the meaning of quantum mechanics, especially by examining some recently developed ways of thinking about it. It may then be easier to understand why our intuition seems to pay no attention to something so important.

The universe consists of matter, and matter is composed of many different kinds of elementary particles, such as electrons and photons. These particles lack individuality—every electron in the universe is identical to every other, and all photons are likewise interchangeable. However, any particle can occupy one of an infinite number of different "quantum states." There are two broad classes of particle. Fermions, such as electrons, obey the Pauli exclusion principle: no two particles of the same kind can occupy the same state at the same time. Bosons, such as photons, obey a kind of antiexclusion principle: two or more particles of the same kind exhibit a preference for being in the same state at the same time. (That property of photons makes possible the operation of the laser, where photons in a given state stimulate the emission of more photons in that same state. All those photons have the same frequency and travel in the same direction, forming the laser beam. "LASER" was originally an acronym for "light amplification by stimulated emission of radiation.")

Bosons, because of their proclivity for crowding together in the same quantum state, can build up their densities so that they behave almost like classical fields, such as those of electromagnetism and gravitation. Particles that are bosons can therefore be regarded as quanta —quantized packets of energy—of those fields. The quantum of the electromagnetic field is the photon. Similarly, theory requires the existence of the quantum of the gravitational field, a boson called the graviton. In fact, any fundamental force must be associated with an elementary particle that is the quantum of the corresponding field. Sometimes the quantum is said to "carry" the corresponding force.

When matter is described as being composed of elementary particles—that is, of fermions and bosons—it should be emphasized that under certain conditions some of the bosons may behave more like a field than like particles (for example, in the electric field surrounding a charge). Fermions too can be described in terms of fields; although those fields do not ever behave classically, they are nevertheless associated, in a sense, with forces.

All matter possesses energy, and all energy is associated with matter. When people refer carelessly to matter being converted into energy (or vice versa), they mean simply that certain kinds of matter and energy are being converted into other kinds. For example, an electron and a related (but oppositely charged) particle called a positron can come together and turn into two photons, a process often described as "annihilation" or even "annihilation of matter to give energy." However, it is merely the transformation of matter into other matter, of certain forms of energy into other forms.

The Standard Model

All the known elementary particles (except the graviton demanded by theoretical considerations) are provisionally described today by a theory that has come to be called the standard model. We shall look into it in some detail a little later on. It seems to be in excellent agreement with observation, although there are a few features that have not yet been confirmed by experiment. Physicists had hoped that those would be tested (along with exciting newer ideas that go beyond the standard model) at the high energy particle accelerator that was partially completed in Texas (the "superconducting supercollider" or "SSC"). But that was scrapped by the U.S. House of Representatives in a conspicuous setback for human civilization. Now the only hope for verification of fundamental theoretical ideas lies in the lower energy accelerator facility being created (by converting an existing machine) at CERN near Geneva, Switzerland. Unfortunately, its energy may be too low.

Those of us who helped put together the standard model are naturally rather proud of it, since it brought a good deal of simplicity out of a bewildering variety of phenomena. However, there are a number of reasons why it cannot be the ultimate theory of the elementary particles.

First, the forces have very similar forms and cry out for unification by means of a theory in which they appear as different manifestations of the same underlying interaction; yet in the standard model they are treated as different from one another and not unified (contrary to claims that are sometimes made). Second, the model is not yet simple enough; it contains more than sixty kinds of elementary particles and a number of interactions among them, but no explanation for all that variety. Third, the model contains more than a dozen arbitrary constants describing those interactions (including the constants that yield the various masses of the different kinds of particles); it is hard to accept as fundamental a theory in which so many important numbers are uncalculable in principle. Finally, gravitation is not included, and any attempt to incorporate it in a straightforward manner leads to disastrous difficulties: the results of calculations of physical quantities turn out to include infinite corrections, rendering them meaningless.

So-called Grand Unified Theories

Elementary particle theorists have tried to cope with these defects in two ways. The more straightforward way involves generalizing the standard model to what some have called a "grand unified theory," although there is little justification for that name. Let us see what such a generalization does about the four problems listed.

First, the interactions in the standard model that required unification are in fact seen to be unified, along with new ones, at very high energies, with a natural description of how, at the lower energies of today's experiments, the interactions appear to be separate. Second, all the elementary particles of the theory are grouped into just a few sets, with the members of each set closely related; thus a good deal of simplification is achieved, even though the number of kinds of particles is substantially increased (with some of the new ones having masses so high that they cannot be observed in the foreseeable future). Third, the theory contains even more arbitrary constants than the standard model, still uncalculable in principle. Finally, gravitation is still omitted, and it is just as difficult as before to incorporate it.

Such a theory may possibly be approximately valid over a wide range of energies. The third and fourth points, however, make clear that it is not credible as the fundamental theory of the elementary particles.

Einstein's Dream

The search for that fundamental unified theory leads to the second way of transcending the standard model. The quest recalls Einstein's dream of a field theory that would unify in a natural way his general-relativistic theory of gravitation and Maxwell's theory of electromagnetism. In his old age, Einstein published a set of equations that claimed to accomplish that task, but unfortunately their appeal was purely mathematical—they did not describe plausible physical interactions of gravitation and electromagnetism. The greatest physicist of modern times had lost his powers. In 1979, at the celebration in Jerusalem of the hundredth anniversary of Einstein's birth, I deplored the fact that a special commemorative coin had been struck with those wrong equations on the reverse—what a shame for a scientist who had produced such beautiful, correct, and crucially important equations when he was younger. I am likewise disturbed that so many pictures and statues of Einstein (such as the sculpture at the headquarters of the National Academy of Sciences in Washington) show him in old age when he was no longer making important contributions and not as the rather handsome, well-dressed younger man who had made all the remarkable discoveries.

Einstein's attempt at a unified field theory was doomed not only by the general decline of his skills, but also by specific flaws in his approach. Among other things, he ignored three important features of the problem:

The existence of other fields besides the gravitational and electromagnetic (Einstein knew in a general way that there must be other forces but did not attempt to describe them).

The need to discuss not only the fields that quantum theory reveals as being composed of bosons like the photon and graviton, but also fermions (Einstein believed that the electron, for example, would somehow emerge from the equations).

The need to construct a unified theory within the framework of quantum mechanics (Einstein never accepted quantum mechanics, although he had helped to lay the groundwork for it).

Still, we theoretical physicists have been inspired by Einstein's dream, in a modern form: a unified quantum field theory embracing not only the photon, the graviton, and all the other fundamental bosons, with their associated electromagnetic, gravitational, and other fields, but also the fermions such as the electron. Such a theory would be contained in a simple formula explaining the great multiplicity of elementary particles and their interactions and yielding, in the appropriate approximations, Einstein's equation for general-relativistic gravitation and Maxwell's equations for electromagnetism.

Superstring Theory—The Dream Perhaps Realized

Now that dream may have been realized. A new type of theory called "superstring" theory seems to have the right properties for accomplishing the unification. In particular, "heterotic superstring theory" is the first viable candidate ever for a unified quantum field theory of all the elementary particles and their interactions.

Superstring theory grew out of an idea called the bootstrap principle after the old saw about the man who could pull himself up by his own bootstraps. The notion was that a set of elementary particles could be treated as if composed in a self-consistent manner of combinations of those same particles. All the particles would serve as constituents, all the particles (even the fermions in a certain sense) would serve as quanta for force fields binding the constituents together, and all the particles would appear as bound states of the constituents. When I tried, many years ago, to describe the idea to an audience at the Hughes Aircraft Company, the engineer who then headed the synchronous satellite program, Harold Rosen, asked me whether it was anything like what he and his team had found when they tried to explain an intrusive signal in the circuits they were building. They finally succeeded by assuming the signal was there and showing that it would then produce itself. I agreed that the bootstrap idea was indeed something of the same sort: the particles, if assumed to exist, produce forces binding them to one another; the resulting bound states are the same particles, and they are the same as the ones carrying the forces. Such a particle system, if it exists, gives rise to itself.

The earliest form of superstring theory was proposed by John Schwarz and André Neveu in 1971, with support from some ideas of Pierre Ramond. Although the theory seemed far-fetched at the time, I invited both Schwarz and Ramond to Caltech, believing that superstrings were so beautiful that they had to be good for something. Schwarz and various collaborators, especially Joël Scherk and Michael Green, developed the theory further over the next fifteen years or so.

At first the theory was applied to just a subset of particles, the same ones that theorists had attempted to describe using the bootstrap principle. It was only in 1974 that Scherk and Schwarz suggested that superstring theory could describe *all* the elementary particles. What convinced them was the discovery that the theory predicted the existence of the graviton, and thus of Einsteinian gravitation. Almost ten years later four physicists at Princeton, known collectively as the "Princeton string quartet," found the particular form called the heterotic superstring theory.

Superstring theory, in particular the heterotic form, may really be the long-sought unified quantum field theory. In a suitable approximation, it implies, as it should, Einstein's theory of gravitation. Moreover, it incorporates Einsteinian gravitation and the other fields into a quantum field theory without running into the usual difficulties with infinities. It also explains why there is a great multiplicity of elementary particles: the number of different kinds is actually infinite, but only a finite number (some hundreds, probably) have sufficiently low masses to be discoverable in the laboratory. Also, the theory does not, at least on the face of it, contain any arbitrary numbers or lists of particles and interactions, although some arbitrariness may reappear on closer examination. Finally, superstring theory emerges from a simple and beautiful principle of self-consistency, originally presented as the bootstrap idea.

Not the Theory of Everything

Of all the important questions about the heterotic superstring theory, the one that particularly interests us here is the following: Assuming that it is correct, is it really the theory of everything? Some people have used that expression, and even the abbreviation TOE, in describing it. However, that is a misleading characterization unless "everything" is taken to mean only the description of the elementary particles and their interactions. The theory cannot, by itself, tell us all that is knowable about the universe and the matter it contains. Other kinds of information are needed as well.

The Initial Condition and the Arrow(s) of Time

One of these pieces of additional information is the initial condition of the universe at or near the beginning of its expansion. We know that the universe has been expanding for around ten billion years. That expansion appears in dramatic form to astronomers using powerful telescopes to observe distant clusters of galaxies, but it is not at all obvious when we look closer at hand. The solar system is not expanding, nor is our galaxy or the cluster of galaxies to which it belongs. The other galaxies and clusters are not expanding either. But the different clusters *are* receding from one another, and that is what reveals the expansion of the universe, which has been compared to an idealized picture of the baking of raisin bread. Under the influence of the yeast, the bread (the universe) expands but the raisins (the clusters of galaxies) do not, although they move apart.

The behavior of the universe since the beginning of the expansion obviously depends not only on the laws that govern the behavior of the particles composing the universe but also on the initial condition. Nor is that initial condition something that shows up only in abstruse problems in physics and astronomy. Far from it. It has an enormous influence on what we observe around us every day. In particular, it determines the arrow (or arrows) of time.

Imagine a film of a meteorite speeding into the Earth's atmosphere, glowing with heat as it streaks across the sky, most of its substance consumed by fire, and then, much diminished in size and weight, crashing into the Earth. If we were to run that film backwards, we would see a rock, partially buried in the earth, rising of its own accord into the air, growing in size and weight as it arcs through the atmosphere, collecting material on the way, and finally, large and cold, flying off into space. The time-reversed film is clearly an impossible sequence of events—we can identify it immediately as a film run backwards.

That asymmetry of the behavior of the universe between forward and backward time is known as the arrow of time. Sometimes various aspects of the asymmetry are discussed separately and labeled as different arrows of time. However, they are all related; they all have the same ultimate origin. Now, what is that origin?

Could the explanation for the arrow or arrows of time be found in the fundamental laws of the elementary particles? If changing the sign of the time variable leaves the form of the equations describing those laws unchanged, the equations are said to be symmetrical between forward and backward time. If reversing the time variable alters the form of the equations, an asymmetry between forward and backward time, or a violation of time symmetry, is said to exist. Such a violation could, in principle, account for the arrow of time. In fact a small violation of that kind is known to exist, but it is much too special an effect to give rise to such a general phenomenon as the arrow of time.

Instead, the explanation is that if we look in both directions of time, we find that about ten or fifteen billion years away in one of those directions, the universe is in a very special state. That direction of time has an arbitrary name: the past. The other direction is called the future. In the state corresponding to the initial condition, the universe was tiny, but that tininess by no means completely characterizes the state, which was also especially simple. If, in the very distant future, the universe stops expanding and starts to recontract, eventually becoming tiny again, there is every reason to believe that its resulting final state will be quite different from the initial one. The asymmetry between past and future will thus be maintained.

A Candidate for the Initial Condition

Since a viable candidate has emerged for the unified law of the elementary particles, it is reasonable to ask whether we also have a plausible theory for the initial condition of the universe. Actually there is one. It was suggested by James Hartle and Stephen Hawking in about 1980. Hawking likes to call it the "no-boundary boundary condition." This is an apt name but it doesn't convey what is particularly interesting about the proposal with respect to "following the information." If the elementary particles are indeed described by a unified theory (which Hartle and Hawking did not explicitly assume), then the appropriately modified version of their initial condition can be calculated in principle from that unified theory, and the two fundamental laws of physics, for the elementary particles and the universe, become a single law.

Instead of Everything, Just Probabilities for Histories

Whether or not the Hartle–Hawking idea is correct, we can still ask the following: If we specify both the unified theory of the elementary particles *and* the initial condition of the universe, can we, in principle, predict the behavior of the universe and everything in it? The answer is no, because the laws of physics are quantum-mechanical, and quantum mechanics is not deterministic. It permits theories to predict only probabilities. The fundamental laws of physics allow, in principle, only the calculation of probabilities for various alternative histories of the uni-

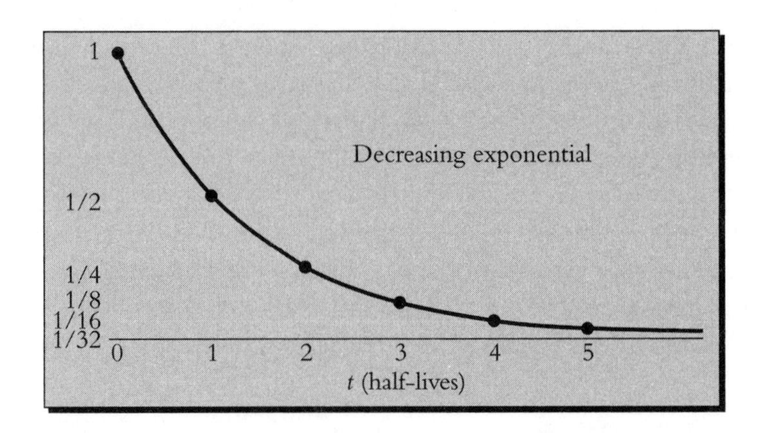

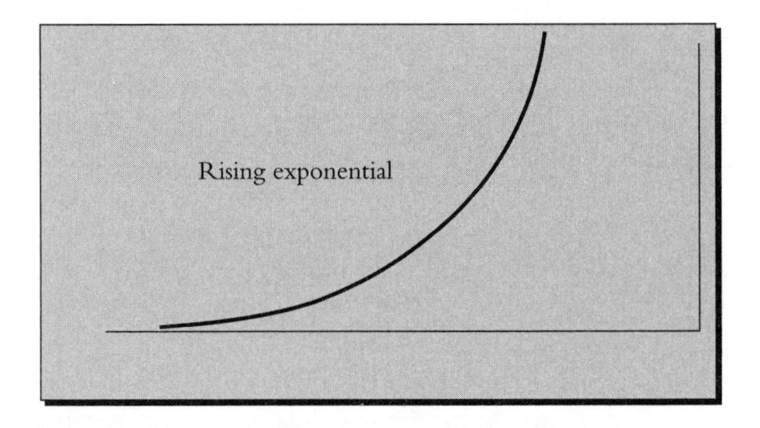

Top: The decreasing exponential curve for the fraction of radioactive nuclei remaining undecayed after time *t*. *Bottom:* A rising exponential.

verse that describe different ways events could play themselves out given the initial condition. Information about which of those sequences of events is actually occurring can be gathered only from observation, and is supplementary to the fundamental laws themselves. Thus there is no way that the fundamental laws can supply a theory of everything.

The probabilistic nature of quantum theory can be illustrated by a simple example. A radioactive atomic nucleus has what is called a "half-life," the time during which it has a 50% chance of disintegrating. For example, the half-life of Pu^{239}, the usual isotope of plutonium, is around 25,000 years. The chance that a Pu^{239} nucleus in existence today will still exist after 25,000 years is 50 percent; after 50,000 years, the chance is only 25 percent; after 75,000 years, 12.5 percent; and so on. The quantum-mechanical character of nature means that for a given Pu^{239} nucleus, that kind of information is all we can know about when it will decay; there is no way to predict the exact moment of disintegration, only a curve of probability against time such as the one on the facing page. (That curve is what is known as a decreasing exponential; the reverse curve, a rising exponential, is also shown. Any exponential curve gives, at equal time intervals, a geometric progression, such as 1/2, 1/4, 1/8, 1/16 . . . , for the decreasing case or 2, 4, 8, 16 . . . for the rising one.)

While the moment of radioactive disintegration cannot be predicted accurately, the direction of decay is completely unpredictable. Suppose the Pu^{239} nucleus is at rest and will decay into two electrically charged fragments, one much larger than the other, traveling in opposite directions. All directions are then equally likely for the motion of one of the fragments, say the smaller one. There is no way to tell which way the fragment will go.

If so much is unknowable in advance about one atomic nucleus, imagine how much is fundamentally unpredictable about the entire universe, even given the unified theory of the elementary particles and the initial condition of the universe. Above and beyond those presumably simple principles, each alternative history of the universe depends on the results of an inconceivably large number of accidents.

Regularities and Effective Complexity from Frozen Accidents

Those accidents have chance outcomes, as required by quantum mechanics. The outcomes have helped to determine the character of individual galaxies (such as our Milky Way), of particular stars and planets (such as the sun and the Earth), of terrestrial life and the particular species that evolved on our planet, of individual organisms such as ourselves, and of the events of human history and our personal lives. The genotype of any human being has been influenced by numerous quantum accidents, not only mutations in the ancestral germ plasm, but even events affecting the fertilization of a particular egg by a particular sperm.

The algorithmic information content of each alternative history of the universe evidently receives a tiny contribution from the simple fundamental laws, along with a gigantic contribution from all the quantum accidents that arise along the way. But it is not only the AIC of the universe that is dominated by those accidents. Although they are chance events, their effects contribute heavily to effective complexity as well.

The effective complexity of the universe is the length of a concise description of its regularities. Like the algorithmic information content, the effective complexity receives only a small contribution from the fundamental laws. The rest comes from the numerous regularities resulting from "frozen accidents." Those are chance events of which the particular outcomes have a multiplicity of long-term consequences, all related by their common ancestry.

The consequences of some such accidents can be far-reaching. The character of the whole universe was affected by accidents occurring near the beginning of its expansion. The nature of life on Earth depends on chance events that took place around four billion years ago. Once the outcome is specified, the long-term consequences of such an event may take on the character of a law, at any but the most fundamental level. A law of geology, biology, or human psychology may stem from one or more amplified quantum events, each of which could have turned out differently. The amplifications can occur through a variety of mechanisms, including the phenomenon of chaos, which introduces, in certain situations, indefinitely large sensitivities of outcome to input.

To understand fully the significance of chance events, it is necessary to look deeper into the meaning of quantum mechanics, which teaches us that chance plays a fundamental role in the description of nature.

CHAPTER

11

A CONTEMPORARY VIEW OF QUANTUM MECHANICS

Quantum Mechanics and the Classical Approximation

When quantum mechanics was first discovered, people were struck most forcibly by the contrast between its probabilistic character and the certainties of the older classical physics, in which exact and complete information about an initial situation would in principle permit, given the correct theory, exact and complete specification of the outcome. Determinism of that kind is never perfectly applicable in quantum mechanics, but it often does apply approximately under the frequently encountered conditions—what may be called the quasiclassical domain—where classical physics is nearly correct. That domain may be crudely characterized as involving the behavior of heavy objects. For instance, the motion of the planets around the sun can be calculated, for any practical purpose, without quantum corrections, which are utterly negligible for such a problem. If the quasiclassical domain were not so relevant, physicists would never have developed and used classical physics in the first place, and classical theories such as those of Maxwell and Einstein would not have achieved their wonderful successes in predicting the results of observations. This is another case where the old paradigm (as Kuhn would call it) is not discarded when a new one is adopted, but remains a valid approximation in a suitable limit (like Newton's theory of gravitation, which remains immmensely useful as an approximation to that of Einstein when speeds are slow relative to the velocity of light). Still, classical physics is only an approximation, while quantum mechanics is, as far as we know, exactly correct. Although many decades have elapsed since the discovery of quantum mechanics, physicists are only now approaching a really satisfactory interpretation, one that affords a deep understanding of how the quasi-classical domain of everyday experience arises from the underlying quantum-mechanical character of nature.

The Approximate Quantum Mechanics of Measured Systems

When first formulated by its discoverers, quantum mechanics was often presented in a curiously restrictive and anthropocentric fashion, and it is frequently so presented to this day. It is assumed, more or less, that some experimental situation (such as the radioactive decay of a particular kind of nucleus) is replicated identically over and over again. The outcome of the experiment is observed each time, preferably by a physicist using some apparatus. It is supposed to be important that the physicist and the apparatus be external to the system being studied. The physicist records the fractions of occurrence of the different possible outcomes of the experiment (such as the times of the decays). As the number of trials gets larger without limit, these fractions tend to approximate the probabilities of the various outcomes, which are predicted by the quantum-mechanical theory. (The probability of radioactive decay as a function of time is closely related to the proportion of nuclei remaining undecayed after various elapsed time intervals, as shown in the curve on page 132. The probability of decay follows a similar curve.)

This original interpretation of quantum mechanics, restricted to repeated experiments performed by external observers, is far too special to be acceptable today as the fundamental characterization, especially since it has become increasingly clear that quantum mechanics must apply to the whole universe. The original interpretation is not wrong, but it is applicable only to the situations it was developed to describe. Moreover, in a wider context that interpretation must be regarded as not only special but approximate. We can refer to it as the "approximate quantum mechanics of measured systems."

The Modern Approach

For describing the universe, a more general interpretation of quantum mechanics is clearly necessary, since no external experimenter or apparatus exists and there is no opportunity for repetition, for observing many copies of the universe. (In any case the universe presumably couldn't care less whether human beings have evolved on some obscure planet to study its history; it goes on obeying the quantum-mechanical laws of physics irrespective of observation by physicists.) That is one reason why what I call the modern interpretation of quantum mechanics has been developed over the last few decades. The other principal reason is the need for a clearer understanding of the relationship between quantum mechanics and the approximate classical description of the world around us.

In early discussions of quantum mechanics it was often implied, and sometimes explicitly stated, that there was a classical domain *apart from* quantum mechanics, so that basic physical theory somehow required classical laws in addition to quantum-mechanical ones. To a generation brought up on classical physics, that arrangement may have appeared satisfactory, but to many of us today it seems bizarre as well as unnecessary. In the modern interpretation of quantum mechanics, it is proposed that the quasiclassical domain emerges from the quantum-mechanical laws, including the initial condition at the beginning of the expansion of the universe. Understanding how that emergence comes about presents a major challenge.

The modern approach was pioneered by the late Hugh Everett III, a graduate student of John A. Wheeler at Princeton and later a member of the Weapons Systems Evaluation Group at the Pentagon. A number of theoretical physicists have worked on it since, including James Hartle and me. Hartle (of the University of California at Santa Barbara and the Santa Fe Institute) is a distinguished theoretical cosmologist and an expert on Einstein's general-relativistic theory of gravitation. In the early 1960s, when he was my doctoral student at Caltech, his dissertation was on elementary particle theory. Later, he and Stephen Hawking wrote the seminal paper entitled "The Wave Function of the Universe," which played a vital role in shaping the field of quantum cosmology. Since 1986, Jim and I have been working together to help clarify how quantum mechanics should be conceived, particularly in relation to the quasiclassical domain.

We consider Everett's work to be useful and important, but we believe that there is much more to be done. In some cases too, his choice of vocabulary and that of subsequent commentators on his work have created confusion. For example, his interpretation is often described in terms of "many worlds," whereas we believe that "many alternative histories of the universe" is what is really meant. Furthermore, the many worlds are described as being "all equally real," whereas we believe it is less confusing to speak of "many histories, all treated alike by the theory except for their different probabilities." To use the language we recommend is to address the familiar notion that a given system can have different possible histories, each with its own probability; it is not necessary to become queasy trying to conceive of many "parallel universes," all equally real. (One distinguished physicist, well versed in quantum mechanics, inferred from certain commentaries on Everett's interpretation that anyone who accepts it should want to play Russian roulette for high stakes, because in some of the "equally real" worlds the player would survive and be rich.)

Another linguistic problem is that Everett avoided the word "probability" in most connections, using instead the less familiar but mathematically equivalent notion of "measure"; Hartle and I see no advantage in that. Words aside, however, Everett left a number of important questions unanswered, and the main challenge is not a matter of language but of filling those gaps in our understanding of quantum mechanics.

Jim Hartle and I are part of an international group of theorists trying in various ways to construct the modern interpretation of quantum mechanics. Among those who have made especially valuable contributions are Robert Griffiths and Roland Omnès, whose belief in the importance of histories we share, as well as Erich Joos, Dieter Zeh, and Wojciech ("Wojtek") Żurek, who have somewhat different points of view. The formulation of quantum mechanics in terms of histories was developed by Dick Feynman, who built on earlier work by Paul Dirac.

That formulation not only helps to clarify the modern interpretation; it is also particularly useful in describing quantum mechanics whenever Einsteinian gravitation is taken into account, as it must be in quantum cosmology. The geometry of space-time is then seen to be subject to quantum-mechanical indeterminacy, and the method based on histories handles that situation particularly well.

The Quantum State of the Universe

Basic to any treatment of quantum mechanics is the notion of a quantum state. Let us consider a somewhat simplified picture of the universe in which each particle has no attributes other than position and momentum and the indistinguishability of all particles of a given type (the interchangeability of all electrons, for example) is set aside. Then what is meant by a quantum state for the whole universe? It is best to start by discussing the quantum state of a single particle and then of two particles before tackling the whole universe.

In classical physics it would have been legitimate to specify exactly both the position and the momentum of a given particle at the same time, but in quantum mechanics that is forbidden, as is well known, by the uncertainty, or indeterminacy, principle. The position of a particle can be specified exactly, but its momentum will then be completely undetermined; that situation characterizes a particular kind of quantum state for a single particle, a state of definite position. In another kind of quantum state the momentum is specified but the position is completely undetermined. There is also an infinite variety of other possible quantum states for a single particle, in which neither position nor momentum is exactly specified, only a smeared-out probability distribution for each. For example, in a hydrogen atom, which consists of a single (negatively charged) electron in the electric field of a single (positively charged) proton, the electron may find itself in the quantum state of lowest energy, in which its position is smeared out over a region of atomic size and its momentum is distributed as well.

Now consider a "universe" of two electrons. It is technically possible for their quantum state to be such that each electron is in a definite quantum state. However, that does not often occur in reality, because the two electrons interact, especially through the electrical repulsion between them. The helium atom, for example, consists of two electrons in the field of a central nucleus with a double positive charge. In the lowest energy state of the helium atom, it is not true that each of the two electrons is in a definite quantum state of its own, although that situation is sometimes discussed as an approximation. Instead, as a result of the interaction between the electrons, their joint quantum state is one in which the states of the two electrons are entangled (correlated) with each other. If you are interested in just one of the electrons, you can "sum over" all the positions (or momenta or values of any other attribute) of the second electron, and your electron is then not in a definite ("pure") quantum state but instead has a set of probabilities for various pure single-electron quantum states. Your electron is said to be in a "mixed quantum state."

We can now proceed directly to the consideration of the whole universe. If the universe is in a pure quantum state, it is a quantum state such that the states of all the individual particles it contains are entangled with one another. If we sum over all the situations in some parts of the universe, then the rest of the universe (what is "followed," what is not summed over) is in a mixed quantum state.

The universe as a whole may be in a pure quantum state. Hartle and Hawking, making that assumption, have proposed a particular form for the pure state that existed near the beginning of the expansion of the universe. As remarked earlier, their hypothesis—suitably generalized to a unified theory of elementary particles—specifies that initial quantum state of the universe in terms of the theory. Moreover, the same unified theory determines how the quantum state varies with time. But a complete specification of the quantum state of the whole universe, not only initially but at all times, still does not supply an interpretation of quantum mechanics.

The quantum state of the universe is like a book that contains the answers to an infinite variety of questions. Such a book is not really useful without a list of those questions to be asked of it. The modern interpretation of quantum mechanics is being constructed by means of a discussion of the appropriate questions to ask of the quantum state of the universe.

Since quantum mechanics is probabilistic rather than deterministic, those questions are inevitably about probabilities. Hartle and I, like Griffiths and Omnès, make use of the fact that the questions always relate ultimately to alternative histories of the universe. (By "history" we do not mean to emphasize the past at the expense of the future; nor do we refer mainly to written records as in human history. A history is merely a narrative of a time sequence of events—past, present, or future.) The questions about alternative histories may be of the type "What is the probability of occurrence of this particular history of the universe rather than those others?" or else "Given these assertions about a history of the universe, what is the probability of those additional statements being true?" Often the latter type of question takes the familiar form, "Given these assertions about the past or the present, what is the probability of those statements about the future coming true?"

Alternative Histories at the Race Track

One place to encounter probabilities is at the race track, where they are related to what we may call true odds. If the true odds against a horse's winning a race are 3 to 1, then that horse's probability of winning is 1/4; if the true odds are 2 to 1, the probability is 1/3, and so on. (Of course the odds actually quoted at a race track are not true odds and do not correspond in this way to true probabilities. We shall return to that point.) If there are ten horses in the race, each one has some positive probability of winning (or zero probability in a really desperate case!), and those ten probabilities add up to 1 if there is to be exactly one winner among the horses. The ten alternative outcomes are then *mutually exclusive* (only one can occur) and *exhaustive* (one of them must occur). An obvious property of the ten probabilities is that they are *additive*: the probability that either the third or the fourth horse will win, for example, is just the sum of the two individual probabilities of victory of the third and the fourth horses.

A closer parallel between the race track experience and histories of the universe can be drawn by considering a sequence of races, say eight races with ten horses in each. Suppose for the sake of simplicity that only winning matters (not "place" or "show") and that there is just one winner of each race (no dead heats). Each list of eight winners is then a kind of history, and those histories are mutually exclusive and exhaustive, as in the case of a single race. The number of alternative histories is the product of eight factors of ten, one for each race, or a hundred million all together.

The probabilities of the different sequences of victories have the same additive property as do the probabilities of individual horses winning a single race: the probability that one or another particular sequence of victories will take place is the sum of the individual probabilities for the two sequences. A situation in which either one sequence or another happens may be called a "combined history."

Let us label two individual alternative histories A and B. The additive property dictates that the probability of the combined history "A or B" is the probability of A plus the probability of B. In other words, the probability that I will go to Paris tomorrow or stay home is the sum of the probabilities of my going to Paris and my staying home. A quantity that doesn't obey this rule is not a probability.

Alternative Histories in Quantum Mechanics

Suppose a set of alternative histories of the universe is specified, and that those histories are exhaustive and mutually exclusive. Does quantum mechanics always assign a probability to each one? Surprisingly, it does not always do so. Instead, it assigns to every *pair* of such histories a quantity called *D*, and it supplies a rule for calculating *D* in terms of the quantum state of the universe. The two histories in a given pair may be different, like the alternatives A and B, or they may be the same, say A and A. The value of *D* will be indicated by an expression such as *D*(A, B), pronounced *D* of A and B. If the two histories in the pair are both A, then we have *D*(A, A). If both of them are the combined history A or B, then the value of *D* is designated *D*(A or B, A or B).

When the two histories in the pair are the same, *D* is a number between zero and one, like a probability. In fact, it can, under certain conditions, be interpreted as the probability of the history. To see what those conditions are, let us examine the relationship among the following quantities:

D(A or B, A or B).

D(A, A).

D(B, B).

D(A, B) plus *D*(B, A).

The first three quantities are numbers between zero and one and thus resemble probabilities. The last quantity can be positive or negative or zero and is not a probability. The rule for calculating *D* in quantum mechanics is such that the first quantity is the sum of the other three. But, if the last quantity is always zero when A and B are different, then *D*(A or B, A or B) is just equal to *D*(A, A) plus *D*(B, B). In other words, if *D* is always zero when the two histories are different, then *D* of a history and that same history always possesses the additive property and can therefore be interpreted as the probability of that history.

The fourth quantity in the list is called the interference term between the histories A and B. If it is *not* zero for every pair of different histories in the set, then those histories cannot be assigned probabilities in quantum mechanics. They "interfere" with each other.

Since the best that quantum mechanics can do in any situation is to predict a probability, it can do nothing in the case of histories that interfere with each other. Such histories are useful only for constructing combined histories that do not interfere.

Fine-Grained Histories of the Universe

Completely fine-grained histories of the universe are histories that give as complete a description as possible of the entire universe at every moment of time. What does quantum mechanics have to say about them?

Let us continue to use the simplified picture of the universe in which particles have no attributes other than their positions and momenta and the indistinguishability between particles of a given type is put aside. If classical deterministic physics were exactly correct, the positions and momenta of all the particles in the universe could be specified exactly at any given time. Classical dynamics could then, in principle, predict with certainty the positions and momenta of all the particles at all future times. (The phenomenon of chaos produces situations in which the slightest imprecision in initial positions or momenta can lead to arbitrarily large uncertainties in future predictions, but in classical theory perfect determinism would still be correct in principle, given perfect information.)

What is the corresponding situation in quantum mechanics, to which classical physics is only an approximation? For one thing, it is no longer meaningful to specify both the exact position and the exact momentum of a particle at the same time; that is part of the celebrated uncertainty principle. In quantum mechanics, therefore, the condition of the simplified universe at a given time could be characterized by specifying just the positions of all the particles (or by the positions of some and the momenta of others, or by the momenta of all, or in an infinite number of other ways). One kind of completely fine-grained history of the simplified universe in quantum mechanics would consist of the positions of all the particles at all times.

Since quantum mechanics is probabilistic rather than deterministic, one might expect it to supply a probability for each fine-grained history. However, that is not the case. The interference terms between fine-grained histories do not usually vanish, and probabilities cannot therefore be assigned to such histories.

Yet at the race track, the bettor does not have to worry about any interference term between one sequence of winners and another. Why not? How is it that the bettor deals with true probabilities that add up properly, while quantum mechanics supplies, at the fine-grained level, only quantities for which the addition is encumbered by interference terms? The answer is that in order to have actual probabilities, it is necessary to consider histories that are sufficiently coarse-grained.

Coarse-Grained Histories

The sequence of eight horse races serves not only as a metaphor, but also as an actual example of a very coarse-grained history of the universe. Since only the list of winners is considered, the coarse graining consists of the following:

1. Ignoring all times in the history of the universe except those at which the races are won.
2. At the times considered, following only the horses entered in the races and ignoring all other objects in the universe.
3. Of those horses, following only the one that wins each race; every part of the horse is ignored except the tip of its nose.

For histories of the universe in quantum mechanics, coarse graining typically means following only certain things at certain times and only to a certain level of detail. A coarse-grained history may be regarded as a class of alternative fine-grained histories, all of which agree on a particular account of what is followed, but vary over all possible behaviors of what is not followed, what is summed over. In the case of the horse races, each coarse-grained history is the class of all fine-grained histories that share the same sequence of eight winning horses on that particular afternoon at that particular track, although the fine-grained histories in the class vary over all possible alternatives for what happens to any other feature of the history of the universe!

All the fine-grained histories of the universe are thus grouped into classes such that each one belongs to one and only one class. Those exhaustive and mutually exclusive classes are coarse-grained histories (such as the different possible sequences of winners of eight races when there are no ties). Suppose that a given class comprises just two fine-grained histories, J and K; the coarse-grained history will then be "J or K," meaning that either J or K occurs. Likewise, if the class comprises many fine-grained histories, the coarse-grained history will be the combined history in which any of those fine-grained ones occurs.

Mathematicians would call these coarse-grained histories "equivalence classes" of fine-grained histories. Each fine-grained history belongs to one and only one equivalence class, and the members of the class are treated as equivalent.

Imagine that the only things in the universe are the horses in the eight races and a certain number of horse flies, and that all each horse can do is either win or not win. Each fine-grained history in this ludicrously oversimplified world consists of the sequence of winning horses and some particular story about the flies. If the coarse-grained histories follow only the horses and their victories and ignore the flies, then each such history will be the set of fine-grained histories in which there is a particular sequence of winning horses and any fate whatsoever for the flies. In general, each coarse-grained history is an equivalence class of fine-grained histories characterized by a particular narrative describing the phenomena followed and any of the possible alternative narratives describing everything that is ignored.

Coarse Graining Can Wash Out Interference Terms

For the quantum-mechanical histories of the universe, how can the grouping of fine-grained histories into equivalence classes yield coarse-grained histories with true probabilities? How is it that suitably coarse-grained histories have no interference terms between them? The answer is that the interference term between two coarse-grained histories is the sum of all the interference terms between pairs of fine-grained histories belonging to those two coarse-grained ones. The sum of all those terms, with their positive and negative signs, can produce a great deal of cancellation and give a small result of either sign, or zero. (Recall that D of a history and itself is always between zero and one, like a real probability; when such quantities are added, they cannot cancel out.)

Any behavior of anything in the universe that is ignored in the coarse-grained histories can be said to have been "summed over" in this summation process. All the details that are left out of coarse-grained histories, all the times and places and objects that are not followed, are summed over. For instance, equivalence classes could group together all the fine-grained histories in which certain particles have specified positions at every moment while all the other particles in the simplified universe can be anywhere at all. We would then say that the positions of the first set of particles are followed at every moment, while those of the second set of particles are ignored or summed over.

Further coarse graining might consist of following the positions of the first set of particles only at certain times, so that everything that happens at all other times is summed over.

Decoherence of Coarse-Grained Histories—True Probabilities

If the interference term between each pair of coarse-grained histories is zero, either exactly or to an exceedingly good approximation, then all the coarse-grained histories are said to *decohere.* The quantity D of each coarse-grained history and itself is then a true probability, with the additive property. In practice, quantum mechanics is always applied to sets of decohering coarse-grained histories, and that is why it is able to predict probabilities. (D, by the way, is called the *decoherence functional*; the word "functional" indicates that it depends on histories.)

In the case of the afternoon at the races, the coarse graining employed can be summarized as follows: the fate of everything in the universe is summed over except the winners of races at a particular track, and events at all times are summed over except the moments at which victories in the eight races occur on a particular day. The resulting coarse-grained histories decohere and have true probabilities. Because of our everyday experience it does not surprise us that things work out that way, but we should be curious about how it happens.

Entanglement and Mechanisms of Decoherence

What is the underlying explanation for decoherence, the mechanism that makes interference terms sum to zero and permits the assignment of probabilities? It is the entanglement of what is followed in the coarse-grained histories with what is ignored or summed over. The horses and jockeys in the races are in contact with air molecules, bits of sand and horse dung on the track, photons from the sun, and horse flies, all of them summed over in the coarse-grained histories of the races. The different possible outcomes of the races are correlated with different fates for everything ignored in the coarse-grained histories. But those fates are summed over, and quantum mechanics tells us that in the summation, under suitable conditions, interference terms vanish between histories involving different fates for what is ignored. Because of the entanglement, the interference terms between different results of the races also give zero.

It is mind-boggling to consider, instead of those decohering coarse-grained histories, an extreme case of fine-grained histories with non-zero interference terms and no true probabilities. Such histories might follow, over the whole period of the races, every elementary particle contained in every horse and in everything that came into contact with each horse. We do not have to go to extremes, however, to find histories that are sufficiently free of entanglement that they interfere with each other. Take the famous experiment in which a photon from a tiny source can pass freely through either of two slits in a screen on its way to a given point on a detector—those two histories interfere and cannot be assigned probabilities. It is then meaningless to say what slit the photon came through.

Probabilities and Quoted Odds

It should be stressed once more for the sake of clarity that the probabilities given, for sufficiently coarse-grained histories, by quantum mechanics together with a correct physical theory are the best probabilities that can be calculated. For a sequence of races, they correspond to what we have called true odds. The odds actually quoted at a race track, however, are quite different in character. They merely reflect bettors' opinions about forthcoming races. Moreover, the corresponding probabilities do not even add up to 1, since the track needs to make a profit.

Decoherence for an Object in Orbit

To illustrate the generality of decoherence, we can pass from the mundane to the celestial for another example: an approximate description of the orbit of an object in the solar system. The object can range in size from a large molecule to a planet; in between, it may be a dust grain, a comet, or an asteroid. Consider coarse-grained histories in which the fates of all other things in the universe are summed over, as are all the internal properties of the object itself, leaving only its center-of-mass position at all times. In addition, suppose that position itself is treated only approximately, so that only small regions of space are considered and all the possibilities for the position inside each region are summed over. Finally, suppose the coarse-grained history sums over what happens at most times, following the approximate position of the object only at a discrete sequence of times, with short intervals between them.

Say the object in orbit has mass M, the linear dimensions of the small regions of space are of order X, and the time intervals are of order T. The different possible coarse-grained histories of the object in the solar system will decohere to a high degree of accuracy over wide ranges of values of the quantities M, X, and T. The mechanism responsible for that decoherence is again frequent interaction with objects the fates of which are being summed over. In one famous example, those objects are the photons composing the background electromagnetic radiation left over from the initial expansion of the universe (the so-called big bang). Our orbiting object will repeatedly encounter such photons and scatter off them. Each time that happens, object and photon will emerge from the collision with altered motions. But the different directions and energies of all the photons are summed over, and that washes out the interference terms between such directions and energies, and consequently the interference terms among different coarse-grained histories of the object in orbit.

The histories (specifying the successive approximate positions of the center of mass of the object in the solar system at the particular instants of time) decohere because of the repeated interactions of the object with things that are summed over, like the photons of the background radiation.

This process answers a question Enrico Fermi often put to me in the early 1950s, when we were colleagues at the University of Chicago: Since quantum mechanics is correct, why is the planet Mars not all spread out in its orbit? An old answer, that Mars is in a definite place at each time because human beings are looking at it, was familiar to both of us, but seemed just as silly to him as to me. The actual explanation came long after his death, with the work of such theorists as Dieter Zeh, Erich Joos, and Wojtek Żurek on mechanisms of decoherence, such as the one involving the photons of the background radiation.

Photons from the sun that scatter off Mars are also summed over, contributing to the decoherence of different positions of the planet, and it is just such photons that permit human beings to see Mars. So, while the human observation of Mars is a red herring, the physical process that makes that observation possible is not a red herring at all, and can be regarded as being partially responsible for the decoherence of different coarse-grained histories of the motion of the planet around the sun.

Decoherent Histories Form a Branching Tree

Such decoherence mechanisms make possible the existence of the quasiclassical domain that includes ordinary experience. That domain consists of decoherent coarse-grained histories, which can be envisaged as forming a tree-like structure. Jorge Luis Borges, in one of his brilliant short stories, described a representation of such a structure as a "garden of forking paths." At each branching, there are mutually exclusive alternatives. A pair of such alternatives has often been likened to a fork in a road, as in Robert Frost's poem "The Road Not Taken."

The structure first branches into alternative possibilities right at, or just after, the beginning of the expansion of the universe. Each branch then splits again a short time later into further alternatives, and so on for all time. At each branching, there are well-defined probabilities for the alternatives. There is no quantum interference between them.

This is well illustrated by what happens at the race track. Each race involves a branching into ten alternatives for the different winners, and for each winner there is a further branching into ten alternatives for the winner of the next race.

At the track, there is not usually a great deal of influence exerted by the outcome of one race on the probabilities for the next one (for example, a jockey's becoming depressed over having lost a previous race). However, in the branching tree of alternative histories of the universe, the outcome at one branching may profoundly affect the probabilities at subsequent branchings, and may even affect the nature of the alternatives in subsequent branchings. For example, the condensation of material to form the planet Mars may have depended on a quantum accident billions of years ago; in those branches where no such planet appeared, it follows that further branchings explicitly related to alternative fates of Mars would not occur.

The tree-like structure of alternative decohering coarse-grained histories of the universe is different from evolutionary trees like those for human languages or for biological species. In the case of evolutionary trees, all the branches are present in the same historical record. For example, the Romance languages all branch off from a late version of Latin, but they are not alternatives. French, Spanish, Portuguese, Italian, Catalan, and others are spoken today, and even the now extinct Romance languages, such as Dalmatian, were actually spoken at one time. By contrast, the branches of the tree of alternative decohering histories are mutually exclusive, and only one branch is accessible to an observer. Even the interpreters of Hugh Everett's work who speak of many worlds, equally real, do not claim to have observed more than one of those branching worlds.

High Inertia and Nearly Classical Behavior

Decoherence alone (giving rise to branching of histories into distinct alternatives with well-defined probabilities) is not the only important property of the quasiclassical domain that includes everyday experience. The domain also exhibits largely classical behavior—hence "quasiclassical." Not only do the successive positions of the planet Mars at a sequence of closely spaced times have true probabilities. Those positions at those times are also very highly correlated with one another (probabilities extremely close to one) and they correspond, in an excellent approximation, to a well-defined classical orbit around the sun. That orbit obeys Newton's classical equations for motion in the gravitational field of the sun and the other planets, with tiny corrections for Einstein's improved (general-relativistic) classical theory and a very small frictional force from collisions with light objects like the background photons. Recall that those objects are ignored and thus summed over in the coarse-grained histories that follow the motion of Mars, which is the reason why the coarse-grained histories decohere.

How can the planet follow a deterministic classical orbit when it is continually being buffeted by random blows from the photons it encounters? The answer is that the heavier an object in an orbit, the less it will exhibit erratic behavior and the more it will peacefully follow its orbit. It is the mass *M* of the planet, its inertia, that resists the buffeting and permits it to behave classically to a very good approximation. An atom or small molecule is too light to follow an orbit with any degree of consistency in the presence of all the objects in the solar system with which it could collide. A large dust grain is heavy enough to follow an orbit fairly well, and a small spacecraft does so still better. But even such a spacecraft is knocked around a little by the solar wind, composed of electrons emitted by the sun. Collisions of the craft with those electrons would be sufficient to disturb certain very delicate experiments used to test Einsteinian gravitation; for this reason, it would be desirable for those experiments to make use of a radar transponder on Mars instead of on a space probe.

Although we have ascribed quasiclassical behavior to the heaviness of objects, it would be more accurate to ascribe it to motions associated with sufficiently high inertia. A batch of very cold liquid helium can be both large and heavy and nevertheless, because some of its internal motions have low inertia, exhibit bizarre quantum effects such as creeping over the edge of an open container.

Fluctuations

Physicists sometimes try to distinguish between quantum and classical fluctuations, where the latter could be, for example, thermal fluctuations associated with the motions of molecules in a hot gas. The coarse graining required to achieve decoherence in quantum mechanics implies that many variables must be summed over, and those variables may easily include some of the ones that describe such molecular motions. Thus classical thermal fluctuations tend to get lumped together with quantum fluctuations. A heavy object that follows a classical orbit fairly well is resisting the effects of both kinds of fluctuations at once. Likewise a lighter object may be significantly affected by both.

Erratic motion caused by repeated collisions with tiny things was noticed in the early nineteenth century by the botanist Robert Brown, after whom the phenomenon is named Brownian motion. It can easily be observed by putting a drop of ink into water and watching the ink granules under a microscope. Their jerky movements were explained quantitatively by Einstein as being caused by fluctuations in collisions with water molecules, thus making molecules effectively susceptible to observation for the first time.

Schrödinger's Cat

In a quasiclassical domain, objects approximately obey classical laws. They are subject to fluctuations, but those are individual events superimposed on a pattern of fairly classical behavior. Once it occurs, however, a fluctuation in the history of an otherwise classical object can be amplified to an arbitrary degree. A microscope can enlarge the image of an ink particle struck by a molecule and a photograph can preserve the magnified image indefinitely.

This brings to mind the famous thought experiment involving Schrödinger's cat, in which a quantum event is amplified so as to control whether or not a cat is poisoned. Such amplification is perfectly possible, if not very nice. A device can be hooked up that makes the life or death of the cat depend, for example, on the direction of motion of a nuclear fragment emitted in a radioactive decay. (Using thermonuclear weapons, one could nowadays arrange for the fate of a city to be determined in the same way.)

The usual discussion of Schrödinger's cat goes on to describe alleged quantum interference between the live and dead cat scenarios. However, the live cat has considerable interaction with the rest of the world, through breathing, for example, and even the dead cat interacts with the air to some extent. It doesn't help to have the cat placed in a box, because the box will interact with the outside world as well as with the cat. Thus there is plenty of opportunity for decoherence between coarse-grained histories in which the cat lives and coarse-grained histories in which it dies. The live and dead cat scenarios decohere; there is no interference between them.

Perhaps it is the interference aspect of the cat story that makes Stephen Hawking exclaim, "When I hear about Schrödinger's cat, I reach for my gun." He is, in any case, parodying the remark (often attributed to one or another Nazi leader, but actually occurring in the early pro-Nazi play *Schlageter* by Hanns Johst) "When I hear the word 'Kultur,' I release the safety catch on my Browning."

Suppose the quantum event that determines the fate of the cat has already occurred, but we don't know what happened until we open a box containing the cat. Since the two outcomes decohere, this situation is no different from a classical one where we open a box inside of which the poor animal, arriving after a long airplane voyage, may be either dead or alive, with some probability for each. Yet reams of paper have been wasted on the supposedly weird quantum-mechanical state of the cat, both dead and alive at the same time. No real quasiclassical object can exhibit such behavior because interaction with the rest of the universe will lead to decoherence of the alternatives.

Additional Coarse Graining for Inertia and Quasiclassicality

A quasiclassical domain naturally requires histories that are sufficiently coarse-grained to decohere to an excellent approximation; it also requires that they be even further coarse-grained so that what is followed in the histories has enough inertia to resist, to a considerable extent, the fluctuations inevitably associated with what is summed over. There then remain continual small excursions from classical behavior and occasional large ones.

The reason additional coarse graining is required for high inertia is that sizable chunks of matter can then be followed, and those chunks can have large masses. (If some stable or nearly stable elementary particles with huge masses were available, they would provide a different source of high inertia. Such particles have not been encountered in our experience, although they might exist and, if so, could have played an important role in the earliest moments of the expansion of the universe.)

Measurement Situations and Measurements

A quantum event may become fully correlated with something in the quasiclassical domain. That is what happens in the sensible part of the cat story, where such an event becomes correlated with the fate of the animal. A simpler and less fanciful example would be a radioactive nucleus that occurs as an impurity in a mica crystal and decays, say, into just two electrically charged fragments moving in opposite directions. The direction of motion of one fragment is completely undetermined until the decay occurs, but then it correlates perfectly with a track left in the mica. Quasiclassical histories, which sum over things like the soft radiation emitted when the track was formed, leave the different directions, with some small spread in each one, decoherent. Such a track, at ordinary temperatures, lasts for tens of thousands of years or more, and of course mere persistence is an example (albeit a trivial one) of a classical history. The radioactive decay has made contact with the quasiclassical domain.

The accumulation in minerals of tracks left by the disintegration products of spontaneous nuclear fission is sometimes used to date those minerals; the method is known as fission-track dating, and it can be applied to rocks that are hundreds of thousands of years old. Suppose a physicist engaged in such research looks at a particular track. While pursuing the work on dating, he or she can also be said to have made a measurement of the direction of decay of the radioactive nucleus. The track, however, has been there ever since it was formed; it does not come into existence when the physicist looks at it (as some clumsy descriptions of quantum mechanics might suggest). A measurement situation has existed since the nucleus decayed and the track was formed; that is when a strong correlation was established with the quasiclassical domain. The actual measurement could have been carried out by a cockroach or any other complex adaptive system. It consists of "noticing" that a particular alternative has occurred out of a set of decoherent alternatives with various probabilities. Exactly the same thing occurs at the racetrack when a particular horse is "observed" to win one of the races. A record of the victory, already present somewhere in the quasiclassical domain, is further registered in the memory of the observer, whether that observer is of high or low intelligence. However, many sensible, even brilliant commentators have written about the alleged importance of human consciousness in the measurement process. Is it really so important? What do noticing and observing really mean?

An IGUS—A Complex Adaptive System as Observer

An observation in this context means a kind of pruning of the tree of branching histories. At a particular branching, only one of the branches is preserved (more precisely, on each branch, only that branch is preserved!). The branches that are pruned are thrown away, along with all the parts of the tree that grow out of the branches that are pruned.

In a sense, the mica with the fission tracks has already performed a pruning operation by registering the actual direction of motion of the fission fragment and thus discarding all other directions. But a complex adaptive system observing the track prunes in a more explicit way, by including the observation in the data stream that gives rise to the evolution of its schemata. The subsequent behavior of the system can then reflect its observation of the particular track direction.

A complex adaptive system acting as an observer probably deserves a special name. Jim Hartle and I call it an IGUS, for information gathering and utilizing system. If the IGUS possesses consciousness or self-awareness to a significant degree (so that it notices itself noticing the direction of a fission track), so much the better. But why is that necessary? Does a measurement made by an arbitrary human being, even one of very low intelligence, really have any greater significance than one made by a gorilla or a chimpanzee? And if not, then why not substitute a chinchilla or a cockroach for the ape?

When it comes to pruning the branching tree of histories, perhaps a distinction should be made between a human observer who knows something about quantum mechanics (and is therefore aware of the origins of the tree) and one who does not. In a sense, the difference between them is greater than that between a human ignorant of quantum mechanics and a chinchilla.

An IGUS can do something else besides eliminating alternative branches when a particular outcome is known: it can bet on that outcome beforehand, using some approximate version of the probabilities supplied by quantum mechanics. Only a complex adaptive system can do that. Unlike a piece of mica, an IGUS can incorporate its estimated probabilities of future events into a schema and base its future behavior on that schema. A desert-dwelling mammal, for instance, may walk a long way to a deep water hole some days after the last rain, but not to a shallow one, since the probability is greater that there will still be water remaining in the deep hole.

The pruning of branches replaces what, in the traditional interpretation of quantum mechanics, is usually called the "collapse of the wave function." The two descriptions are not unrelated mathematically, but the collapse is often presented as if it were a mysterious phenomenon peculiar to quantum mechanics. Since pruning, however, is just the recognition that one or another of a set of *decohering* alternatives has occurred, it is quite familiar. It is exemplified by noticing, for example, that I have not gone to Paris after all but stayed home. All the branches of history that depended on my going to Paris have been discarded; their probabilities are now zero no matter what they were before.

The point often left unclear in discussions of the so-called collapse is that even if the pruning involves the measurement of a quantum event, it is still an ordinary discrimination among decohering alternatives. Quantum events can be detected only at the level of the quasiclassical domain. There the situation is just one of classical probabilities, as in throws of dice or tosses of a coin, with the probabilities changing to one and zero when the outcome is known. The quasiclassical domain admits the possibility of reasonably persistent records of the outcome, records that can be amplified or copied over and over in a quasiclassical chain of near-certain agreement of each record with the previous one. Once a quantum event is correlated with the quasiclassical domain (creating a measurement situation), the particular outcome in a given branch of history becomes a fact.

Self-awareness and Free Will

Since the issue of consciousness has been raised, let us digress briefly and explore it a bit further. The human brain has greatly enlarged frontal lobes compared with those of our close relatives the great apes. Neurobiologists have identified areas of the frontal lobes that seem to be associated with self-awareness and intention, thought to be especially well-developed in human beings.

In conjunction with the many parallel processing strands in human thought, consciousness or attention seems to refer to a sequential process, a kind of spotlight that can be turned from one idea or sensory input to another in rapid succession. When we believe we are attending to many different things at once, we may really be employing the spotlight in a time-sharing mode, moving it around among the various objects of our attention. The parallel processing strands differ in their accessibility to consciousness, and some sources of human behavior lie buried in layers of thought that are difficult to bring to conscious awareness.

Nevertheless, we do say that utterances and other actions are to a considerable degree under conscious control, and that statement reflects not only the recognition of the spotlight of awareness but also the strong belief that we have a degree of free will, that we can choose among alternatives. The possibility of choice is an important feature, for example, of "The Road Not Taken."

What objective phenomena give rise to that subjective impression of free will? To say a decision is taken freely means that it is not strictly determined by what has gone before. What is the source of that apparent indeterminacy?

A tempting explanation is that it is connected with fundamental indeterminacies, presumably those of quantum mechanics enhanced by classical phenomena such as chaos. A human decision would then have unpredictable features, which could be labeled retrospectively as freely chosen. One might wonder, however, what feature of the human brain cortex makes the contribution of quantum fluctuations and chaos particularly prominent there.

Instead of invoking only those straightforwardly physical effects, we might also consider processes more directly associated with the brain and the mind. Recall that, for a given coarse graining, *all* the phenomena that are summed over (not followed) can contribute apparent indeterminacies (for instance thermal fluctuations) that are lumped in with quantum fluctuations. Since there are always many strands of thought not illuminated by the searchlight of consciousness, those strands are being summed over in the extremely coarse-grained histories that are consciously remembered. The resulting indeterminacies would seem more likely to contribute to the subjective impression of free will than the indeterminacies narrowly associated with physics. In other words, human beings probably act on hidden motives more often than they use the results of an internal random or pseudorandom number generator. But the whole matter is poorly understood, and for the time being we can only speculate. (Speculation about such questions is far from new, and is usually rather vague. Nonetheless, I see no reason why the subject should not be pursued in terms of scientific inquiry into the possible role of various indeterminacies in the functioning of the human brain cortex and the corresponding mental processes.)

What Characterizes the Familiar Quasiclassical Domain?

In the coarse-grained histories of the quasiclassical domain that incorporates familiar experience, certain kinds of variables are followed while the rest are summed over, which means that they are ignored. Which kinds are followed? Roughly speaking, the usual quasiclassical domain follows gravitational and electromagnetic fields and exactly conserved quantities such as energy or momentum or electric charge, as well as quantities that are approximately conserved, like the number of dislocations (irregularities) in a crystal produced by the passage of a charged particle. A quantity is said to be conserved when the total amount existing in a closed system remains unchanged through time; it is approximately conserved when the total amount in a closed system varies only a little as time passes. A conserved quantity like energy cannot be created or destroyed, only transformed. The dislocations in a crystal, however, obviously can be created, by the passage of a charged particle, for example; still, they can last for tens or hundreds of thousands of years, and in that sense they are nearly conserved.

The familiar quasiclassical domain involves summing over everything except ranges of values of these fields and conserved and nearly conserved quantities within small volumes of space, but volumes large enough to have the inertia necessary to resist the fluctuations associated with the effects of all the variables that are summed over. That is to say, the fluctuations are resisted sufficiently so that the quantities followed exhibit quasiclassical behavior.

Those quantities must be followed at time intervals that are not too close together, so the alternative coarse-grained histories can decohere. In general, if the graining becomes too fine (because the time intervals are too short, the volumes too small, or the ranges of values of the quantities followed too narrow), the danger of interference between histories rears its head.

Let us consider a set of alternative coarse-grained histories that are maximally refined, so that any further fine graining would ruin either the decoherence or the nearly classical character of the histories or both. The small volumes in which the conserved and nearly conserved quantities are followed at suitable intervals of time can then cover the whole universe, but with a coarse graining in space and time (and in ranges of values of the quantities) that is just adequate to yield decoherent and nearly classical alternative histories.

The experience of human beings and of the systems with which we are in contact is of a domain that is very much more coarse-grained than such a maximal quasiclassical domain. A huge amount of additional coarse graining is required to go from that maximal quasiclassical domain to the domain accessible to actual observation. The accessible domain follows only very limited regions of space-time, and the coverage of the variables in those regions is very spotty. (The interiors of stars and of other planets are almost entirely inaccessible, for instance, and what happens on their surfaces can be detected only in a very coarse-grained manner.)

In contrast, the coarse-grained histories of the maximal quasiclassical domain need not sum over, and so ignore, all the variables inaccessible to human observation. Instead, those histories can include descriptions of alternative outcomes of processes arbitrarily remote in space and time. They can even cover events near the beginning of the expansion of the universe, when there were presumably no complex adaptive systems anywhere to act as observers.

In summary, a maximal quasiclassical domain is an exhaustive set of mutually exclusive coarse-grained histories of the universe that cover all of space-time, that are decoherent with one another and nearly classical most of the time, and that are maximally refined consistent with the other conditions. In this particular kind of maximal quasiclassical domain we are discussing, the quantities followed are ranges of values of conserved and nearly conserved quantities over small volumes. The actual domain of familiar human experience is obtained from such a maximal domain by the application of an extreme amount of additional coarse graining, corresponding to the capabilities of our senses and instruments.

The Branch Dependence of Followed Quantities

It is important to reemphasize that the specific quantities followed at a given time may depend on the outcome of a previous branching of histories. For instance, the distribution of mass in the Earth, as represented by the amount of energy contained in each of a huge number of small volumes within the planet, will presumably be followed by coarse-grained histories as long as the Earth exists. But what if the Earth is one day blown to smithereens by some presently unforeseen catastrophe? What if that catastrophe vaporizes the planet, as in some B movie? Presumably, for the histories in which that happens, the quantities subsequently followed by the coarse-grained histories will be different from what they were before the catastrophe. In other words, what is followed in the case of a given coarse graining of histories may be branch-dependent.

Individual Objects

We have discussed the quasiclassical domain that includes familiar experience in terms of ranges of values of fields and of exactly or approximately conserved quantities in small volumes of space. But how do individual objects like a planet come into the story?

Early in the history of the universe, masses of material began to condense under the influence of gravitational attraction. The narratives of the various alternative coarse-grained histories after that time are much more concise when described in terms of the objects thus formed. It is much simpler to record the motion of a galaxy than to list separately all the coordinated changes in the densities of matter in a trillion trillion small volumes of space as the galaxy moves.

As galaxies gave rise to stars, planets, rocks and in some places to complex adaptive systems like the living things on Earth, the existence of individual objects became a more and more striking feature of the quasiclassical domain. Many of the regularities of the universe are most concisely described in terms of such objects; thus the properties of individual things represent a great deal of the effective complexity of the universe.

In most cases the description of individual objects is simplest when the definition allows for the accretion or loss of comparatively small amounts of matter. When a planet absorbs a meteorite or a cat breathes, the identity of the planet or the cat is not altered.

But how is individuality to be measured? One way is to look at a set of comparable objects and, for a given coarse graining, to describe as briefly as possible the properties that distinguish them (such as the lost feathers of the eleven California condors I saw feasting on a calf). The number of bits in the description of a typical individual can then be compared with the number of bits necessary to count the individuals in the set. If, for the particular coarse graining involved, the description contains many more bits than the enumeration, the objects in the set are exhibiting individuality.

Consider the set of all human beings, now numbering about five and a half billion. Assigning a different number to each person takes around 32 bits, because 2 multiplied by itself 32 times is 4,294,967,296. But even a cursory look at each person close up, accompanied by a brief interview, can easily reveal many more than 32 bits of information. When studied more closely, the people will exhibit far more individuality. And imagine how much additional information will be available when their individual genomes can be read.

The stars in our galaxy, not counting possible dark kinds that astronomers may some day discover, number some 100 billion. Assigning each one a serial number would take about 37 bits. For the sun, which is close by, much more information than that is available to astronomers, but the graining for other stars is much coarser. Position in the sky, the brightness, the spectrum of light emitted, and motion can all be measured to some extent, more or less accurately depending on distance. The total number of bits of information is not typically very much greater than 37, and in some cases it may be less. As viewed today by astronomers, stars other than the sun do exhibit some individuality, but not a great deal.

The particular coarse graining characteristic of today's observations can be avoided by switching to a maximal quasiclassical domain, which consists of alternative histories, covering all of space-time, that are not only decoherent and nearly classical, but also in some sense maximally fine-grained given their decoherence and quasiclassicality. Where appropriate, those histories may be expressed in terms of individual objects, which are followed in extraordinary detail and exhibit a correspondingly high degree of individuality.

In the usual maximal quasiclassical domain, the information about any star is enormously greater than what we now know about the sun. Likewise, the information about any human being is much richer than what is available to us today. In fact, no complex adaptive system observing a star or a human being could possibly make use of such a gigantic quantity of information. Moreover, much of the data would refer to random or pseudorandom fluctuations of matter densities in a stellar interior or inside some bone or muscle. It is hard to imagine what use a complex adaptive system could make of the bulk of such information. Yet regularities in the data could be very useful; in fact, physicians make use of just such regularities when they employ magnetic resonance imaging (MRI) or X-ray computer-assisted tomography (CAT scans) to diagnose illness. As usual, a descriptive schema formulated by an observing complex adaptive system is a concise listing of regularities, and the length of such a list is a measure of the effective complexity of the thing observed.

The Protean Character of Quantum Mechanics

Like classical probabilistic situations such as a series of horse races, the coarse-grained alternative histories of the universe that constitute the maximal quasiclassical domain form a branching tree-like structure, with well-defined probabilities for the different possibilities at each branching. How then does quantum mechanics differ from classical mechanics? One obvious difference is that in quantum mechanics the coarse graining is necessary for the theory to yield anything useful, whereas coarse graining is introduced in classical mechanics only because of the imprecision of measurements or some other practical limitation. But another difference may account, more than anything else, for the counter-intuitive nature of quantum mechanics—it is protean. Recall that Proteus, in classical mythology, was a reluctant prophet who could transform himself into one type of creature after another. To get him to make predictions, it was necessary to hold him fast as he went though a great many changes of form.

Say we return to our simplified fine-grained histories of the universe, which specified the position of every particle in the universe at every instant. In quantum mechanics position is an arbitrary choice. While the Heisenberg uncertainty principle makes it impossible to specify both the position and the momentum of a given particle at the same time with arbitrary accuracy, it does not prevent momentum instead of position from being specified at some of those instants. Consequently, fine-grained histories can be chosen in a great many different ways, with each particle being characterized at certain times by its momentum and at the remaining times by its position. Moreover, there is an infinite variety of other, subtler ways of constructing fine-grained histories of the universe.

Are There Many Inequivalent Quasiclassical Domains?

For each of those sets of fine-grained histories it is possible to consider many different coarse grainings and to ask which ones, if any, lead to a maximal quasiclassical domain characterized by decoherent coarse-grained histories that exhibit nearly classical behavior, with continual small excursions and occasional large ones. Furthermore, one can ask whether there are really significant distinctions among the domains or whether they are all more or less the same.

Jim Hartle and I, among others, are trying to answer that question. Unless the contrary is demonstrated, it will remain conceivable that there is a large set of inequivalent maximal quasiclassical domains, of which the familiar one is just a single example. If that is true, what distinguishes the familiar quasiclassical domain from all the others?

Those who espouse the early view of quantum mechanics might say that human beings have chosen to measure certain quantities and that our choice helps to determine the quasiclassical domain with which we deal. Or, a little more generally, they might say that human beings are capable of measuring only certain kinds of quantities and that the quasiclassical domain must be based, at least in part, on such quantities.

Home for Complex Adaptive Systems

It is true that quasiclassicality guarantees to all human beings and to all systems in contact with us the possibility of comparing records, so that we are all dealing with the same domain. But do we collectively *select* that domain? Such a point of view may be needlessly anthropocentric, like other aspects of the old-fashioned interpretation of quantum mechanics.

Another, less subjective approach is to start with a maximal quasiclassical domain and note that along certain branches, during certain epochs of time, and in certain regions of space, it can exhibit just the kind of mixture of regularity and randomness that favors the evolution of complex adaptive systems. The nearly classical behavior supplies the regularity, while the excursions from determinism—the fluctuations—supply the element of chance. Mechanisms of amplification, including ones that involve chaos, permit some of those chance fluctuations to come into correlation with the quasiclassical domain and give rise to branchings. Hence, when complex adaptive systems evolve, they do so in connection with a particular maximal quasiclassical domain, which need not be regarded as having somehow been chosen by those systems according to their capabilities. Instead, the location and capabilities of the systems determine the degree of additional coarse graining (in our case, very coarse indeed) that is applied to the particular maximal quasiclassical domain in order to arrive at the domain perceived by the systems.

Suppose that the quantum mechanics of the universe allows mathematically for various possible maximal quasiclassical domains that are genuinely inequivalent. Suppose, too, that complex adaptive systems actually evolved to exploit some coarse graining of each of those maximal quasiclassical domains. Each domain then provides a set of alternative coarse-grained histories of the universe, and information gathering and utilizing systems (IGUSes) record in each case the outcomes of various probabilistic branchings in the tree of possible histories, a tree that would be quite different in the two cases!

If there is some degree of agreement in the phenomena followed by the two otherwise distinct quasiclassical domains, the two IGUSes might become aware of each other and even communicate to some extent. But a great deal of what is followed by one IGUS could not be apprehended directly by the other. Only through a quantum-mechanical calculation or measurement might one IGUS achieve any appreciation of the full range of phenomena perceived by the other (this may remind some people of the relation between men and women).

Could an observer utilizing one domain really become aware that other domains, with their own sets of branching histories and their own observers, were available as alternative descriptions of the possible histories of the universe? This fascinating issue has been raised by science fiction writers (who sometimes use the expression "goblin worlds," according to the Russian theorist Starobinsky), but it is only now receiving enough attention from theorists of quantum mechanics to be properly studied.

Those of us working to construct the modern interpretation of quantum mechanics aim to bring to an end the era in which Niels Bohr's remark applies: "If someone says that he can think about quantum physics without becoming dizzy, that shows only that he has not understood anything whatever about it."

CHAPTER

12

QUANTUM MECHANICS AND FLAPDOODLE

While many questions about quantum mechanics are still not fully resolved, there is no point in introducing needless mystification where in fact no problem exists. Yet a great deal of recent writing about quantum mechanics has done just that.

Because quantum mechanics predicts only probabilities, it has gained a reputation in some circles of permitting just about anything. Is it true that in quantum mechanics anything goes? That depends on whether events of very, very low probability are included. I remember as an undergraduate being assigned the problem of calculating the probability that some heavy macroscopic object would, during a certain time interval, jump a foot into the air as a result of a quantum fluctuation. The answer was around one divided by the number written as one followed by sixty-two zeroes. The point of the problem was to teach us that there is no practical difference between that kind of probability and zero. Anything that improbable is effectively impossible.

When we look at what can really happen with significant probability, we find that many phenomena that were impossible in classical physics are still effectively impossible in quantum mechanics. However, public understanding of this has been hampered in recent years by a rash of misleading references in books and articles to some elegant theoretical work done by the late John Bell and to the results of a related experiment. Some accounts of the experiment, which involves two photons moving in opposite directions, have given readers the false impression that measuring the properties of one photon instantaneously affects the other. Then the conclusion has been drawn that quantum mechanics permits faster-than-light communication, and even that claimed "paranormal" phenomena like precognition are thereby made respectable! How can this have happened?

Einstein's Objections to Quantum Mechanics

The story begins, in a way, with the attitude of Albert Einstein toward quantum mechanics. Although he had helped to prepare the way for it early in the twentieth century with his brilliant work on photons, in which he took seriously the original quantum hypothesis of Max Planck, Einstein never liked quantum mechanics itself. At the Solvay Conference in Brussels in 1930, Einstein gave what purported to be a demonstration that quantum mechanics was inconsistent. Niels Bohr and his allies worked frantically for the next few days to find the flaw in the great man's argument. Sure enough, before the conference ended, they were able to show that Einstein had omitted something; remarkably, it was the effect of general relativity that he had forgotten. Once that was included, the alleged inconsistency disappeared.

After that, Einstein gave up trying to show that quantum mechanics was internally inconsistent. Instead, he concentrated on identifying the principle it violated that he believed a correct theoretical framework should obey. In 1935, together with two young associates, Podolsky and Rosen, he published a paper describing that principle and a hypothetical experimental situation in which quantum mechanics would fail to conform to it. The principle, which he called "completeness," challenged the essential nature of quantum mechanics.

What Einstein required was roughly the following. If, by means of a certain measurement, the value of a particular quantity Q could be predicted with certainty, and if, by an alternative, quite different measurement, the value of another quantity R could be predicted with certainty, then, according to the notion of completeness, one should be able to assign exact values simultaneously to both of the quantities Q and R. Einstein and his colleagues succeeded in choosing the quantities to be ones that cannot simultaneously be assigned exact values in quantum mechanics, namely the position and momentum of the same object. Thus a direct contradiction was set up between quantum mechanics and completeness.

What is the actual relationship in quantum mechanics between a measurement that permits the assignment of an exact value to a particle's position at a given time and another measurement that permits its momentum at the same time to be exactly specified? Those measurements take place on two different branches, decoherent with each other (like a branch of history in which one horse wins a given race and another branch in which a different horse wins). Einstein's requirement amounts to saying that the results from the two alternative branches must be accepted *together.* That clearly demands the abandonment of quantum mechanics.

Hidden Variables

In fact, Einstein did want to replace quantum mechanics with a different kind of theoretical framework. In remarks made elsewhere, he indicated his belief that the successes of quantum mechanics stemmed from theoretical results that were only approximately correct and that represented a kind of statistical average over the predictions of another sort of theory.

Einstein's idea assumed a more definite form when various theorists, at different times, suggested that quantum mechanics might be replaced by a deterministic, classical framework—but one in which there is a very large number of "hidden variables." Those variables may be imagined as describing invisible flies buzzing about everywhere in the universe, more or less at random, interacting with the elementary particles and affecting their behavior. As long as the flies are undetectable, the best the theorist can do in making predictions is to take statistical averages over their motions. But the unseen flies will cause unpredictable fluctuations, creating indeterminacies. The hope was that the indeterminacies would somehow match those of quantum mechanics, so that the predictions of the scheme would agree with quantum-mechanical predictions in the many cases where observation confirms the latter.

Bohm and Einstein

I knew one theorist who vacillated, at least for a time, between believing in quantum mechanics and thinking that it might have to be replaced by something like a "hidden variable" approach. That was David Bohm, who was preoccupied during his whole career with understanding the meaning of quantum mechanics.

In 1951, when I was a fresh Ph.D. and a post-doc at the Institute for Advanced Study in Princeton, David was an assistant professor at Princeton University. We were both bachelors and sometimes spent the evening walking around Princeton together discussing physics. David told me that as a Marxist he had had difficulty believing in quantum mechanics. (Marxists tend to prefer their theories to be fully deterministic.) Since quantum mechanics was immensely successful and not contradicted by any observation, he had tried to convince himself that it was, after all, philosophically acceptable. In attempting to reconcile quantum mechanics with his Marxist convictions, he had written an elementary textbook on quantum theory, emphasizing the problem of interpretation. That book was about to appear, and David was anxious to show Einstein the relevant chapters and see if he could overcome the great man's objections. David asked me to arrange an appointment. I replied that I was not the best person to do so, since I hardly knew Einstein, but that I would talk with Miss Dukas, Einstein's formidable secretary, and see what could be done.

When I met David a day or two later and started to tell him that I was working on his appointment, he interrupted me excitedly to report that it was unnecessary. His book had appeared and Einstein had already read it and telephoned him to say that David's was the best presentation he had ever seen of the case against him and that they should meet to discuss it. Naturally, when next I saw David I was dying to know how their conversation had gone, and I asked him about it. He looked rather sheepish and said, "He talked me out of it. I'm back where I was before I wrote the book." From then on, for more than forty years, David tried to reformulate and reinterpret quantum mechanics so as to overcome his doubts. Very recently, I learned with great sadness that he had died.

The EPRB Experiment

Many years ago David Bohm proposed replacing the hypothetical "completeness" experiment of Einstein, Podolsky, and Rosen (which need not be described here) by a modified and more practical version. Bohm's experiment (called EPRB after the four physicists) involves the decay of a particle into two photons. If the particle is at rest and has no internal "spin," then the photons travel in opposite directions, have equal energy, and have identical circular polarizations. If one of the photons is left-circularly-polarized (spinning to the left), so is the other; likewise if one is right-circularly-polarized (spinning to the right), so is the other. Furthermore, if one is plane-polarized along a particular axis (that is, has its electric field vibrating along that axis), then the other one is plane-polarized along a definite axis. There are two cases, depending on the character of the spinless particle. In one case the plane polarization axes of the two photons are the same. In the other they are perpendicular. For simplicity let us take the former case, even though in the practical situation (where the decaying particle is a neutral pi meson) the latter case applies.

The setup is assumed to be such that nothing disturbs either photon until it enters a detector. If the circular polarization of one of the photons is measured by the detector, the circular polarization of the other is certain—it is the same. Similarly, if the plane polarization of one of the photons is measured, that of the other photon is certain—again, it is the same as that of the first photon. Einstein's completeness would imply that both the circular and plane polarization of the second photon could then be assigned definite values. But the values of the circular polarization and the plane polarization of a photon cannot be exactly specified at the same time (any more than the position and momentum of a particle can be so specified). Consequently, the requirement of completeness is just as unreasonable in this case, from the point of view of quantum mechanics, as in the case discussed by Einstein and his colleagues. The two measurements, one of circular and the other of plane polarization, are alternatives; they take place on different branches of history and there is no reason for the results of both to be considered together.

EPRB and the Hidden Variable Alternative

Later on, some theoretical work of John Bell revealed that the EPRB experimental setup could be used to distinguish quantum mechanics from hypothetical hidden variable theories, by means of certain polarization measurements on both photons. Bell's theorem (also called Bell's inequalities) concerns a particular quantity that specifies the correlation between the polarizations of the two photons. In quantum mechanics, that quantity can attain values that are not allowed in a classical hidden variable theory.

After the publication of Bell's work, various teams of experimental physicists carried out the EPRB experiment. The result was eagerly awaited, although virtually all physicists were betting on the correctness of quantum mechanics, which was, in fact, vindicated by the outcome. One might have expected that interested people all over the world would heave a sigh of relief on hearing the news and then get on with their lives. Instead, a wave of reports began to spread alleging that quantum mechanics had been shown to have weird and disturbing properties. Of course, it was the same old quantum mechanics. Nothing was new except its confirmation and the subsequent flurry of flapdoodle.

The Story Distorted

The principal distortion disseminated in the news media and in various books is the implication, or even the explicit claim, that measuring the polarization, circular or plane, of one of the photons somehow affects the other photon. In fact, the measurement does not cause any physical effect to propagate from one photon to the other. Then what does happen? If, on a particular branch of history, the plane polarization of one photon is measured and thereby specified with certainty, then on the same branch of history the plane polarization of the other photon is also specified with certainty. On a different branch of history the circular polarization of one of the photons may be measured, in which case the circular polarization of both photons is specified with certainty. On each branch, the situation is like that of Bertlmann's socks, described by John Bell in one of his papers. Bertlmann is a mathematician who always wears one pink and one green sock. If you see just one of his feet and spot a green sock, you know immediately that his other foot sports a pink sock. Yet no signal is propagated from one foot to the other. Likewise no signal passes from one photon to the other in the experiment that confirms quantum mechanics. No action at a distance takes place. (The label "nonlocal" applied by some physicists to quantum-mechanical phenomena like the EPRB effect is thus an abuse of language. What they mean is that *if interpreted classically in terms of hidden variables,* the result would indicate nonlocality, but of course such a classical interpretation is wrong.)

The false report that measuring one of the photons immediately affects the other leads to all sorts of unfortunate conclusions. First of all, the alleged effect, being instantaneous, would violate the requirement of relativity theory that no signal—no physical effect—can travel faster than the speed of light. If a signal were to do so, it would appear to observers in some states of motion that the signal was traveling backward in time. Hence the limerick:

> There was a young lady named Bright
> Who could travel much faster than light.
> She set out one day, in a relative way,
> And returned home the previous night.

Next, certain writers have claimed acceptability in quantum mechanics for alleged "paranormal" phenomena like precognition, in which the results of chance processes are supposed to be known in advance to "psychic" individuals. Needless to say, such phenomena would be just as upsetting in quantum mechanics as in classical physics; if genuine, they would require a complete revamping of the laws of nature as we know them.

A third manifestation of flapdoodle is the submission of proposals, for example to the U. S. Department of Defense, for using quantum mechanics to achieve faster-than-light communication in military contexts. One wonders whether the advent of this new category of far-out requests means a declining number in more old-fashioned areas, like antigravity and perpetual motion. If not, the bureaucracy that deals with them must be growing.

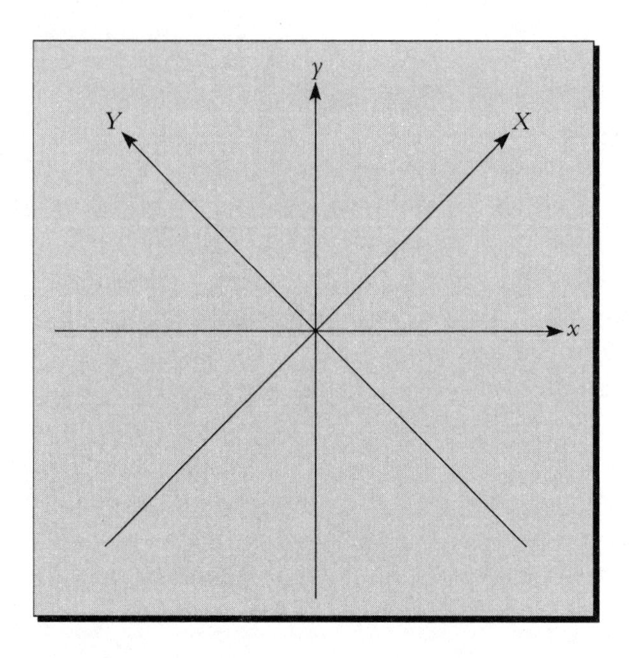

Axes for plane polarizations used in quantum cryptography.

Serious Potential Applications of EPRB

Meanwhile, serious researchers have begun to think about ways in which the EPRB effect might actually be put to use. Instead of crackpot ideas, they have come up with some fascinating potential applications. For example, Charlie Bennett, Gilles Brassard, and Artur Ekert have been working on a form of quantum cryptography in which the EPRB effect is used over and over again to generate a string of randomly generated bits known to two people and to no one else. That string can then be used as the basis of an unbreakable cipher for transmitting messages secretly between the two individuals.

The method works roughly as follows. Suppose a steady supply of EPRB photon pairs is available to Alice and Bob. Of each pair, one photon goes to Alice and one to Bob. They agree in advance to make a long series of plane polarization measurements of their respective photons, on half the occasions distinguishing between two perpendicular directions called x and y, and on the other half distinguishing between two other mutually perpendicular directions (halfway between x and y) called X and Y. (The X and Y axes are rotated 45 degrees from the x and y axes, as shown in the illustration above.) Alice chooses at random, for each of her photons, whether it will be subjected to an x versus y or an X versus Y measurement. Bob does the same, separately and independently.

Once the work is finished, Alice tells Bob which kind of measurement she made on each of her photons, x versus y or X versus Y, and Bob gives Alice the analogous information. (The conversation can take place over a public telephone and be overheard by spies without doing any harm.) They learn on which occasions they both made the same kind of measurement (that will have happened about half the time). For each such common measurement, the results obtained by Alice and Bob must be identical, because of the EPRB effect. The results of those common measurements are then known to both of them and to no one else (assuming each made the measurements in secret and did not divulge the results). Those results can be represented as a string of 1s (standing for x or X) and 0s (standing for y or Y) known only to Alice and Bob. That string can then serve as the basis for an unbreakable secret cipher to be used between them.

If Alice and Bob are especially worried about security, they can waste the results of a few of the measurements they made in common, comparing the corresponding 1s and 0s over an open telephone line to make sure they are really identical (while using the rest of the 1s and 0s for their secret messages). Any spy who had somehow been making his or her own measurements on the photons would thereby have destroyed the perfect agreement between Alice's and Bob's results. The comparison of some of those results would reveal the work of the spy.

Quantum cryptography does not actually require the EPRB effect. Subsequently, a group of six physicists (including Bennett) has invented a clever procedure, in which EPRB *is* essential, for destroying a photon and creating one in the same state of polarization but elsewhere (i.e., with a different probability distribution in space).

As we have continued to learn more and more about the elementary particle system, a remarkable interplay has developed between the apparent complexities revealed by experiment and the simplification achieved by theory. The discovery of many different kinds of particles and several different types of interaction among them has created and reinforced the impression that particle physics is complicated. At the same time, on the theoretical side, progress toward unification in the description of particles and interactions has uncovered more and more of the underlying simplicity. Although elementary particle physics is much less than a century old, we may already be at the stage where the unity of the subject is beginning to reveal itself, in the form of a single principle that is expected to predict the existence of the observed diversity of elementary particles.

CHAPTER

13

QUARKS AND ALL THAT: THE STANDARD MODEL

All respectable theorizing about the elementary particles is carried out within the framework of quantum field theory, which includes both the standard model and superstring theory. Quantum field theory is based on three fundamental assumptions: the validity of quantum mechanics, the validity of Einstein's relativity principle (special relativity when gravitation is not included, and general relativity otherwise), and locality (meaning that all fundamental forces arise from local processes and not by means of action at a distance). Those local processes involve the emission and absorption of particles.

QED—Quantum Electrodynamics

The first successful example of quantum field theory was quantum electrodynamics (QED), the theory of the electron and the photon. The electron is a fermion (that is, it obeys the Pauli exclusion principle) and it possesses one fundamental unit of electric charge (labeled "negative" according to a convention that dates back to Benjamin Franklin). The photon is a boson (in other words, it obeys the antiexclusion principle) and it is electrically neutral.

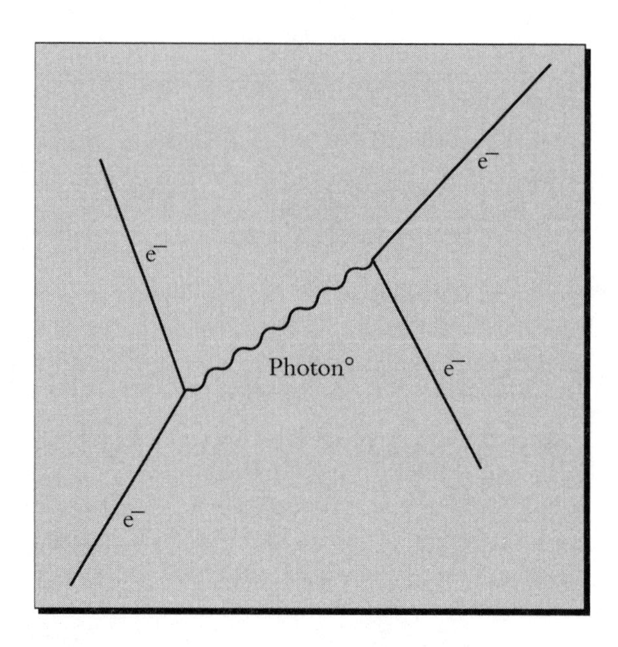

Two electrons exchanging a virtual photon, which gives rise to the electromagnetic force between them.

In quantum electrodynamics the electromagnetic force between two electrons comes about through the emission of a photon by one electron and its absorption by the other. If you know some classical physics, you may object that for an electron to emit a photon (that is, to turn into an electron plus a photon) violates the principle of the conservation of energy, the conservation of momentum, or both; likewise for the absorption of a photon. If you know some quantum physics, however, you are probably aware that conservation of energy need not hold over finite time intervals, only in the long run. That property of quantum mechanics can be regarded as a manifestation of Heisenberg's uncertainty principle applied to energy and time. The system can borrow some energy for a while to permit the first electron to emit a photon, and the energy can then be returned when the other electron absorbs the photon. Such a process is called the "virtual" exchange of a photon between electrons. The photon is emitted and absorbed only in the Pickwickian sense of quantum mechanics.

For any quantum field theory, we can draw funny little pictures, invented by my late colleague Dick Feynman, which give the illusion of understanding what is going on. In the one on the facing page electrons are virtually exchanging a photon to give the electromagnetic force between them. Each electron is labeled "e," with a minus sign attached to indicate its single unit of negative electric charge. The photon similarly carries a nought superscript to indicate its electrical neutrality. An "e" with a positive sign would represent the positron, the antiparticle of the electron. But what is an antiparticle?

Particle–Antiparticle Symmetry

Quantum field theory turns out to imply a fundamental symmetry of the elementary particle system between particles and their "antiparticles." For every particle there is a corresponding antiparticle, which behaves like the particle moving backwards in space and time. The antiparticle of the antiparticle is the particle itself. If two particles are antiparticles of each other, they have opposite electrical charges (that is, charges of equal magnitude but opposite sign) and the same mass. The antiparticle of the electron is called the positron because of its positive electric charge. Some electrically neutral particles, such as the photon, are their own antiparticles.

When Dirac published his relativistic equation for the electron in 1928, he opened the way for the discovery of quantum electrodynamics, which followed soon afterwards. The interpretation of the Dirac equation pointed to the necessity of the positron, but initially Dirac did not actually predict the existence of that particle. Instead, he indicated that somehow the expected positively charged object might be identified with the proton, which was well known experimentally but is almost two thousand times heavier than the electron (from which it differs in other important ways as well). When I asked him, many decades later, why he had not immediately predicted the positron, Dirac replied in his usual pithy manner, "Pure cowardice."

It was left to experimentalists to make the discovery. The positron turned up in 1932 in the laboratories of my late colleague Carl Anderson at Caltech and Patrick Blackett in England; they shared a Nobel prize in physics a few years later. Their experiments established that the particle–antiparticle symmetry of quantum field theory is a real phenomenon.

To a great extent the standard model can be regarded as a generalization of quantum electrodynamics. The electron and positron are supplemented by many other particle-antiparticle pairs of fermions, and the photons are supplemented by other quanta. Just as the photon is the carrier or quantum of the electromagnetic force, so the other quanta carry other fundamental forces.

Quarks

For a long time it was thought that among the particles accompanying the electron on the list of fundamental fermions would be the neutron and proton, the constituents of atomic nuclei. However, that turned out to be false; the neutron and proton are not elementary. Physicists have learned on other occasions as well that objects originally thought to be fundamental turn out to be made of smaller things. Molecules are composed of atoms. Atoms, although named from the Greek for uncuttable, are made of nuclei with electrons around them. Nuclei in turn are composed of neutrons and protons, as physicists began to understand around 1932, when the neutron was discovered. Now we know that the neutron and proton are themselves composite: they are made of quarks. Theorists are now quite sure that it is the quarks that are analogues of the electron. (If, as seems unlikely today, the quarks should turn out to be composite, then the electron would have to be composite as well.)

In 1963, when I assigned the name "quark" to the fundamental constituents of the nucleon, I had the sound first, without the spelling, which could have been "kwork." Then, in one of my occasional perusals of *Finnegans Wake,* by James Joyce, I came across the word "quark" in the phrase "Three quarks for Muster Mark." Since "quark" (meaning, for one thing, the cry of a gull) was clearly intended to rhyme with "Mark," as well as "bark" and other such words, I had to find an excuse to pronounce it as "kwork." But the book represents the dream of a publican named Humphrey Chimpden Earwicker. Words in the text are typically drawn from several sources at once, like the "portmanteau words" in *Through the Looking Glass*. From time to time, phrases occur in the book that are partially determined by calls for drinks at the bar. I argued, therefore, that perhaps one of the multiple sources of the cry "Three quarks for Muster Mark" might be "Three quarts for Mister

Mark," in which case the pronunciation "kwork" would not be totally unjustified. In any case, the number three fitted perfectly the way quarks occur in nature.

The recipe for making a neutron or proton out of quarks is, roughly speaking, "Take three quarks." The proton is composed of two "*u* quarks" and one "*d* quark," while the neutron contains two "*d* quarks" and one "*u* quark." The *u* and *d* quarks have different values of the electric charge. In the same units in which the electron has an electric charge of −1, the proton has a charge of +1, while the neutron has charge 0. The charge of the *u* quark in those same units is 2/3 and that of the *d* quark −1/3. Sure enough, if we add 2/3, 2/3, and −1/3, we get 1 for the charge of the proton; and if we add −1/3, −1/3, and 2/3, we get 0 for the charge of the neutron.

The *u* and *d* are said to be different "flavors" of quark. Besides flavor, the quarks have another, even more important property that is called "color," although it has no more to do with real color than flavor in this context has to do with the flavors of frozen yoghurt. While the name color is mostly a joke, it also serves as a kind of metaphor. There are three colors, labeled red, green, and blue after the three basic colors of light in a simple theory of human color vision. (In the case of paints, the three primary colors are often taken to be red, yellow, and blue, but for mixing lights instead of paints for their effect on human observers, yellow is replaced by green.) The recipe for a neutron or proton is to take one quark of each color, that is, a red quark, a green quark, and a blue quark, in such a way that color averages out. Since, in vision, white can be regarded as a mixture of red, green, and blue, we can use the metaphor to say that the neutron and proton are white.

Quarks Confined

Quarks have the remarkable property of being permanently trapped inside "white" particles such as the neutron and proton. It is only white particles that are directly observable in the laboratory. Color is averaged out in the observable particles and only inside them can colored objects exist. Likewise the electric charge of an observable object is always a whole number (such as 0, 1, −1, or 2), and fractionally charged particles can exist only inside.

When I proposed the existence of quarks, I believed from the beginning that they were permanently confined in some way. I referred to such quarks as "mathematical," explaining carefully what I meant by the term, and contrasted them with what I called "real quarks," which would be capable of emerging so that they could be detected singly. The reason for the choice of language is that I didn't want to face arguments with philosophically inclined critics demanding to know how I could call quarks "real" if they were always hidden. The terminology proved unfortunate, however. Numerous authors, ignoring my explanation of the terms "mathematical" and "real," as well as the fact that the situation I was describing is the one now generally accepted as correct, have claimed that I didn't really believe the quarks were there! Once such a misunderstanding becomes established in popular literature, it tends to perpetuate itself, because the various writers often simply copy one another.

Colorful Gluons

For quarks to be confined as they are, the forces between them must be very different from familiar forces like electromagnetism. How does that difference arise?

Just as the electromagnetic force between electrons is generated by the virtual exchange of photons, so the quarks are bound to one another by a force that comes from the exchange of other quanta, called gluons because they glue the quarks together to make observable white objects like the neutron and the proton. The gluons pay no attention to flavor—we might say they are "flavor-blind." However, they are very sensitive to color. In fact, color plays the same kind of role for them that electric charge plays for the photon: the gluons interact with color much as the photon interacts with electric charge.

The triple nature of color requires the gluons to have a property not shared by the photon: for different color situations, there are different gluons. In the top sketch on the facing page, a red quark is shown turning into a blue one with the virtual emission of a red-blue gluon, which is virtually absorbed by a blue quark, turning it into a red one. Another color situation is exhibited in the lower sketch, where a blue quark turns into a green one, virtually emitting a blue-green gluon,

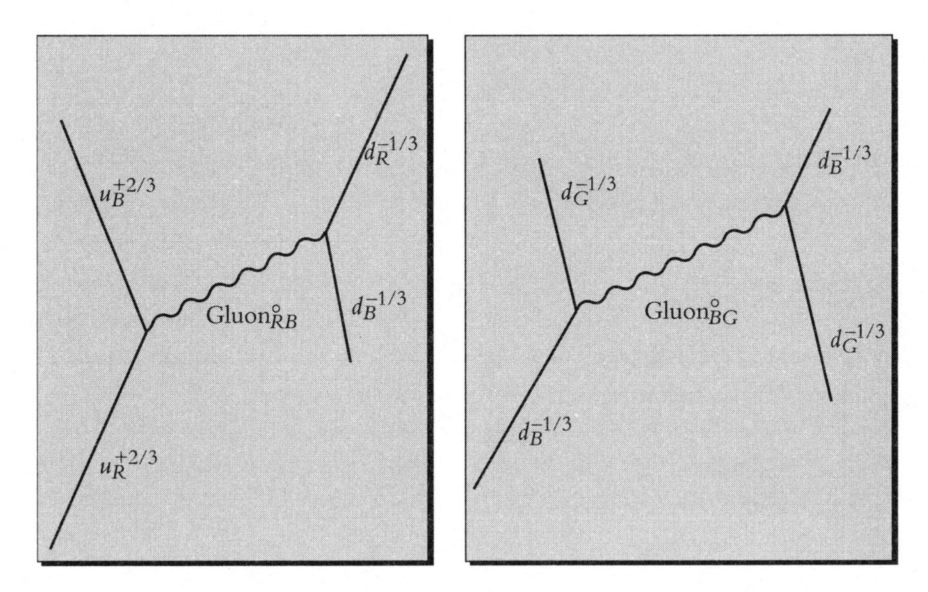

Forces between quarks from gluon exchange.

the virtual absorption of which turns a green quark into a blue one. (Note, by the way, that the antiparticle of a gluon is also a gluon; for example, blue-green and green-blue gluons are antiparticles of each other.) The flavor labels have been chosen to be different in the two sketches to illustrate the irrelevance of flavor to the color processes mediated by gluons.

Quantum Chromodynamics

Around 1972, a number of us contributed to the formulation of a definite quantum field theory of quarks and gluons. I named it quantum chromodynamics, using the Greek root *chromo* meaning color. It seems to be the correct theory and is generally recognized as such, although a considerable amount of mathematical work is still required before we can be sure that its detailed quantitative predictions agree with observation—confirming that the quarks, antiquarks, and gluons (of which all nuclear objects such as the neutron and proton are composed) truly behave according to the rules of quantum chromodynamics.

	QED	QCD		
		$u_R^{+2/3}$	$d_R^{-1/3}$	....
Fermions	e^-	$u_G^{+2/3}$	$d_G^{-1/3}$	....
		$u_B^{+2/3}$	$d_B^{-1/3}$	....
Quanta (bosons)	Photon°	Colorful gluons°		
		$\bar{u}_R^{-2/3}$	$\bar{d}_R^{+1/3}$	....
Fermions	e^+	$\bar{u}_G^{-2/3}$	$\bar{d}_G^{+1/3}$	....
		$\bar{u}_B^{-2/3}$	$\bar{d}_B^{+1/3}$	....

Antiparticles of each other

Comparison of QED and QCD. The quarks and antiquarks are coupled to the photon through their electric charges, but the electron and positron are not coupled to the gluons.

To compare quantum electrodynamics (QED) and quantum chromodynamics (QCD), we can construct the sort of dictionary shown above. In QED, electrons and positrons interact through the virtual exchange of photons, while in QCD quarks and antiquarks similarly interact through the virtual exchange of gluons. The electromagnetic force arises from electric charges; we can think of the color force as arising from color charges. Both electric and color charges are perfectly conserved—just as electric charge cannot be created or destroyed, neither can color charge.

However, there is a crucial difference between the two theories: in QED the photon, which carries the electromagnetic interaction, is electrically neutral, while in QCD the gluons, which carry the color force, are themselves colorful. Because the gluons are not without color, they interact among themselves in a way that photons do not, and that gives rise to terms in the equations of QCD that have no analogue in QED. As a result, the color force turns out to behave very differently from electromagnetism or any other previously known force: at large distances, it does not die away. That property of QCD explains why colored quarks and antiquarks and colorful gluons are permanently confined inside white objects such as the neutron and proton. The color force acts something like a spring holding them together.

Although quarks are trapped forever and cannot be directly detected in the laboratory, beautiful experiments have been performed that confirm their existence inside the proton. For example, a beam of high-energy electrons can be used to make a kind of electron micrograph of the proton's interior, and sure enough the quark structure is revealed. I was delighted when my friends Dick Taylor, Henry Kendall, and Jerry Friedman shared the Nobel prize in physics for such an experiment. (I only wish I had noticed in advance that this would be a good way to confirm the existence of quarks.)

Simplicity Revealed by QCD

In an atomic nucleus, neutrons and protons are bound together (unlike quarks, they are not confined and can be extracted individually). Now that those particles are known to be composed of quarks, how are the nuclear forces between them to be described? When I was a graduate student, the character of those forces was one of the great mysteries we hoped some day to solve. Most theorists now believe that QCD has supplied the solution, although the relevant calculations are by no means complete. The situation is analogous to that of the forces between atoms or molecules, which were explained in the late 1920s after the discovery of quantum mechanics. Those forces are not in any way fundamental, but only indirect consequences of the electromagnetic force treated quantum-mechanically. Similarly, the nuclear force is not fundamental but arises as an indirect effect of the color force, which in turn comes from the quark–gluon interaction.

The neutron and proton are not the only observable (white) nuclear particles, although they are the best known. Hundreds of other nuclear particle states have been discovered since the late 1940s in high energy collisions, first in cosmic ray experiments and then at high energy particle accelerators. They have all now been explained as composites of quarks, antiquarks, and gluons. The quark scheme, embodied in the explicit dynamical theory of quantum chromodynamics, has thus exposed the simplicity underlying an apparently very complicated pattern of states. Moreover, those states all interact with one another through the "strong interaction," which includes the nuclear force. The many manifestations of the strong interaction are all thought to be describable as indirect consequences of the fundamental quark–gluon interaction. Quantum chromodynamics has thus revealed the simplicity of the strong interaction as well as that of the nuclear particle states, which are the participants in that interaction.

Electron and Electron Neutrino—The Weak Force

Important as they are, we know that there is more to matter than the nuclear particles and their fundamental constituents. The electron, for example, does not possess color and pays no attention to the color force or to the resulting nuclear force. In a heavy atom, the innermost electrons that surround the nucleus actually spend a good deal of their time inside, but do not feel the nuclear force, although naturally they are susceptible to electromagnetic effects such as the electrical pull of the protons.

Although the electron does not possess color, it does have flavor. Just as the *d* quark has the *u* quark as its flavor partner, so the electron has the electron neutrino as its partner. The electron neutrino is something of a silent partner because, being electrically neutral, it ignores not only the nuclear force (as the electron does), but also the electromagnetic force. It can pass through the earth, for example, with very little probability of interacting. The neutrinos produced by the thermonuclear reactions in the center of the sun reach the surface of the earth by raining down on us during the day, but at night they come up at us through the earth. When the writer John Updike read about that aspect of neutrino behavior, he was inspired to write the following, entitled "Cosmic Gall":

Neutrinos, they are very small.
    They have no charge and have no mass
And do not interact at all.
The earth is just a silly ball
    To them, through which they simply pass,
Like dustmaids down a drafty hall
    Or photons through a sheet of glass.
    They snub the most exquisite gas,
Ignore the most substantial wall,
    Cold-shoulder steel and sounding brass,
Insult the stallion in his stall,
    And, scorning barriers of class,
Infiltrate you and me! Like tall
And painless guillotines, they fall
    Down through our heads into the grass.
At night, they enter at Nepal
    And pierce the lover and his lass
From underneath the bed—you call
    It wonderful; I call it crass.

(In the third line it is tempting to employ scientific license and alter "do not" to "scarcely.")

Unfortunately detection of solar neutrinos is still fraught with many problems. The rate of detection seems to be lower than predicted, leading physicists to propose various explanations of varying degrees of plausibility. My colleague Willy Fowler once went so far as to suggest that maybe the nuclear furnace at the center of the sun went out some time ago, but the energy transfer mechanisms in the sun are so slow that the news has not yet reached the surface. Not many people believe that is the correct explanation, but if it is, then we are heading for a real energy crisis some day.

How can neutrinos be produced in the center of the sun and how can they be detected in laboratories here on earth if they are subject neither to the strong force nor to the electromagnetic one? Another force, the so-called weak force, is responsible. The electron neutrino does participate in that interaction, along with the electron. Hence the suggested revision of John Updike's phrase "do not interact at all."

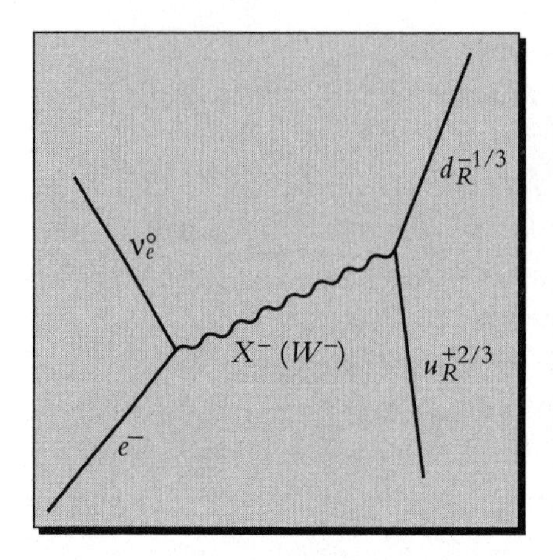

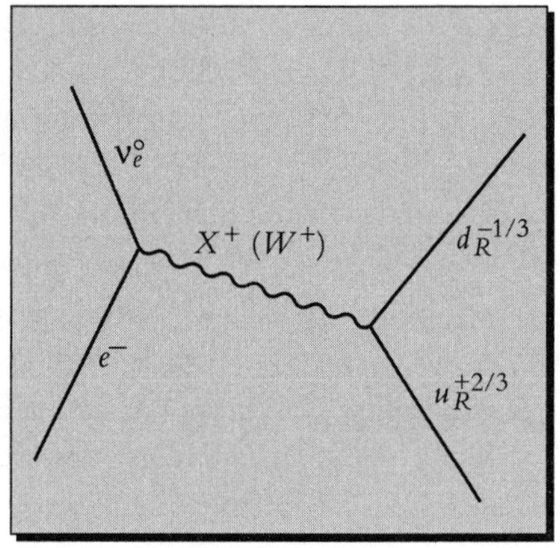

An electron turning into an electron neutrino while a *u* quark turns into a *d* quark. Two versions of the same Feynman diagram.

The weak force gives rise to reactions like the following:

1. An electron turns into an electron neutrino while a proton turns into a neutron. This reaction is an example of how neutrinos can be produced; the proton involved is part of a heavy nucleus and the electron is in one of the innermost electron states around that nucleus, inside of which it spends a sizable fraction of its time.
2. The inverse process, in which an electron neutrino turns into an electron while a neutron turns into a proton. This illustrates how a neutrino can be detected, with the target neutron being inside a nucleus.

Since neither the neutron nor the proton is elementary, however, such reactions are not the basic processes, which involve quarks instead:

1. An electron turns into an electron neutrino while a *u* quark turns into a *d* quark.

2. An electron neutrino turns into an electron while a *d* quark turns into a *u* quark.

These reactions involve a change of flavor, both for the electron turning into an electron neutrino (or vice versa) and for the *u* quark turning into a *d* quark (or vice versa). As in any such process in quantum field theory, a quantum is being exchanged. For each of the two reactions, (the first of which is illustrated on the facing page) there are two versions of the same Feynman diagram, one version involving the exchange of a positively charged quantum and the other the exchange of a negatively charged quantum. The existence of such quanta was first discussed by some of us in the late 1950s, and they were discovered at CERN twenty-five years later, in experiments that procured a Nobel Prize for Carlo Rubbia and Simon van der Meer. The quanta are usually called W^+ and W^-, as they were designated in a celebrated paper by T. D. Lee and C. N. Yang, but I still often refer to them by the names X^+ and X^- that Dick Feynman and I used to employ.

Quantum Flavor Dynamics and the Neutral Weak Force

Both the electromagnetic and weak forces can be regarded as flavor forces, since the electric charge varies with flavor and the weak force involves the changing of flavor. During the 1950s and 1960s, a sort of quantum flavor dynamics was formulated, incorporating both quantum electrodynamics and a theory of the weak force. Quantum flavor dynamics (associated particularly with the names of Sheldon Glashow, Steven Weinberg, and Abdus Salam) successfully predicted, among other things, the existence of a new flavor force that causes simple scattering of electron neutrinos off neutrons or protons, without change of flavor.

More basically, in terms of quarks, the new force causes scattering of electron neutrinos off *u* and *d* quarks, again without any change of flavor. The scattering takes place through the exchange of a new, electrically neutral quantum called Z^0, as illustrated on page 190. The existence of that quantum, too, was confirmed by Rubbia, van der Meer, and their colleagues.

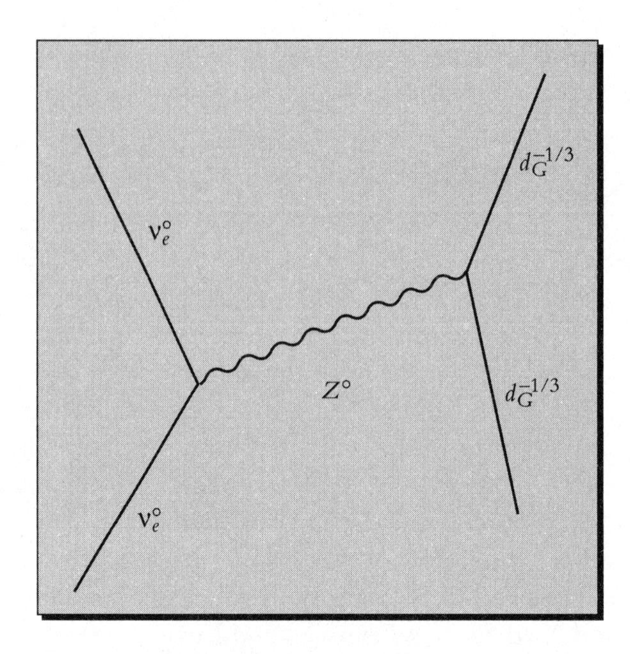

The scattering of an electron neutrino off a *d* quark.

Families of Fermions

The diagram on page 191 summarizes what we have said so far about particles and forces. There is a family of fermions composed of the electron and its neutrino and two flavors of tricolored quarks; a corresponding antifamily consists of the positron and the anti–electron–neutrino and two flavors of tricolored antiquarks. Coupled to the color variable (which does not exist for the electron and its neutrino and their antiparticles) are the colorful gluons of quantum chromodynamics. Coupled to the flavor variable, which does exist for the entire family and for the entire antifamily, are the four quanta of quantum flavor dynamics.

That fermion family, it turns out, does not stand alone. There are two more such families, with very similar structure. Each consists of an electron-like particle, a corresponding neutrino, and two flavors of quark with the electric charges –1/3 and +2/3, like the *d* and *u* respectively.

The electron-like particle in the second family is the muon, discovered by Carl Anderson and Seth Neddermeyer at Caltech in 1937. It is a heavy version of the electron, about two hundred times heavier, and it has its own neutrino—the muon neutrino. The quarks in the second fermion family are the "strange" *s* quark (analogous to *d*) and the "charmed" *c* quark, (analogous to *u*). Like the muon, they are heavier than their analogues in the first family.

A third family of fermions is also known, consisting of the tauon (about twenty times as heavy as the muon); the tauon neutrino; the *b* (or "bottom") quark, with a charge of −1/3; and the *t* (or "top") quark, with a charge of +2/3, for which two independent teams of experimentalists have recently provided convincing evidence. If they had not confirmed the existence of the top quark, we theorists would have had to "fall on our fountain pens," as my former colleague Marvin "Murph" Goldberger used to put it. These days, though, fountain pens are scarce. Besides, the ancient Roman hero who wanted to kill himself after a defeat had a trusty retainer to hold his sword—it is not clear if a pen could be held steady enough by a graduate student.

Can there be additional fermion families besides the three that are known? Light was thrown on that question by a recent experiment on the rate of disintegration of the Z^0 quantum. The result was in agreement with theoretical predictions that allow for the decay of Z^0 into any of three different kinds of neutrino–antineutrino pairs, corresponding precisely to electron, muon, and tauon neutrinos. There is no room for a fourth kind of neutrino unless it has a gigantic mass, unlike the

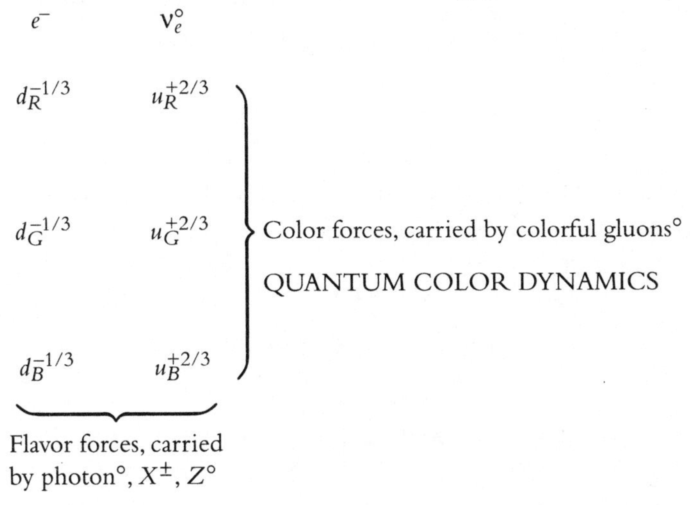

Elementary particles and forces discussed so far. (The antiparticles of the fermions are omitted from the diagram for the sake of simplicity.)

other three, which are very light. A fourth family is thus excluded, unless its neutrino is very different from the others.

With the three families of fermions, their antiparticles, and the quanta of the electromagnetic, weak, and gluonic interactions we have come nearly to the end of our description of the standard model, and it is still a fairly straightforward generalization of QED. The photon is accompanied by other quanta and the electron by other fermions. The patterns of those quanta and fermions, including their various masses and the strengths of the forces mediated by the quanta, show some apparent complexity. But the standard model is not yet the fundamental theory, and only at the fundamental level should the full simplicity of the underlying theory be revealed.

The Zero-Mass Approximation

One way to bring out the simplicity that the standard model does possess is to consider an approximation in which all the particles mentioned so far are assigned zero mass, meaning that they always travel at the speed of light and can never be brought to rest. When the quanta of the weak interaction are treated as massless, the fundamental similarity of the three interactions is made manifest. Quantum flavor dynamics and quantum chromodynamics have related mathematical structures; they belong to the same class of theories—so-called gauge theories, or Yang–Mills theories (as generalized by Shelly Glashow and me long ago).

When the fermions are assigned zero mass as well, a great deal of symmetry appears in the system of fermions. In particular, the three families then have identical properties.

The question that immediately arises is how the zero-mass approximation gets broken. But before describing the mechanism that induces nonzero masses, we should take a look at the actual values.

Masses (or Energies) Large and Small

In dealing with masses and energies, it is essential to make use of Einstein's celebrated relation between mass and energy, which states that a particle with nonzero mass, when at rest, has an energy equal to its mass times *c* squared, where *c* is the velocity of light. This equivalence of mass and rest energy can be used to assign an energy equivalent to any mass. The neutron and proton masses, when thus converted to energy, are both rather close to the energy unit known as the GeV or giga-electron-volt. The prefix giga- is used to indicate a billion; a GeV is thus the energy that would be acquired by an electron accelerated from rest by an electrical potential difference of one billion volts. It is a convenient unit in which to measure the energy equivalents of particle masses.

The nonzero masses of the elementary particles in the standard model are mostly quite different from one another. The electron mass is around one two-thousandth of a GeV. The neutrino masses, if they are not actually zero, are at most on the order of a hundred-millionth of a GeV. The tauon mass is approximately 2 GeV. The *X* (or *W*) plus and minus and Z^0 bosons have masses in the vicinity of 100 GeV. The heaviest quark, the *t* quark, is expected to have a mass around 170 GeV. All these masses violate the special symmetries of the zero-mass approximation.

Spontaneous Symmetry Violation

What makes those masses nonzero, and what makes them so different from one another? The mechanism that operates in the standard model is at least partially understood. It is connected with the existence of a new kind (or new kinds) of boson. At least one such type of boson may be observable at energies available now or soon to be available at the new accelerator at CERN. Such a particle is called a Higgs boson (or higgson). It was actually discussed not only by Peter Higgs of Edinburgh (in a beautiful piece of theoretical work), but also, in various ways, by several other elementary particle physicists, including Tom Kibble, Gerald Guralnik and C. R. Hagen, and Robert Brout and François Englert. In addition, it was proposed in general terms even earlier by my friend Philip Anderson, a theoretical condensed matter physicist and currently Vice-Chairman of the Science Board of the Santa Fe Institute. He won a Nobel prize for his work in condensed matter physics, but his anticipation of the general idea of the Higgs boson has not been widely recognized by elementary particle physicists. I cannot avoid a sneaking suspicion that if his contribution had been more generally acknowledged we would be spared some of his eloquent public arguments against the construction of new particle accelerators. Is there any way he could strongly oppose a machine that is built, in part, to try to find the Anderson-Higgs boson, or even the Higgs-Anderson boson?

For the sake of fairness, I suggest we retain the term "higgson" but use the name "Anderson-Higgs" to label the mechanism that breaks the symmetry of the zero-mass approximation and is responsible for the various different non-zero particle masses in the standard model. The Anderson-Higgs mechanism is a special case of a more general process called spontaneous symmetry breaking.

For a familiar example of that process, think of an ordinary magnet, in which all the tiny atomic magnets are aligned with one another. The equations for the elementary particles composing the magnet, in interaction with one another but subject to no outside influence, are perfectly symmetrical with respect to directions in space: they are indifferent, so to speak, to the direction in which the magnet points. But then any external influence, however feeble, (say a very weak external magnetic field), can determine the magnet's orientation, which would otherwise be totally arbitrary.

The equations for the particles composing the magnet possess a symmetry because they treat all directions alike, but each individual solution of the equations violates the symmetry by pointing in a definite direction. The set of all those unsymmetrical solutions does, however, possess the symmetry because every direction corresponds to a solution and the set of all directions is perfectly symmetrical.

The essence of spontaneous symmetry breaking lies in this very circumstance: equations with a particular symmetry can have solutions that individually violate that symmetry, although the set of all solutions is symmetrical.

The greatest virtue of the Anderson-Higgs mechanism of spontaneous symmetry breaking is that it permits the fermions and the quanta of the weak interaction to acquire nonzero masses without introducing disastrous infinities into the calculations of quantum flavor dynamics. Particle theorists had searched for some time for such a "soft" mechanism for producing nonzero masses before it was shown that the higgson would do the job.

Violation of Time Symmetry

The Anderson-Higgs mechanism may be responsible not only for the nonzero masses in the standard model but also for the observed small deviation from symmetry under reversal of time in elementary particle physics. The equations of the underlying fundamental theory would then be symmetrical under time reversal. (Indeed, heterotic superstring theory, the only serious candidate for the unified theory of elementary particles, does possess that symmetry.) The violation would represent another instance of symmetrical equations having a symmetrical set of unsymmetrical solutions, just one of which is found in nature. In this case there would be two solutions, differing in the direction of time.

This violation of time symmetry at the elementary particle level does not in any case seem capable of explaining the arrow (or arrows) of time—the conspicuous differences that we continually observe between events moving forward in time and the corresponding time-reversed version of those events. Those conspicuous differences arise instead from the special initial condition at the beginning of the expansion of the universe, as we have already mentioned and as we shall discuss in more detail later on.

Violation of Matter–Antimatter Symmetry

If the mathematical operation that interchanges forward and backward time is combined with the interchange of left and right and also with the interchange of matter and antimatter, the resulting operation (called CPT) is an exact symmetry of quantum field theory. Thus it should not come as a complete surprise that the spontaneous violation of time symmetry also violates the symmetry between matter and antimatter. Can that violation be responsible for the gross asymmetry of the world around us, where just about everything is composed of matter, and antimatter is produced only in rare collisions at high energy?

That proposal was made years ago by Andrei Sakharov, the late Russian physicist well known for his coming up (along with Ya. B. Zel'dovich) with the crucial idea for the Soviet hydrogen bomb and later for his struggles on behalf of peace and human rights in the Soviet

Union. Sakharov put together a package of ideas that has undergone considerable modification at the hands of other theoretical physicists but has always included the following point. The very early universe was symmetrical between matter and antimatter, but soon produced the present asymmetrical situation through the same effect that produces spontaneous violation of time symmetry. Sakharov's proposal appeared quite strange at first, but in its successive transformations has looked better and better. It does seem that spontaneous symmetry violation is responsible for the predominance of matter over antimatter.

Spin

The higgson involved in the Anderson–Higgs mechanism of spontaneous symmetry violation is a different kind of boson from the quanta of the gluonic, weak, and electromagnetic interactions. A very important distinction lies in the value of the spin angular momentum (spin for short), which quantifies how much the particle rotates about its own axis. Quantum mechanics supplies a natural unit for spin, and in terms of that unit a boson can have a spin of 0 or 1 or 2, and so on, while a fermion can have a spin of 1/2 or 3/2 or 5/2, and so on.

All the elementary fermions of the standard model have spin 1/2. All the quanta of quantum chromodynamics and quantum flavor dynamics have spin 1. The higgson, however, must have spin 0.

How Can There Be So Many Elementary Particles?

The huge multiplicity of observed nuclear particle states was explained by the discovery that they are composites, formed according to the rules of quantum chromodynamics, from elementary quarks, antiquarks, and gluons. But the quarks, with their three colors and six flavors, and the gluons, which come in eight (rather than nine) color combinations, are already rather numerous. Moreover, outside the realm of the strongly interacting nuclear particles, we encounter in addition the electron and its neutrino, the muon and its neutrino, and the tauon and its neutrino.

And all the fermions have antiparticles distinct from themselves. In addition, we have the photon, and the three intermediate bosons of the weak interaction. The higgson completes our list of elementary particles demanded by the standard model.

Let us add up the total number. We have eighteen quarks, three electron-like particles, and three neutrinos, making twenty-four fermions in all. With their antiparticles, we have forty-eight. Then there are the known quanta: the eight gluons, the photon, and the three intermediate bosons for the weak interaction, bringing the total to sixty. With the higgson (if there is only one), we get sixty-one.

To a lay observer, it seems crazy to suppose that the basic law of all matter in the universe could rest on such a large and heterogeneous collection of fundamental objects. The expert in elementary particle physics can only agree. The solution to the puzzle must lie in the incorporation of the standard model into a larger theory containing less arbitrariness, preferably a unified theory of all the particles and all their interactions. While the standard model is supported by copious evidence from experiment, any unified theory must at present, in the absence of direct support from observation, be regarded as speculative. A unified theory must of course be testable, that is, able to make predictions that can be verified by observation. But how can such a theory deal with the profusion of elementary particles with which we are now faced in the standard model?

There would seem to be three ways out. The first is to suppose that today's elementary particles are really composite, and that the ultimate description of matter involves a smaller number of new, truly elementary constituents. I do not believe there is any theoretical or experimental evidence today that points in such a direction. Moreover, the hypothetical new constituents would themselves have to be rather numerous in order to explain the great variety of properties of the known elementary particles. Hence the reduction achieved in the number of elementary objects would not be dramatic.

A related idea is that the process just discussed, of explaining the apparently elementary objects at one level as composites of still more elementary objects at the next level down, will go on forever. Such an endless chain of composition was advocated in the People's Republic of China by the late Chairman Mao (if he seems an unlikely recruit to this field of endeavor, remember that Lenin wrote about the electron and

Stalin intervened in numerous controversies in the sciences, humanities, and arts, sometimes with most unfortunate consequences for those he attacked). In accordance with the Chairman's ideas, the quark was for a time called "layer child" in Chinese, recalling the coinage "foundation child" for atom. Under the rule of Mao and the Gang of Four, it was no doubt inadvisable for Chinese scientists to disagree too violently with the idea of an infinite sequence of layers. Under the milder regimes that have followed, however, the late Chairman's foray into theoretical physics has largely been ignored.

The third possibility is that a simple theory underlies the elementary particle system, according to which the number of such particles can be regarded as infinite, with only a finite number accessible to experimental detection at available energies. Superstring theory falls into this category of explanation.

CHAPTER

14

SUPERSTRING THEORY: UNIFICATION AT LAST?

For the first time in history, we have in superstring theory—and specifically in heterotic superstring theory—a serious proposal for a unified theory of all the elementary particles and their interactions and, therefore, of all the forces of nature. The next step is to extract predictions from the theory and compare them with what is known, as well as with what will soon be measurable, about the elementary particles. A striking feature of that comparison is the occurrence in the equations of a characteristic energy or mass (the Planck mass), close to which the full unity of superstring theory starts to manifest itself directly. But the energy equivalent of the Planck mass is enormous compared to the energy scale of phenomena detectable in the laboratory. Therefore, the elementary particles that can be studied more or less directly in the laboratory all belong to the "low-mass sector" of the theory.

The Low-Mass Sector

A large but finite number of particles (say between one and two hundred) have masses low enough that they will appear in accelerator experiments of the foreseeable future. Those particles and their interactions constitute the low-mass sector of superstring theory.

All the other elementary particles (an infinite number of them) are enormously more massive, so that their presence can be verified only through virtual effects (such as the generation of forces by the virtual exchange of quanta). Some of those virtual effects may be of critical importance, such as the ones that enable Einsteinian gravitation to be part of the theory without engendering crippling infinities.

The standard model, including the three fermion families, their antiparticles, and the twelve known quanta, forms part of the low-mass sector of the unified theory. The graviton, with zero mass, obviously belongs to the low-mass sector as well, as do other predicted particles.

Renormalizability of the Standard Model

The standard model is distinguished from, say, the theory of the graviton by a wonderful property called renormalizability. This means it can be separated off in an excellent approximation from the rest of the unified theory without encountering infinities that make calculations meaningless. A renormalizable portion of the unified theory can be used on its own, almost as if it were the final theory. However, there is a price to be paid for that separation, which in the case of the standard model is the occurrence of more than a dozen arbitrary numbers that cannot be calculated and must be taken instead from experiment. Those numbers represent the dependence of the model on the rest of the fundamental unified theory, including the infinite set of high-mass states.

Comparison with Observation Not at All Impossible

Because the ratios of the Planck mass to the nonzero masses of the low-mass sector are so large, a few theoretical physicists and a number of lay authors have claimed that the predictions of the theory are difficult or impossible to check against observation. However, such arguments are not correct. There are many possible ways of confronting the theory with experimental results.

1. To begin with, superstring theory already predicts, in a suitable limit, Einstein's general-relativistic theory of gravitation. The automatic incorporation of Einsteinian gravitation into a unified quantum field theory without encountering the usual difficulties (with infinities) is already a major triumph.
2. The next challenge is to determine whether superstring theory can predict, in an appropriate approximation, the standard model.
3. But recall that the standard model has a great many arbitrary constants (parameters), the values of which the theory should be capable of predicting.
4. The low mass sector of superstring theory contains additional, new particles, the predicted properties of which can be checked against observation.
5. In particular, the standard model is embedded in a larger renormalizable model that is part of the low-mass sector. The properties of that larger theory, including its particle content and the constants describing the masses and interactions of the particles, can all be compared with the results of experiments.
6. In addition, the virtual effects of the high-mass sector may introduce some observable corrections to the physics of the low-mass sector.
7. Finally, superstring theory may have consequences for cosmology that are verifiable by astronomical observation.

So we need not despair of finding ways to compare the predictions of the theory with the facts of nature, but theoreticians must proceed with the difficult task of extracting those predictions.

Basic Units of Energy and Other Quantities

What is the huge energy that characterizes superstring theory and where does it come from? It is the basic unit of energy, derived from the fundamental, universal constants of nature. There are three such constants: c, the velocity of light in empty space; h, the quantum constant of Max Planck; and G, the gravitational constant of Isaac Newton.

The constant *h* is the universal ratio of the energy of any quantum of radiation to the vibration frequency of that radiation. In practice, it is usually employed in the form $\hbar$, which means *h* divided by two pi, where two pi is the familiar ratio of the circumference of any circle to its radius. (Werner Heisenberg used to wear a tie pin in the form of $\hbar$ as a way of showing his pride in having discovered quantum mechanics. That symbol is so familiar to physical scientists that my late friend, the brilliant and amusing mathematician Stanisław Ulam, used to describe the *ł*, the "dark" Polish *l* in his first name, as *l* divided by two pi.

G is the universal constant in Newton's formula for the gravitational force between any two point particles, which is equal to *G* times the product of the two masses divided by the square of the distance between them. (Newton showed that the same formula applies to two spherically symmetrical bodies if the distance used is the distance between their centers, so that the formula can be employed approximately for the sun and the planets and for satellites like the moon.)

By multiplying and dividing suitable powers of the three universal constants, *c*, $\hbar$, and *G*, one can construct the fundamental unit of any physical quantity, such as length, time, energy, or force. The fundamental length is about a centimeter divided by a number written as 1 followed by thirty-three zeroes. Dividing that length by the velocity of light yields the fundamental unit of time, on the order of a second divided by a number written as 1 followed by forty-four zeroes.

By way of contrast, the units to which most people are accustomed are arbitrary; they are not constructed from the universal constants of nature. Although the foot is no longer (if it ever was) the average length of the shod feet of the first ten men to leave church on Sunday, it is still not fundamental. Neither is the meter, defined formerly as the length of a specific bar of metal in a vault near Paris and nowadays as a certain multiple of the wavelength of the light emitted by a krypton atom in a particular state of excitation.

Particle Masses and the Basic Unit

The fundamental unit of mass, the Planck mass, is about a hundred thousandth of a gram. It may not seem enormous on a human scale, but on the scale of the neutron and proton masses (both approximately equal to a GeV) it is large indeed—about twenty billion billion times as big. Turning the relation around, we can say that the masses of neutron and proton are extremely tiny in terms of fundamental units. The mass of the electron is about two thousand times smaller still. Why do these very small numbers occur? The short answer is that we do not yet know. Heterotic superstring theory does not explicitly contain any adjustable parameters. If it is indeed the correct theory, it must somehow generate by itself the tiny ratios of the masses of the known particles to the fundamental unit of mass. Verifying by calculation that it does so will be one of the most severe tests of the theory.

So far, there are only hints of how the small numbers may arise in the theory. A fairly obvious guess is that there is a useful approximation in which all the particles of the low-mass sector are assigned a mass of zero. Symmetry-breaking corrections to that approximation (including those induced by the Anderson–Higgs mechanism) then become responsible for the actual tiny but nonzero masses. A few mass values, including those of the photon and graviton, receive no correction at all and remain zero.

Energies available in current experiments are on the order of a thousand GeV; soon they may be something like ten times greater, but no more. They are still very small compared with the fundamental unit of energy, around twenty billion billion GeV. Since experimentalists cannot produce in the laboratory particles with masses higher than the energies available at their accelerators, they will be dealing directly only with particles that belong to the low-mass sector.

The Meaning of the Term Superstring

What general observations can be made about the particle content of heterotic superstring theory? The answer to that question is connected with the meaning of the word string and of the prefix super-.

String indicates that the theory can be regarded as describing particles in terms of tiny loops instead of points; the typical dimension of each loop is approximately the fundamental unit of length, around a billionth of a trillionth of a trillionth of a centimeter. Those loops are so small that for many purposes there is an equivalent description in terms of point particles, actually an infinity of kinds of point particles. How are those different particles related to one another? In particular, how are those of the low-mass sector related to those that have masses comparable to the Planck mass or greater?

A good analogy is with a violin string, which has a lowest mode of vibration and an infinite number of other modes (harmonics) of higher and higher musical frequency. But in quantum mechanics energy is like frequency multiplied by Planck's constant *h*. The particles of the low-mass sector can be visualized as the lowest modes of the various sorts of loops of string occurring in superstring theory, while particles with masses comparable to the fundamental unit of mass represent the next-lowest modes, and still heavier particles represent higher modes, and so on forever.

The prefix super- indicates that the theory has approximate "supersymmetry," which means in turn that for every fermion in the list of particles there is a corresponding boson and vice versa. If the supersymmetry of the elementary particle system were exact, each fermion would have precisely the same mass as its related boson. However, supersymmetry manages to get itself "broken," in a way that is still only dimly understood, causing the masses of the corresponding fermions and bosons to be separated by what I like to call a "supergap." This is not exactly the same for each fermion–boson couple, but it is probably always of the same general order of magnitude. If the supergap is anything like the fundamental unit of mass, we can despair of observing directly in the laboratory the superpartners of the known elementary particles.

Superpartners and New Accelerators

If, however, the energy equivalent of the supergap is only hundreds or even thousands of GeV, superpartners may be observable in the next few years, provided the new CERN accelerator is built. (The chance of observing them would have been greater if the higher energy SSC had been completed.) The theoretical analysis of certain experimental results indicates that the supergap is probably of the right size to have been bridged by the SSC and may perhaps be bridged by the CERN machine. Assuming those indications are correct, I believe the prospect of discovering superpartners is the most exciting of all the specific motivations for building new accelerators. (There is always in addition the nonspecific goal of exploring the unknown and seeing if some unforeseen phenomenon turns up.)

The names assigned to the hypothetical superpartners follow two different patterns. When the known particle is a boson, the related fermion is given a name ending in the Italian diminutive -ino, first employed (in a different way) by Fermi in naming the neutrino. Thus the expected partner of the photon is the photino, that of the graviton the gravitino, and so forth. Since the electrically charged bosons transmitting the weak interaction are often called *W* plus and *W* minus, the corresponding predicted fermions have acquired the bizarre appellation "winos." In the case where the known particle is a fermion, the boson partner is called by the same name as the fermion but with the letter "s" prefixed (presumably standing for super). Thus one gets rather weird terms like squarks and selectrons. (I should like to emphasize that I am not responsible for these names, although I must admit, reluctantly, to having been present when the -ino suffix was chosen for the fermion partners of known bosons.)

Since the superpartner of a boson is a fermion and vice versa, the spins of the two superpartners must always be different, one a whole number and the other a whole number plus 1/2. In fact, the two spins must differ by 1/2. The Higgs bosons (or higgsons) have spin 0 and their partners (higgsinos) have spin 1/2. The three families of fermions have spin 1/2 and their partners (squarks, selectrons, and so on) have spin 0. The quanta (gluons, photons, X or W bosons, and Z^0) have spin 1 and their partners (gluinos, photinos, and so on) have spin 1/2. The graviton has spin 2 and its partner, the gravitino, spin 3/2.

In superstring theory, the standard model is incorporated into a larger renormalizable theory, which we may call the superstandard model, containing the twelve quanta, the usual fermions, and some higgsons, along with the superpartners of all those particles. The prediction of the validity of the superstandard model provides a great many experimental tests of superstring theory.

The Approach to the Planck Mass

As energy increases and the low mass sector—the sector directly accessible to experiment—is left behind, superstring theory predicts that the gluonic, the electromagnetic, and the weak interactions all approach one another in strength and reveal their close relationship. (Extrapolation of present experimental results to high energies yields the same conclusion provided broken supersymmetry is assumed, with a supergap that is not too large. Thus there is already some indirect evidence for supersymmetry.) At the same time, the symmetries among the fermions also assert themselves.

Now let energy continue to increase. It may or may not turn out, at an energy interval just below the Planck mass, that the superstandard model gives way temporarily to a supersymmetric version of a "grand unified theory" before exhibiting, in the vicinity of the Planck mass, the first excited modes of the superstring.

Although none of us will live to see energies comparable to the Planck mass produced in the laboratory, such energies were available in the infant universe as it began to expand. The fundamental unit of time, about a hundred millionth of a trillionth of a trillionth of a trillionth of a second, measures the period during which the tiny universe experienced the full physical effects of the superstring theory. Is there any cosmic evidence remaining today that could test the validity of superstring theory through its effects at that crucial but remote moment of time?

Theorists are not sure whether such evidence remains or not. That brief interval was almost certainly followed by a period of inflation, an explosive expansion of the universe that was succeeded by the more gradual expansion still going on. Inflation nearly wiped out many features of the very early universe, and may thus have suppressed numerous consequences of superstring theory. But the constraints it imposes on the character of the inflation may permit the theory to be tested after all by cosmological means.

The same kind of reasoning applies to the initial condition of the universe, which, according to the proposal of Hartle and Hawking, is tied to the unified quantum field theory. Assuming their idea and superstring theory are both correct, the initial condition is uniquely determined, but its effects on the later universe are filtered through the process of inflation.

Apparent Multiplicity of Solutions

Besides the great disparity between the characteristic energy scale of the superstring theory and the energies available for elementary particle experiments, there is another reason why a few physicists have expressed doubts about whether the theory will be testable. This has to do with the discovery of numerous approximate solutions of the equations of heterotic superstring theory.

Each such solution supplies, among other things, a list of the particles that have zero mass in the approximation used. It is plausible to assume that those particles are the same ones that make up the low-mass sector of the theory when the small nonzero mass corrections are included. The zero-mass particle content of each approximate solution can then be compared with the content of the superstandard model. For certain solutions, there is indeed agreement: the low-mass sector contains the superstandard model and a few additional particles, including the graviton and gravitino.

The trouble is that thousands of other approximate solutions have turned up, and it looks as if many more will be found. It is therefore not at all impossible that the observed situation is compatible with a solution of the superstring theory, but what is to be done with all the other solutions?

There are several possible answers. One, of course, is that the theory is wrong, but I see no reason whatever to draw such a drastic conclusion from the existence of a plethora of approximate solutions. A second possibility is that the difficulty stems entirely from the approximation (which is by no means fully justified, but merely convenient), and that when the approximation is improved, all the solutions but one will be seen to be spurious and can be thrown away. (A modified version of that possibility is that only a few genuine solutions will survive.)

Action

To discuss the remaining possible answers to the problem of multiple solutions, it is helpful to bring up a most important quantity called action, usually designated by the symbol *S*. It was introduced into classical Newtonian physics ages ago and proved quite useful, but with the advent of quantum mechanics it became not only useful but essential. (Action has the dimensions of energy times time; $\hbar$, Planck's constant divided by two pi, has the same dimensions and can be regarded as the fundamental unit of action.) Recall that the probabilities for coarse-grained histories in quantum mechanics are sums over values of the quantity D for pairs of completely fine-grained histories. A theory in quantum mechanics assigns to each fine-grained history a particular value of the action S, and it is those values of the action (along with the initial condition) that determine the values of D.

It is clearly highly desirable to find the formula for the action S in heterotic superstring theory. So far, however, that has proved an elusive goal. What seems to be within reach today—as a result of work by my former student Barton Zwiebach, by Michio Kaku, and by a group in Kyoto—is to express the action as the sum of an infinite series, but the summation of that series remains a formidable task.

It may be illuminating to compare the situation to an exercise that my late colleague Dick Feynman went through in 1954. (He discussed his project with me at the time I visited Caltech back in 1954, when I was offered and accepted a job there. I had, in fact, started a similar project myself.) Feynman began by imagining that Einstein had never had his brilliant insight into the nature of gravitation around 1914, the realization that it would have to obey the invariance principle of general relativity and relate to the geometry of spacetime. Dick asked whether it would be possible, without that insight, to construct the theory by brute force. He found that it could be done. However, the action came out in the form of an infinite series, and summation of that series was virtually impossible in the absence of the geometrical point of view and the invariance principle. That principle, general relativity, yields the answer directly, without any need for the brute force method or the infinite series. In fact, once Einstein understood, on the basis of general relativity, what kind of formula he needed in order to describe gravitation, he was able to learn the relevant mathematics from an old classmate, Marcel Grossmann, and write down the formula for the action, from which the equation on page 87 can be derived.

Perhaps the situation in superstring theory is similar. If theorists understood the invariance principle of superstring theory, they might be able to write down the formula for the action in short order, without resorting to the summation of the infinite series. (While we are waiting for it to be discovered, what should we call that principle? Field marshal relativity? Generalissimo relativity? Certainly it goes far beyond general relativity.) In any case, a deep understanding of superstring theory will go hand in hand with finding the formula for the action S.

As noted earlier, the idea motivating the work that led to superstring theory in the first place was the "bootstrap" principle of self-consistency, a simple and powerful idea but not yet formulated in the right language for discovering the action or the full symmetry principle underlying the action. When superstring theory is expressed in the language of quantum field theory and when its action and its symmetries have been discovered, it will truly have come of age.

Effective Action

Starting from the action, one can in principle calculate a related quantity, which I denote by the symbol $\overline{S}$. A theorist might call it something like "quantum-corrected Euclideanized action." It is a kind of average of a modified version of the action $\overline{S}$, where the modification involves altering the character of the time variable. We can refer to the quantity $\overline{S}$ as the "effective action." It plays a most important role in the interpretation of the theory. In the first place, it is in terms of $\overline{S}$ that Hartle and Hawking would express their initial condition for the universe. In the second place, it is to the quantity $\overline{S}$ that we must look for guidance if there really are many solutions to the superstring theory. Somehow that quantity, calculated for the different solutions, must discriminate among them.

Reasoning from classical physics, where the principle of least action provides an elegant means of formulating classical dynamics, some theorists might argue that the correct solution for physical purposes—the one that characterizes the real universe—would have to be the one with the smallest value of the effective action $\overline{S}$. That could indeed be the right criterion for picking out the correct solution.

Since we are dealing with quantum mechanics, however, it may turn out that there is no single correct solution for the universe, but rather a probabilistic situation in which all the true solutions are possible alternatives, each with its own probability, which gets smaller as the value of $\overline{S}$ gets larger. In fact, the formula for the probability in terms of $\overline{S}$ would be a decreasing exponential, described by a curve like the one on page 132. The solution with the least value of $\overline{S}$ would then have the highest probability of characterizing the universe, but other solutions would have some chance as well.

A Particular Solution Determined by Chance?

The particular solution that applies to the universe would then determine the structure of the system of elementary particles. In fact, it would do more than that. Remarkably enough, it would even determine the number of dimensions of space.

One way to think of the spatial aspect of the heterotic superstring is that the theory starts out describing a spacetime with one time dimension and nine spatial dimensions; various solutions then correspond to the collapse of some of the spatial dimensions, leaving only the remaining ones observable. If the probabilistic interpretation of $\overline{S}$ applies, then the three-dimensional character of space in our universe is a consequence of the chance occurrence of a particular solution to the superstring equations (as is, for instance, the existence of a particular number of fermion families containing particular sets of particles).

Such a probabilistic situation is the most intriguing possible outcome of the current attempts to solve the puzzle posed by the many apparent solutions to the equations of superstring theory. Suppose it is the right outcome. We can then think of the branching tree of alternative coarse-grained histories of the universe, with a probability for each one, as beginning with a first branching that selects a particular solution of the superstring equations.

The predictions of superstring theory, whether or not they depend on such a probabilistic "choice" of solution, will have to be compared with our experience of three-dimensional space as well as all the properties of the elementary particle system.

If heterotic superstring theory does turn out to make correct predictions in all the cases where it is feasible to test it, the problem of the fundamental theory of the elementary particles will presumably have been solved. The underlying dynamics for the evolution of the state of the universe will be known. But the description of the history of the universe depends on the initial condition as well and also on the chance outcomes of all the branchings of the tree of universal history.

Multiple Universes?

So far the quantum cosmology we have discussed has referred to alternative histories of the universe, treated as a single entity embracing all matter everywhere. But quantum cosmology is in a state of flux and abounds in speculative ideas of great interest, the status of which is still in question; some of those ideas refer, in one way or another, to multiple universes. Since uni- means one, that sounds like a contradiction in terms, and perhaps a new word will help to avoid at least the linguistic confusion that could arise if some of those ideas turn out to be even partially correct. We can employ the term "multiverse" to refer to the entire ensemble of universes, of which our familiar universe is a member.

The introduction of multiple universes is pointless unless our universe is largely autonomous. One proposal is that the other universes are "baby universes" virtually created and destroyed in quantum processes, much like the virtual quanta that carry forces in quantum field theory. As viewed by Stephen Hawking and others, the virtual creation and destruction of baby universes alter the results of calculations in elementary particle theory, but do not necessarily challenge in an essential way the concept of a branching tree of history for our universe.

Another suggested possibility is that numerous universes exist, many of them comparable in size with our own, but that our universe has only limited contact, if any at all, with the others, even if such contact might have occurred in the distant past or might become possible in the far future. In one such picture, the universes are like bubbles in the multiverse, bubbles that separated from one another very long ago, thus inaugurating an era with no communication among the universes, an era that would last for an exceedingly long time. If this kind of multiple universe picture turns out to have any validity, one may try to identify what goes on in the various bubbles with different possible branches of the history of the universe. The door is then opened to the notion that fantastically many branches of the tree of coarse-grained histories are actually all realized, but in different bubbles. The probability of each history would then be essentially a statistical probability, the fraction of the various "universes" in which the particular history takes place.

Now suppose the probabilistic interpretation of the many approximate solutions of superstring theory is right, so that there are many true solutions associated with various elementary particle patterns and various dimensions of space. Then, if numerous universes really exist as bubbles in a multiverse, sets of them could be characterized by different solutions of superstring theory, with numbers of occurrences declining exponentially with the value of the effective action $\overline{S}$.

Even if such theoretical speculations are shown to be without foundation, the notion of multiple, largely independent universes still provides a nice (if rather abstract) way of *thinking* about probabilities in quantum cosmology.

"Anthropic Principles"

Some quantum cosmologists like to talk about a so-called anthropic principle that requires conditions in the universe to be compatible with the existence of human beings. A weak form of the principle states merely that the particular branch of history on which we find ourselves possesses the characteristics necessary for our planet to exist and for life, including human life, to flourish here. In that form, the anthropic principle is obvious.

Suppose the correct cosmology really involves a multiverse with a probabilistic distribution of universes displaying various solutions of the fundamental equations. Perhaps those who like to invoke an anthropic principle believe that the solution corresponding to our universe is an improbable one, while the more probable ones do not admit human life in any of their branching histories. But in that case we would still have to say that our universe must be compatible with the existence of everything observed, including human life, whether probable or not. Again, such a weak anthropic principle is a triviality.

In its strongest form, however, such a principle would supposedly apply to the theory of the elementary particles and the initial condition of the universe, somehow shaping those fundamental laws so as to produce human beings. That idea seems to me so ridiculous as to merit no further discussion.

Nevertheless, I have tried to find some version of the idea of an anthropic principle that is neither trivial nor absurd. The best I can come up with is the following. Among the various solutions of the fundamental equations (if there are in fact multiple exact solutions) and among the various branches of history, certain solutions and certain histories create favorable conditions in many places for the evolution of complex adaptive systems, which can act as IGUSes (information gathering and utilizing systems), observers of the outcomes of quantum-mechanical branchings. (Those conditions include the prevalence of a situation suitably intermediate between order and disorder.) The characterization of those solutions and branches poses a theoretical problem of great interest, which could, I suppose, be called the search for IGUSic conditions. A minor incidental feature of those conditions favorable to IGUSes would be that the existence of Earth, of life on Earth, and of human life in particular would be permitted and could occur on certain branches.

One application of such theoretical research would be the refinement of calculations of the probability of receiving signals from intelligent complex adaptive systems on planets orbiting distant stars (as in the SETI project, the search for extraterrestrial intelligence). There are many factors entering into such calculations. One of them is the probable length of time that a technical civilization would last and be capable of and interested in transmitting signals, since catastrophic war or the decline of technology could put an end to them. Another factor is the probability that a planet can harbor complex adaptive systems, for example ones resembling living beings on Earth. Here a number of subtle considerations may enter in. For instance, Harold Morowitz, investigating the requirements on the Earth's atmosphere at the time of the prebiotic chemical reactions that initiated life, has concluded that some fairly restrictive conditions had to apply in order for those reactions to take place. Other experts, however, are not so sure.

It seems that in place of some awesome "anthropic principle" we are faced with a set of fascinating but rather conventional questions in theoretical science about the conditions necessary for the evolution of complex adaptive systems on various branches of history and at various times and places, given the fundamental theory of elementary particles and the initial quantum state of the universe.

The Role of the Initial Condition

We have encountered several times the role of the initial condition in supplying the order in the early universe that made possible the subsequent evolution first of celestial objects like galaxies, stars, and planets and then of complex adaptive systems. We have also discussed one of the most dramatic consequences of the initial condition, that time flows steadily forward throughout the universe. Let us now explore the flow of time in more detail.

CHAPTER

15

TIME'S ARROWS
Forward and Backward Time

Remember the meteorite passing through the atmosphere and landing. If a film of the whole sequence of events were run backwards, we would know instantly that time was being reversed. We understand that the ultimate reason for the unidirectionality of time is that the universe was in a very special state some ten or fifteen billion years ago. Looking through time toward that simple configuration, we are contemplating what we call the past; looking in the other direction, we are seeing what we call the future stretched out before us.

The compactness of the initial state (at the time of what some like to call the big bang) does not fully characterize its simplicity. After all, cosmologists consider it possible, even probable, that at some almost inconceivably distant time in the future the universe will recollapse to a very tiny structure. If so, however, that structure will be quite different from the one that existed in the past. During the period of recollapse, the universe will not be running through its expansion in reverse. The notion that the expansion and contraction would be symmetrical with each other is what Stephen Hawking calls his "greatest mistake."

Radiation and Records

It is easy to think of many ways in which forward time and backward time are different. For example, hot objects like stars and galaxies radiate energy outward. The most familiar form of radiated energy consists of photons—such as those of light, radio waves, and gamma rays—which make possible optical astronomy, radio astronomy, gamma ray astronomy, and so on. In addition to the observation of photons, neutrino astronomy is just coming into its own, and some day soon we will have gravitational wave astronomy. All are based on detecting the outward flow of energy in the form of particles and waves. Likewise, when we see the light coming from a fire or an electric light bulb, our eyes are detecting a stream of emitted photons. If time were reversed, energy in each of these cases would instead be flowing inward. The outward flow of energy can carry signals; if a star turns into a supernova and suddenly becomes enormously brighter for a while, that information is propagated outward at the speed of light.

Another difference between the past and the future is the existence of records of the past, like the tracks left in mica by the charged particles emitted when radioactive nuclei disintegrated long ago. Similar records of future disintegrations are conspicuous by their absence. That asymmetry between past and future is so obvious that we tend to overlook it.

We humans make use of radiation to send signals and of records to learn about the past. We even make and keep records ourselves. But the existence of signals and records in general is quite independent of the existence of complex adaptive systems—like us—that employ some of them.

The Initial Condition and Causality

The time asymmetry of signals and records is part of physical causality, the principle that effects follow their causes. Physical causality can be traced directly to the existence of a simple initial condition of the universe. But how does that initial condition enter into the theory?

The quantum-mechanical formula for the quantity *D*, which yields the probabilities of alternative histories of the universe, already contains the asymmetry between past and future. At one end, corresponding to what we call the past, it contains a specification of a quantum state for the early universe, which we call the initial condition. At the other end, corresponding to the remote future, the formula contains a summation over all possible states of the universe. That summation can be described as a condition of complete indifference as to the state of the universe in the distant future.

If the initial condition were also one of complete indifference, there would be no causality and not much history. Instead, the initial condition is a special and simple one (perhaps even the one described by Hartle and Hawking, which requires no information beyond the dynamical law governing the system of elementary particles).

If the future condition were not one of complete indifference, violations of causality would result and events would occur that were inexplicable (or at least exceedingly improbable) in terms of the past but were required (or nearly so) by the condition specified for the distant future. As the age of the universe increased, more and more such events would occur. There is no evidence for such a situation reflecting predestination and considerable evidence against it; therefore in the absence of any convincing new argument, we can disregard the possibility that the future condition is anything but one of indifference. However, while relegating it to the domain of science fiction or superstition, we can still consider a special condition on the future as an interesting contrary-to-fact case contrasting with the causal situation we strongly believe to be correct.

From the basic quantum-mechanical formula for the probabilities of histories, with a suitable initial condition, it is possible to deduce all the familiar aspects of causality, such as signals and records pointing from the past to the future. All the arrows of time correspond to various features of coarse-grained histories of the universe, and the formula exhibits the tendency of all those arrows to point forward rather than backward everywhere.

Entropy and the Second Law

Of the arrows that mark the distinction between forward and backward time, one of the most famous is the tendency of the quantity called entropy to increase (or at least not to decrease) in a closed system, yielding the principle known as the second law of thermodynamics. (According to an old physics joke, the first law of thermodynamics says you can't win, while the second law says you can't break even. Both laws are frustrating for someone who wants to invent a perpetual motion machine.) The first law merely states the conservation of energy: the total energy of a closed system stays the same. The second law, requiring the increase (or constancy) of entropy, is subtle, and yet entropy is really very familiar to all of us in our daily lives. It is a measure of disorder, and who would deny that disorder tends to increase in a closed system?

If you spend all afternoon at a table sorting pennies according to date or nails according to size, and something knocks over the table, isn't it overwhelmingly likely that the pennies or nails will get mixed up? In a household where children make peanut butter and jelly sandwiches, isn't there a tendency for the peanut butter in the jar to acquire an admixture of jelly and for the jelly jar to get some dollops of peanut butter in it? If a chamber is divided into two parts by a partition, the left-hand side containing oxygen gas and the right-hand side containing an equal amount of nitrogen, isn't the removal of the partition almost certain to yield a mixture of oxygen and nitrogen in both parts?

The explanation is that there are more ways for nails or pennies to be mixed up than sorted. There are more ways for peanut butter and jelly to contaminate each other's containers than to remain completely pure. And there are more ways for oxygen and nitrogen gas molecules to be mixed up than segregated. To the extent that chance is operating, it is likely that a closed system that has some order will move toward disorder, which offers so many more possibilities.

Microstates and Macrostates

How are those possibilities to be counted? An entire closed system, exactly described, can exist in a variety of states, often called microstates. In quantum mechanics, these are understood to be possible quantum states of the system. These microstates are grouped into categories (sometimes called macrostates) according to the various properties that are being distinguished by the coarse graining. The microstates in a given macrostate are then treated as equivalent, so that only their number matters.

Consider the chamber containing equal numbers of nitrogen and oxygen molecules separated by a partition, which is then removed. Now all the possible microstates of the nitrogen and oxygen molecules can be grouped into macrostates such as the following: those in which the left-hand part of the chamber has less than 10 percent nitrogen and the right-hand part less than 10 percent oxygen, those in which the contaminations lie between 10 and 20 percent, those in which they lie between 20 and 30 percent, and so forth. The macrostates in which the contamination lies between 40 and 50 percent (or between 50 and 60 percent) contain the most microstates. Those are also the most disordered macrostates, the ones in which the gases are mixed with each other to the greatest extent.

Actually, counting the number of different ways in which a closed system can be in a particular macrostate is closely related to the technical definition of entropy (as measured in the most convenient unit, called Boltzmann's constant). The entropy of a system in a given macrostate is, roughly speaking, the amount of information—the number of bits—necessary to specify one of the microstates in that macrostate, with the microstates all treated as if they were equally likely.

Recall that the game of twenty questions, played perfectly, is capable of eliciting 20 bits of information beyond whether the unknown is animal, vegetable, or mineral. Twenty bits correspond to the information needed to distinguish 1,048,676 different, equally probable alternatives, where that number is just 2 multiplied by itself 20 times. Likewise 3 bits correspond to 8 equally likely possibilities, because 8 is 2 multiplied by itself 3 times. Four bits correspond to 16 possibilities, 5 bits to 32 possibilities, and so forth. If the number of possibilities lies *between* 16 and 32, then the number of bits lies *between* 4 and 5.

Thus, if the number of microstates in a macrostate is 32, then the entropy of a system in that macrostate is 5 units. If the number is 16, the entropy is 4 units, and so on.

Entropy as Ignorance

Entropy and information are very closely related. In fact, entropy can be regarded as a measure of ignorance. When it is known only that a system is in a given macrostate, the entropy of the macrostate measures the degree of ignorance about which the microstate system is in, by counting the number of bits of additional information needed to specify it, with all the microstates in the macrostate treated as equally probable.

Now suppose the system is not in a definite macrostate, but occupies various macrostates with various probabilities. The entropy of the macrostates is then averaged over them according to their probabilities. In addition, the entropy includes a further contribution from the number of bits of information it would take to fix the macrostate. Thus the entropy can be regarded as the average ignorance of the microstate within a macrostate plus the ignorance of the macrostate itself.

Specification corresponds to order and ignorance to disorder. The second law of thermodynamics tells us merely that, other things being equal, a closed system of low entropy (considerable order) will tend, at least for a very long time, to move toward higher entropy (more disorder). Since there are more ways for disorder to occur than order, the tendency is to move toward disorder.

The Ultimate Explanation: Order in the Past

A deeper question is why the same argument is not applicable when the direction of time is reversed. Why should a film for a system run backwards not show it moving toward probable disorder instead of toward order? The ultimate answer to that question lies in the simple initial condition of the universe at the beginning of its expansion some ten billion years ago, contrasted with the condition of indifference that is applied to the distant future in the probability formula of quantum mechanics. It is not only the causal arrow of time that points from past to future as a result but also the other arrows, including the order–disorder or "thermodynamic" arrow of time. The original condition of the universe leads, later on, to the gravitational condensation of matter and the formation of young galaxies. As galaxies of certain kinds age, young stars and planetary systems are formed inside them. Then the stars and planets age. The arrow of time is communicated from universe to galaxy to star and planet. It points forward in time everywhere in the universe. On Earth it is communicated to the origin of terrestrial life and its evolution and to the birth and aging of every living thing. Most large-scale order in the universe arises from order in the past and ultimately from the initial condition. That is why the transition from order to the statistically much more probable disorder tends to proceed everywhere from past to future and not the other way around.

We can think of the universe metaphorically as an old-fashioned watch that is fully wound at the beginning of its expansion and then gradually runs down while spawning smaller, partially wound watches that slowly run down in their turn, and so on. At every stage, as a new entity is formed, it inherits from existing structures the property of being at least partly wound. We can identify the aging of each approximately isolated entity with the running down of its corresponding watch.

How do galaxies and stars and planets behave as they grow older? Consider what happens to certain familiar categories of stellar objects. At the center of stars like the sun thermonuclear reactions take place, at temperatures of tens of millions of degrees, that convert hydrogen to helium, with the release of energy that finally emerges from the surface in the form of sunlight or starlight. Eventually the star runs out of its nuclear fuel and changes its character, often in dramatic ways. If it is heavy enough, it may turn suddenly into a supernova and then, after shining with great brilliance for a couple of months, collapse into a black hole. Clearly such a process is unidirectional in time!

When we humans set up a pattern of order (with pennies, say) and leave it alone except for an agent that can upset it (a dog, for example), the closed system (pennies on a table plus clumsy dog) will evolve toward disorder because disorder is so probable. That change will occur forward in time because we humans behave causally, like everything else acting forward in time, and we created the pattern of order first and then left it alone with the dog. Such a situation involving an increase of entropy is not so very different from what is going on in the stars and galaxies.

What *is* somewhat distinctive is the setting up of the pattern of order in the first place: sorting the pennies or re-sorting them after the dog knocks them over. That is clearly a decrease of entropy for the set of pennies, although it doesn't violate the second law of thermodynamics, because the set of pennies is not closed. In fact, the second law says that the entropy of the environment and of the person doing the sorting has to increase by at least as much as the entropy of the pennies decreases. How does that work? What are the symptoms of increasing entropy in the person doing the sorting and in the surroundings?

Maxwell's Demon

In trying to answer those questions, it is useful to discuss a hypothetical demon that spends its time sorting, namely Maxwell's demon, dreamt up by the same James Clerk Maxwell who found the equations for electromagnetism. He was treating a very common (and perhaps the earliest) application of the second law of thermodynamics: to a hot body and a cold body in proximity to each other. Imagine a chamber divided into two parts by a removable partition. On one side is a hot sample of gas and on the other side a cold sample of the same gas. The chamber is a closed system with a certain amount of order, because the statistically faster moving molecules of the hot gas on one side of the partition are segregated from statistically slower moving molecules of the cold gas on the other side.

Suppose first that the partition is made of metal, so that it conducts heat. Everyone knows that the hot sample of gas will then tend to get cooler and the cold sample warmer until the two reach the same temperature. That is clearly what the second law requires, since the orderly segregation of hotter and colder gas disappears and entropy thus increases.

Now suppose that the partition fails to conduct heat, so that the segregation of hotter and colder gas is maintained. Entropy will then remain constant, which is also compatible with the second law. But what if there is a demon at work sorting molecules into faster and slower ones? Can it decrease the entropy?

Maxwell's demon guards a trap door in the partition, which is still assumed not to conduct heat. It spots molecules coming from either side and judges their speeds. The molecules of the hot gas are only *statistically* faster than those of the cold gas; each sample of gas contains molecules moving at very different speeds. The perverse demon manipulates the trap door so as to allow passage only to the very slowest molecules of the hot gas and the very fastest molecules of the cold gas. Thus the cold gas receives extremely slow molecules, cooling it further, and the hot gas receives extremely fast molecules, making it even hotter.

In apparent defiance of the second law of thermodynamics, the demon has caused heat to flow from the cold gas to the hot one. What is going on?

Because the law applies only to a closed system, we must include the demon in our calculations. Its increase of entropy must be at least as great as the decrease of entropy in the gas-filled halves of the chamber. What is it like for the demon to increase its entropy?

A New Contribution to Entropy

Leo Szilard began to answer that question back in 1929 by introducing the relation between entropy and information. Later, after the Second World War, Claude Shannon set forth clearly the mathematical notion of information, which was further clarified by the French theoretical physicist Léon Brillouin. In the 1960s, the concept of algorithmic complexity or algorithmic information content was introduced by Kolmogorov, Chaitin, and Solomonoff. Finally, Rolf Landauer and Charlie Bennett of IBM worked out in detail how information and algorithmic information content are connected to the activity of a person, demon, or device that decreases the entropy of a physical system while increasing its own entropy by an equal or a greater amount.

Bennett considered a device acquiring appropriate information about a physical system and then recording it (say on paper or computer tape). The information refers to alternative possible sets of data, with a probability for each set. Bennett made use of those probabilities to study what happens on the average over all the different alternative results. He found that the device can really use the recorded information (about which alternative actually occurred) to make heat flow from a cold object to a hot one, *as long as the device has blank paper or tape available.* The entropy of the system composed of the hot and cold objects is thus decreased, but at the price of using up the paper or tape. Previously, Landauer had shown that erasing the records, leaving no copy, produces an increase of entropy that at least makes up for the decrease. Eventually the device must run out of recording space, and thus in the long run, when records are erased to make room for more, the second law of thermodynamics will be restored.

We have just mentioned that it is the erasure of the *last* copy of the information that *must* produce an increase of entropy at least sufficient to restore the second law. Actually, erasing any copy is likely to lead in practice to a similar entropy increase, but it is only the last copy that *must* do so in principle. The reason is that if at least two copies exist, methods are available under certain conditions for using one of them to "uncopy" another one reversibly, without any increase of conventional entropy.

Meanwhile, it is possible to maintain the second law in some form—even during the period when the records exist—by modifying the definition of entropy of the entire system. This can be done by adding a term equal to the amount of relevant information registered in the surviving records. That amount is not affected by the existence of extra copies of the records. All that matters is whether there exists at least one record of which alternative really occurred.

The usual entropy is a measure of ignorance, and the modified definition adds to it the amount of recorded information. Thus, when new data are obtained and registered, ignorance is reduced—on the average—by a certain amount, while the information in the records is increased by the same amount. When erasure takes place, the information in the records is decreased but ignorance of the situation of the entire closed system is increased by at least as much. The desirability of using a modified definition of entropy in this connection was first pointed out by Wojtek Żurek of Los Alamos National Laboratory and the Santa Fe Institute. The definition discussed here is a slightly different one suggested by Seth Lloyd.

Erasure and Shredding

As the demon performs its task of sorting, it must do something with the information it is acquiring about the individual molecules. To the extent that it stores the information, it will eventually run out of storage space. To the extent that the information is erased, the act of erasure increases the entropy of the demon and its surroundings. What does it mean, though, for the erasure to be completed?

Think of a notation in pencil being effaced by an ordinary gum eraser. The eraser sheds tiny bits of itself, each bearing a small portion of the notation, and these are dispersed all over the desk and even onto the floor. That kind of dispersal of order is itself an increase of entropy. In reality, the messy process of erasure typically produces an entropy increase much larger than the amount of information being rubbed out and much of that entropy production has a quite conventional character (for instance, the generation of ordinary heat). In order to make a point, however, we have been ignoring that extra entropy increase and concentrating on the minimum increase that must accompany the destruction of the information-bearing records.

It matters whether the destruction is irreversible. If the process can be reversed by reconstructing the notation from the fragments of eraser, then the entropy increase specifically associated with erasure has not taken place—but neither has the erasure taken place: a copy of the information still exists in the fragments.

It may be objected that such a reconstruction is always possible in principle, that it is just a practical question whether the information can be recovered from the nasty little bits of eraser. A dramatic example of such a situation was provided when the "students" who invaded and occupied the U.S. Embassy in Teheran in 1979 gathered up the strips of classified documents that had been shredded at the last moment by embassy employees and patiently pieced them together, so that the documents could be read and their contents made public. Although present-day shredders cut up documents in both dimensions, making such reconstruction very much more difficult, it is still not totally impossible in principle. How then can we talk about irreversible erasure or dispersal of information, or indeed about the disruption of any sort of order? Why is the whole idea of increase of entropy, of the conversion of order into disorder, not a fraud?

Entropy Useless Without Coarse Graining

Let us return to the oxygen molecules mixing with the nitrogen molecules. We can ask in what sense the mixing of the gases really increases disorder, since every oxygen and nitrogen molecule is *somewhere* at every moment (at least in the classical approximation), and therefore the situation at any time is just as orderly as at any previous time (provided the position of every molecule is described and not just the distribution of oxygen and nitrogen).

The answer is that entropy, like effective complexity, and algorithmic information content, and other quantities that we have discussed, depends on the coarse graining—the level of detail at which the system is being described. Indeed, it is mathematically correct that the entropy of a system described in perfect detail would not increase; it would remain constant. In fact, however, a system of very many parts is always described in terms of only some of its variables, and any order in those comparatively few variables tends to get dispersed, as time goes on, into other variables where it is no longer counted as order. That is the real significance of the second law of thermodynamics.

A related way to think of coarse graining is in terms of macrostates. A system that is initially described as being in one or a few macrostates will usually find itself later on in a mixture of many, as the macrostates get mixed up with one another by the dynamical evolution of the system. Furthermore, those macrostates that consist of the largest numbers of microstates will tend to predominate in the mixtures. For both those reasons, the later value of entropy will tend to be greater than the initial value.

We can try to connect coarse graining here with coarse graining in quantum mechanics. Recall that a maximal quasiclassical domain consists of alternative coarse-grained histories of the universe that are as fine-grained as possible consistent with being decoherent and nearly classical. Thus the quasiclassical domain that includes familiar experience is associated at each time with a set of highly refined macrostates, which provide a possible coarse graining for defining entropy. The resulting entropy seems to have suitable physical properties, with a tendency to increase as required for the thermodynamic arrow of time.

The Arrows of Time and the Initial Condition

Given an appropriate coarse graining, we can trace back the thermodynamic arrow of time to the simple initial condition of the universe and the final condition of indifference in the quantum-mechanical formula for probabilities of decohering coarse-grained histories of the universe. The same can be said of the arrow of time associated with outward radiation, and likewise of what I call the true cosmological arrow of time, which involves the aging, the running down, of the universe and its component parts. (Stephen Hawking defines his cosmological arrow of time by the expansion of the universe, but that is not a true arrow of time according to my definition. If, after a fantastically long time, the universe contracts, the contraction will take place forward in time as well—the aging will continue, as Hawking himself emphasizes.)

The arrow of time associated with the formation of records also derives ultimately from the simple initial condition of the universe. Finally, the so-called psychological arrow of time, referring to the experience of the forward flow of time on the part of human beings and all other complex adaptive systems, arises from the same condition. Memories are just records, and they obey forward causality just as other records do.

The Emergence of Greater Complexity: Frozen Accidents

The passage of time seems to open up opportunities for complexity to increase. Yet we know that complexity can also decline in a given system, such as a society forced to retreat to simpler social patterns because of severe stress from climate, enemies, or internal strife. Such a society may even disappear altogether. (The Classic Maya collapse certainly involved a reduction of complexity, even if many individuals survived.) Nevertheless, as time goes on, higher and higher social complexity keeps appearing. The same tendency occurs in biological evolution. Although some changes may involve decreases in complexity, the trend is toward higher maximum complexity. Why?

Recall that effective complexity is the length of a concise description of the regularities of a system. Some of those regularities can be traced back to the fundamental physical laws governing the universe. Others arise from the fact that many characteristics of a given part of the universe at a given time are related to one another through their common origin in some past incident. Those characteristics have features in common; they exhibit mutual information. For example, automobiles of a given model resemble one another because they all originate from the same design, which contains many arbitrary features that could have been chosen differently. Such "frozen accidents" can make themselves felt in all sorts of ways. Looking at coins of King Henry VIII of England, we may reflect upon all the references to him not only on coins but in charters, in documents relating to the seizure of abbeys, and in history books and how those would all be different if his elder brother Arthur had survived to mount the throne instead of him. And how much subsequent history may depend on that frozen accident!

We can now throw light on a class of fairly deep questions mentioned near the beginning of this book. If we find a coin bearing the image of Henry VIII, how can we employ the fundamental dynamical equations of physics to deduce that other such coins should turn up? Finding a fossil in a rock, how can we deduce from the fundamental laws that there are probably more fossils of a similar kind? The answer is: only by using the initial condition of the universe as well as the fundamental dynamical laws. We can then utilize the tree of branching histories and argue, starting from the initial condition and the resulting causality, that the existence of the found coin or fossil means that a set of events occurred in the past that produced it, and those events are likely to have produced other such coins or fossils. Without the initial condition of the universe, the dynamical laws of physics would not be able to lead us to such a conclusion.

A frozen accident may also explain, as we discussed earlier, why the four nucleotides abbreviated A, C, G, and T constitute the DNA of all living organisms on Earth. Planets orbiting distant stars may harbor complex adaptive systems that closely resemble terrestrial life but utilize genetic material composed of other molecules. Some theorists of the origin of life on Earth conclude that there may be thousands of such possible alternatives to the set A, C, G, and T. (Others, it should be noted, speculate that the familiar set of nucleotides may be the only possibility.)

An even more likely candidate for a frozen accident is the occurrence of certain right-handed molecules that play important roles in the chemistry of terrestrial life, while the corresponding left-handed molecules are not found in those roles and in some cases may be entirely lacking in life forms on Earth. It is not hard to understand why various kinds of right-handed molecules would be compatible with one another in biochemistry, and the same for left-handed molecules, but what determined the choice of one or the other?

Certain theoretical physicists tried for a long time to connect this left–right asymmetry with the striking behavior of the weak interaction, which exhibits left-handedness in ordinary matter (made of quarks and electrons) and right-handedness in antimatter (made of antiquarks and positrons). Their efforts do not appear to have borne fruit, and so it seems probable that the biochemical left–right asymmetry is a frozen characteristic of the ancestor of all surviving terrestrial life, and that it could just as well have turned out the other way.

The biological left–right asymmetry illustrates in a striking way that many frozen accidents can be regarded as instances of spontaneous symmetry breaking. There may be a symmetrical set of possibilities (in this case right- and left-handed molecules), only one of which actually occurs in a particular part of the universe during a particular time interval. In elementary particle physics, typical instances of spontaneous symmetry breaking are thought to apply to the whole universe. (There may also be others, even in elementary particle physics, that apply only over gigantic regions of the universe; if so, even that subject would to some extent have the character of an environmental science!)

The tree-like structure of branching histories involves a game of chance at every branching. Any individual coarse-grained history consists of a particular outcome of each of those games. As each history continues through time, it registers increasing numbers of such chance outcomes. But some of those accidents become frozen as rules for the future, at least for some portion of the universe. Thus the number of possible regularities keeps increasing with time, and so does the possible complexity.

This effect is by no means restricted to complex adaptive systems. The evolution of physical structures in the universe shows the same trend toward the emergence of more complex forms through the accumulation of frozen accidents. Random fluctuations gave rise to galaxies and clusters of galaxies in the early universe; the existence of each of those objects, with its individual characteristics, has been from the time of its birth a regularity of great importance for its part of the universe. Likewise the condensation of stars, including multiple stars and stars with planetary systems, out of gas clouds in those galaxies provided new regularities of great local importance. As the entropy—the overall disorder—of the universe increases, self-organization can produce local order, as in the arms of a spiral galaxy or the multiplicity of symmetrical forms of snowflakes.

The complexity at a given time of an evolving system (whether a complex adaptive system or a nonadaptive one) does not supply a measure of what levels of complexity it or its descendants (literal or figurative) may attain in the future. To fill that need, we introduced earlier the concept of potential complexity. To define it, we consider the possible future histories of the system and average the system's effective complexity at each future time over those histories, with each one weighted according to its probability. (The natural unit of time for this purpose is the average interval between random changes in the system.) The resulting potential complexity, as a function of time in the future, tells us something about the likelihood that the system will develop into something highly complex by that time, perhaps even by spawning a whole new kind of complex adaptive system. In the example we discussed earlier, potential complexity would distinguish emerging humans from the other great apes, even though their effective complexity at the time was not much different. Likewise, a planetary surface with a significant probability of generating life within a certain time would be distinguished from one on which life was not a serious possibility.

Will the Emergence of Greater Complexity Continue Forever?

After an enormously long time (even by cosmological standards), the universe, as it continues to expand, will become very different. Stars will die; black holes, having become more numerous than they are now, will decay, and probably even protons (and heavier nuclei) will decay as well. All the structures with which we are now familiar will disappear. It may be, therefore, that regularities will become fewer and fewer and the universe will be describable mostly in terms of randomness. The entropy will then be very high and so will the algorithmic information content, so that effective complexity will be low and depth fairly low as well (see pages 59 and 104).

Between now and then, if that picture is correct, the emergence of more and more complex forms will come gradually to a halt and the regression to lower complexity will become the rule. Furthermore, conditions will no longer be conducive to the existence of complex adaptive systems. Even individuality may decline, as well-defined individual objects become progressively scarcer.

This gloomy image is by no means entirely free of controversy. More theoretical research on the very distant future is needed. Although not directly of much practical value, such research will throw light on the significance of the era of complexity in which we find ourselves. Also the universe may be headed, after a very long time indeed, for recollapse, and theorists are investigating that phenomenon as well, attempting to describe what it would be like for entropy to continue increasing in a shrinking universe and what the prospects are for complexity during that possible phase of cosmic evolution.

Meanwhile, here on Earth the characteristics of our planet and our sun have provided frozen accidents that profoundly affect the rules of geology, meteorology, and other "environmental" sciences. In particular, they furnish the background for terrestrial biology. The evolution of the Earth, of the weather on its surface, of the prebiotic chemical reactions that led to the emergence of life, and of life itself, all illustrate the accumulation of frozen accidents that have become regularities for restricted regions of space and time. Biological evolution, especially, has given rise to the emergence of higher and higher effective complexity.

PART III

SELECTION AND FITNESS

CHAPTER

16

SELECTION AT WORK IN BIOLOGICAL EVOLUTION AND ELSEWHERE

All kinds of complex adaptive systems, including biological evolution, operate in accordance with the second law of thermodynamics. However, it is occasionally claimed by anti-evolutionists that biological evolution contradicts the second law, on the grounds that the emergence of more and more complex forms represents an increase of order with the passage of time. There are a number of reasons why that argument is wrong.

First, in the evolution of nonadaptive systems like galaxies, stars, planets, and rocks, more and more complex forms emerge with the passage of time, for reasons that we described earlier, without any conflict with the increase of entropy. The structures all age in accordance with the second law, but as time goes on there is also a broader and broader distribution of complexity, with the maximum complexity gradually increasing.

Second, the second law of thermodynamics applies only to closed (that is, completely self-contained) systems. One crucial mistake made by those who claim a contradiction between that law and biological evolution lies in looking only at what happens to certain organisms and not taking into account the environment of those organisms.

The most obvious way in which living systems fail to be closed arises from the need for sunlight as a direct or indirect source of energy. Strictly speaking, we cannot expect the second law of thermodynamics to hold unless we include that absorption of energy from the sun. Furthermore, energy flows out as well as in; ultimately it leaves in the form of radiation to the sky (think of the thermal radiation transmitted from your house to the cold, dark night sky). The flow of energy through a system can produce local order.

Even apart from that effect, however, the influence of information from the terrestrial environment must be taken into account. To see what happens when that environmental information is included, consider an oversimplified case in which the influence exerted by the environment is steady and the interaction among different kinds of organisms is ignored. A population of a given kind of organism is then evolving in the presence of a consistent environment. As time goes on, the population tends to become better adapted to its surroundings, since different genotypes within the population compete with one another and some are more successful than others in creating phenotypes that survive and reproduce. Consequently, a kind of discrepancy in information between the environment and the organism is gradually reduced. That process is reminiscent of the way that the temperatures of a hot object and a cold object placed in contact with each another approach thermal equilibrium, in conformity with the second law. Biological evolution, far from contradicting that law, can provide instructive metaphors for it. The process of adaptation of a population is a kind of winding down in the presence of its environment.

In hot sulfur springs all over the world (and in the ocean depths where hot vents mark the boundaries between tectonic plates), primitive organisms called extremophiles (or crenarchaeota) thrive in an environment that most living things would find extremely hostile. In the life of extremophiles at the bottom of the ocean, sunlight plays a limited role, confined mostly to processes that supply oxidizing chemicals. For instance, sunlight helps to maintain other life forms, nearer the surface of the water, that keep dropping organic material down to where the extremophiles live.

Strong indirect evidence points to the existence, more than three billion years ago, of organisms that were similar, at least metabolically, to modern extremophiles. No one knows whether the entire underlying genotypes were also very similar, or whether parts of the genome have undergone substantial drift, leaving the practical result in the real world of selection pressures much the same. In either case, we can say that the rather difficult problem of living in that hot, acidic, sulfurous environment was solved when the Earth was young. The extremophiles reached a kind of steady state, something like an evolutionary equilibrium with their surroundings.

Rarely, though, are the surroundings so stable. Most natural situations are more dynamic, with the environment undergoing significant changes as time goes by. For example, the composition of the Earth's atmosphere as we know it owes a great deal to the presence of life. The presence today of significant amounts of oxygen can be attributed, at least in great part, to the plants that have proliferated on the surface of the planet.

Co-evolving Species

Moreover, the environment of any given species of organism includes a huge number of other species, which are themselves evolving. The genotype of each organism, or else the cluster of genotypes that characterizes each species, can be regarded as a schema that includes a description of many of the other species and how they are likely to react to different forms of behavior. An ecological community consists, then, of a great many species all evolving models of other species' habits and how to cope with them.

In some cases it is a useful idealization to consider just two species co-evolving and dealing with the developments in each other's capabilities. For example, in walking through forests in South America, I have often encountered a species of tree that provides nutriment for a particularly nasty species of stinging ant. In return, the ant repels many kinds of animals, including us humans, that might otherwise want to harm the tree. Just as I have learned what that tree looks like, to avoid bumping into it by mistake, so have other mammals learned to recognize it and not munch on it. Such a situation of symbiosis must have been produced by a substantial period of co-evolution.

In the same forest, one may encounter offensive–defensive competitions that also proceed by the evolution of two species adapting to each other. A tree may evolve the capacity to exude a toxic substance that repels a destructive kind of insect. That insect species, in turn, may evolve a means of metabolizing the poison, which is then no longer a threat. Further evolution of the tree may result in modification of the poison so that it again becomes effective, and so on. Such chemical arms races can result in the natural production of chemical agents that are very potent biologically. Some of those can be of great utility to human beings, in medicine, integrated pest management, and other fields.

In the realistic situation, viewed without simplification, many species evolve together in an ecological community, with nonliving surroundings that gradually (or even rapidly) alter with time. That is much more complicated than the idealized examples of symbiosis or competition of two species, just as they are more complicated than the even more idealized case of a single species evolving in a fixed environment. In each case, the process of biological evolution is compatible with the thermodynamic arrow of time, as long as the whole system is taken into account; but only in the simplest situations, like that of extremophiles, does evolution lead to a kind of informational steady state. In general, the process is one of continuing dynamic change, just as it is for a complex physicochemical system like a galaxy, a star, or a planet without life. All are aging, winding down as time goes forward, albeit in a complicated manner.

In an ecological community, the process of mutual adjustment through evolution is an aspect of that aging. Biological evolution is part of the winding-down process by which the informational gap between the potential and the actual tends to be reduced. Once a complex adaptive system exists, the discovery and exploitation of opportunities is not only possible but probable, because that is the direction in which the system is pushed by the selection pressures operating on it.

Punctuated Equilibrium

Biological evolution does not usually proceed at a more or less uniform rate, as some specialists used to imagine. Instead, it often exhibits the phenomenon of "punctuated equilibrium," in which species (and higher groupings or taxa, such as genera, families, and so forth) stay relatively unchanged, at least on the phenotypic level, for long periods of time and then undergo comparatively rapid change over a brief period. Stephen Jay Gould, who proposed the idea in technical publications co-authored with his colleague Niles Eldredge, has also written a great deal about punctuated equilibrium in his engaging popular articles and books.

What causes the comparatively rapid changes that constitute the punctuation? The mechanisms thought to be responsible can be divided into various categories. One comprises alterations, sometimes widespread alterations, in the physicochemical environment. At the end of the Cretaceous Period about sixty-five million years ago, at least one heavy object collided with the Earth, the one that formed the huge crater of Chicxulub on the edge of the Yucatán peninsula. The resulting atmospheric changes helped to produce the Cretaceous extinction, which did away with the large dinosaurs and very many other forms of life. Hundreds of millions of years earlier, during the Cambrian Period, great numbers of ecological niches opened up and were filled by new life forms (somewhat the way a new and popular technology leads to numerous job opportunities). The new life forms created still more new niches, and so on. Some evolutionary theorists have tried to connect that explosion of diversity with an increase in the oxygen content of the atmosphere, but that hypothesis is not generally accepted today.

Another kind of rapid change that may punctuate apparent evolutionary equilibrium is largely biological in character. It does not require dramatic sudden changes in the physical environment. Instead, it results from the tendency of genomes to change gradually with time in ways that do not profoundly affect the viability of the phenotype. As a result of that process of "drift," a cluster of genotypes constituting a species may move toward an unstable situation in which fairly small genetic changes can radically alter the phenotype. It may happen at a certain time that a number of species in an ecological community are approaching that kind of instability, creating a situation that is ripe for the occurrence of mutations that do lead to important phenotypic changes in one or more organisms. Those changes can initiate a series of linked events, in which some organisms become more successful, others die out, the whole community is altered, and new ecological niches open up. Such an upheaval can then provoke change in neighboring communities, as, for example, new kinds of animals migrate there and compete successfully with established species. A temporary apparent equilibrium has been punctuated.

Gateway Events

Particularly dramatic biological events are sometimes responsible for critical instances of punctuated equilibrium in the absence of radical change in the physicochemical surroundings. Harold Morowitz of George Mason University and the Santa Fe Institute points out the great significance of breakthroughs or gateway events that open up whole new realms of possibility, sometimes involving higher levels of organization or higher types of function. Harold has particularly emphasized cases where these gateways are unique or nearly so, and where they are dependent on a biochemical innovation.

To begin with, he theorizes about possible chemical gateways in the course of the prebiotic chemical evolution that led to the origin of life on Earth. Those gateways include:

1. one that led to energy metabolism using sunlight and so to the possibility of a membrane isolating a portion of matter such as that represented later on by the cell;
2. one that supplied the catalysts for the transition from keto acids to amino acids and thence to the production of proteins; and
3. chemical reactions that resulted in molecules called dinitrogen heterocycles and so led to the nucleotides that constitute DNA, thus permitting the genome, the biological schema or information package, to come into existence.

In all these cases Harold emphasizes the narrowness of the gateway. Typically just a few special chemical reactions make possible the entry into a new realm; sometimes just a single reaction is responsible. (The specificity of such reactions does not mean they are necessarily improbable—even a unique reaction may take place easily.)

Analogous gateway events occurred in biological evolution, following the development of the life form ancestral to all organisms alive today. Many of those events opened up new levels of organization. One example is the evolution of eukaryotes, organisms in which the cell possesses a true nucleus (containing the principal genetic material) and also other "organelles"—mitochondria or chloroplasts. The transformation of more primitive organisms into single-celled eukaryotes is thought by many researchers to have come about through their incorporating other organisms, which became endosymbionts (meaning that they lived on inside and in symbiosis with the cell) and then evolved into organelles.

Another example is the evolution of animal-like single-celled eukaryotes (presumably the ancestors of true animals). It is thought that plant-like eukaryotes came first, each engaging in photosynthesis and equipped with a cell wall composed of cellulose, as well as a membrane inside that wall. (The membrane required a biochemical breakthrough, the formation of sterols, related to cholesterol and to human sex hormones.) Evolution then led to organisms possessing the membrane without the wall and thus able to dispense with photosynthesis by devouring photosynthesizing organisms instead. The emergence of that capability was the key to the later appearance of true animals.

The evolution of many-celled organisms from single-celled ones, presumably through aggregation, was made possible by another biochemical innovation—a glue that can hold the cells together.

Harold Morowitz and others believe that, at least in many cases, a small change in the genome, brought about by one mutation or just a few, but coming on top of a long series of earlier changes, can trigger a gateway event and start one of the revolutions that punctuate in major ways the relative stability of evolutionary equilibrium. In entering the realm opened up by the gateway event, an organism acquires new and very significant regularities, which raise it to a higher level of complexity.

As with physical disturbances such as earthquakes (or collisions of the Earth with other objects in the solar system), such major events can be regarded either as individual occurrences of great significance or else as unusual events of great magnitude lying on the tail of a distribution comprising events that are mostly of much smaller magnitude.

Aggregation Resulting in Higher Levels of Organization

In the evolution of an ecological community or an economy or a society, opportunities for increased complexity keep arising, as they do in biological evolution, and result in a tendency for the maximum complexity to drift upward. The most fascinating increases in complexity are those that involve a transition from a level of organization to a higher one, typically through the formation of a composite structure, as in the evolution of many-celled plants and animals from single-celled organisms.

Families or bands of human beings can come together to form a tribe. A number of people can pool their efforts to make a living by constituting a business firm. In the year 1291, three cantons, soon to be joined by a fourth, founded the Swiss Confederation, which grew over time into modern Switzerland. The thirteen North American colonies joined in a confederation and then, by ratifying the constitution of 1787, turned themselves into the federal republic called the United States of America. Cooperation leading to aggregation can be effective.

Although competition among schemata is a characteristic of complex adaptive systems, the systems themselves may indulge in a mixture of competition and cooperation in their interactions with one another. It is often beneficial for complex adaptive systems to join together to form a collective entity that also functions as a complex adaptive system, for instance when individuals and firms in an economy operate under a government that regulates their behavior in order to promote values important to the community as a whole.

Cooperation of Schemata

Even among schemata, competition leavened with cooperation is sometimes both possible and advantageous. In the realm of theories, for instance, competing notions are not always mutually exclusive; sometimes a synthesis of several ideas comes much closer to the truth than any of them does individually. Yet proponents of a particular theoretical point of view can often reap rewards in academic life and elsewhere by claiming their proposal to be completely correct and entirely new, while arguing that competing points of view are wrong and should be discarded. In some fields and in certain cases, that approach may be justified. Often, however, it is counterproductive.

In archaeology, for example, and in other parts of anthropology, disputes have long raged over diffusion of cultural traits versus independent invention. Yet it seems obvious that both occur. The invention of the zero in India (whence it was brought to Europe through the work of al-Khwarizmi) seems overwhelmingly likely to have been independent of its invention in Mesoamerica (where it was employed, for example, by the Classic Maya). If contacts of some kind were responsible instead, how strange that the wheel was almost entirely lacking in the New World in pre-Columbian times (it has been found only on a few toys from Mexico, I understand) when it had been known for so long in the Old World. The bow and arrow seem to have diffused from North America to Mesoamerica, while numerous other cultural developments, such as the domestication of maize, spread in the opposite direction. How can scholars still be berating one another as diffusionists and anti-diffusionists?

Some cultural anthropologists like to point out ecological and economic rationales for tribal customs that may appear at first sight arbitrary or irrational. In doing so, they perform a valuable service, but sometimes they go on to poke fun at the very idea that irrationality and arbitrary choice play important roles in belief systems and patterns of social behavior. Surely that is going too far; a reasonable point of view would temper ecological and economic determinism with a measure of allowance for the vagaries of tribal schemata. For instance, a particular dietary prohibition, say against eating okapis, may make sense for a certain tribe given the nutritional needs of the population and the amount of labor needed to hunt the okapi versus producing other foods in the ecological context of the surrounding forest. However, the restrictions might also stem from a former identification with the okapi as the totem of the tribe; or a mixture of both causes may be at work. Is it wise to insist that one point of view or the other is always the right one?

One of the virtues of the Santa Fe Institute is that an intellectual climate has been created in which scholars and scientists feel drawn to one another's ideas and seek, much more than at their home institutions, to find ways of harmonizing those ideas and creating useful syntheses out of them when that seems indicated. On one occasion, a seminar at the Institute was attended by several professors from the same department of the same university, who found that they could somehow converse constructively in Santa Fe about issues that merely provoked disputes at home.

In biological evolution, the closest thing to cooperation among schemata is probably the genetics of sexual reproduction, in which the genotypes of parent organisms become mixed in their descendants. We shall soon turn to sexual reproduction, but first let us explore further the trend toward higher complexity.

Is There a Drive Toward Higher Complexity?

We have seen that the dynamics of biological evolution can be complicated. Yet it has often been portrayed in simplistic ways. The emergence of ever more complex forms has sometimes been mistaken for a steady progression toward some kind of perfection, which in turn may be identified with the species, and perhaps even the race or gender, of the author. Fortunately, that kind of attitude is on the wane, and today it is possible to look upon evolution as a process rather than regarding it as teleological, a means to an end.

Nevertheless, even now, and even among some biologists, the idea persists to a certain extent that a "drive" toward complexity is inherent in biological evolution. As we have seen, what actually takes place is a bit more subtle. Evolution proceeds by steps, and at each step, complexity can either increase or decrease, but the effect on the whole set of existing species is that the greatest complexity represented has a tendency to grow larger with time. A similar process takes place in a community that is growing wealthier in such a way that any individual family may see its income grow or decline or stay the same, even though the range of incomes is growing wider, so that the largest family income tends to keep increasing.

If we were to ignore any advantages conferred on species by increased complexity, we could regard the changing distribution of complexities as a kind of diffusion, exemplified by a "random walk" on a line. A great many fleas start from the same point and keep jumping, at random, by the same distance each time, either away from the starting point or toward it (initially, of course, they all jump away). At any given moment thereafter, one or more fleas will have jumped the furthest from the starting point. Which particular fleas have gone the greatest distance may keep changing, naturally, depending on which have achieved by chance the greatest number of net jumps away from the

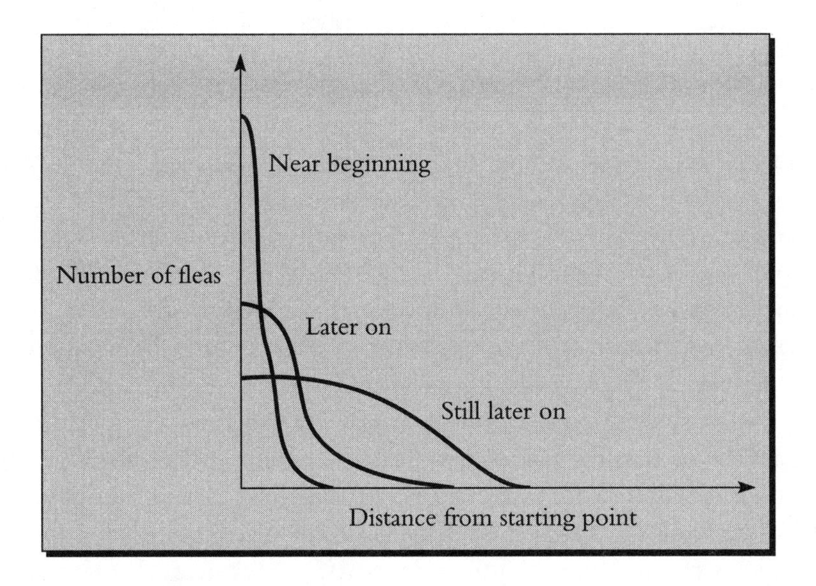

Changing distributions of distances in a random walk.

starting point. The greatest distance that any flea has moved from that point tends to keep increasing as time goes on. The distribution of the distances of the fleas from the starting point spreads out as the fleas undergo *diffusion* toward greater and greater distances, as shown above.

The movement toward higher maximum complexity can proceed in a manner that resembles diffusion, especially in nonadaptive systems like galaxies. In complex adaptive systems such as biological evolution, though, it often happens that selection pressures favor higher complexity in certain situations. The distribution of complexities in such cases will differ, in shape as a function of time, from the result of a random walk. Although there is still no reason to believe in a steady drive toward more complex organisms, the selection pressures that favor higher complexity may often be strong. To characterize those systems and environments in which complexity is highly advantageous presents an important intellectual challenge.

Gateway events in the course of biological evolution typically give rise to large increases in complexity and also to very significant advantages. The opening of a critical gateway results in an explosion of ecological niches, the filling of which may well look as if it were caused by a drive toward greater complexity.

Since we humans are the most complex organisms in the history of biological evolution on Earth, it is understandable that some of us look upon the whole process of evolution as leading up to *Homo sapiens sapiens*. Even though that view is anthropocentric foolishness, there is a sense in which the role of biological evolution does end with us, or at least is now suspended. Our effect on the biosphere is so profound and our ability to transform life (not just by ancient and slow procedures like dog breeding but by modern methods like genetic engineering) will soon be so great that the future of life on Earth really does depend in large part on crucial choices made by our species. Barring some spectacular renunciation of technology (very difficult to accomplish in view of the enormous human population we are already committed to sustaining), or the self-destruction of most of the human race—followed by the reversion of the rest to barbarism—it looks as if the role of natural biological evolution in the foreseeable future will be secondary, for better or for worse, to the role of human culture and *its* evolution.

The Diversity of Ecological Communities

Unfortunately, it will be a long time before human knowledge, understanding, and ingenuity can match—if they ever do—the "cleverness" of several billion years of biological evolution. Not only have individual organisms evolved their own special, intricate patterns and ways of life, but the interactions of huge numbers of species in ecological communities have undergone delicate mutual adjustments over long periods of time.

The various communities consist of sets of species that differ, depending on the region of the globe and, within each region, on the physical environment. On land, the character of the community varies according to factors such as the altitude, the rainfall and its distribution over the year, and the temperature and its pattern of variation. The regional differences are related to species distributions that have in many cases been affected by movements of continents over millions of years and by accidents of ancient migrations and dispersals.

Thus forests tend to differ greatly from one another, even within the tropics. Not all tropical forests are lowland rainforests, as certain news releases would have us believe. Some are lowland dry forests, others are montane cloud forests, and so forth. Furthermore, one can distinguish hundreds of distinct rainforests, all containing significantly different flora and fauna. Brazil, for instance, is home not only to the vast expanse of Amazonian lowland rainforest, itself varying considerably in composition from place to place, but also to the very different Atlantic rainforest, now reduced to only a small percentage of its former territory. At its southern end, the Atlantic forest merges with the Alto Paraná forest of Paraguay and the tropical forest of Misiones province in Argentina. The progressive destruction of the Amazonian forest is now a matter of general concern, although a huge amount of it still remains standing (sometimes, unfortunately, in a degraded condition that is not easily detectable from the air), but the preservation of what is left of the Atlantic forest is even more urgent.

Likewise, deserts differ from one another. In the Namib Desert of Namibia, the flora and fauna are, in great part, different from those of the Sahara Desert at the other end of Africa, and from those of the spiny desert in southern Madagascar.

The Mojave and Colorado deserts of southern California are quite distinct from each other and neither shares many species with, say, the Negev in Israel. (Note, though, that the famous cactus of Israel, the sabra, is just an import from Mexico and California.) On visits to the Colorado Desert and to the Negev, superficial looks will reveal many resemblances in appearance of the flora, but much of that similarity is attributable not to close relationship of species but to evolutionary convergence, the result of the operation of similar selection pressures. Likewise, many euphorbias of the arid high plains of East Africa resemble New World cacti, but only because they have adapted to similar climates; they belong to different families. Evolution has produced a number of distinct but similar solutions, in various parts of the globe, to the problem of a community of organisms living under a given set of conditions.

Faced with such a wide variety of natural communities, will human beings have the collective wisdom to make appropriate policy choices? Have we acquired the power to effect enormous changes before we have matured enough as a species to use that power responsibly?

The Biological Concept of Fitness

Ecological communities made up of many complex individuals, belonging to a large number of species, all evolving schemata for describing and predicting one another's behavior, are not systems likely to reach or even closely approach an ultimate steady state. Each species evolves in the presence of constantly changing congeries of other species. The situation is about as different as it could be from that of the oceanic extremophiles, which evolve in a fairly constant physicochemical environment and interact with other organisms mainly through the organic matter that descends to their level through the water.

Even a comparatively simple and nearly self-contained system such as that of the extremophiles cannot usually be assigned a rigorously defined numerical attribute called "fitness," and certainly not one that continues to increase, as evolution proceeds, until a steady state is achieved. Even in such a simple case, it is much safer to concentrate directly on the selection pressures, the effects that favor one phenotypic character over another and thus affect the competition among the different genotypes. Those selection pressures may not be expressible in terms of a single, well-defined quantity called fitness. They may require a more complicated description, even in the idealized case of a single species adapting to a fixed environment. It is even less likely, then, that a truly meaningful measure of fitness can be assigned to an organism when the environment is changing, and especially when it belongs to a highly interactive ecological community of organisms adapting to one another's peculiarities.

Still, a simplified discussion of biological evolution in terms of fitness can often be instructive. The idea underlying the biological concept of fitness is that propagation of genes from one generation to the next depends on survival of the organism until it reaches the stage of reproduction, followed by the generation of a reasonable number of offspring that in turn survive to reproduce. Differential rates of survival and reproduction can often be roughly described in terms of a fitness quantity, defined so that that there is a general tendency for organisms with higher fitness to propagate their genes more successfully than those with lower fitness. At the extreme, organisms with genetic patterns that are consistently associated with failure to reproduce have very low fitness and tend to disappear.

Fitness Landscapes

One general difficulty becomes apparent when we introduce the crude notion of a "fitness landscape." Imagine the different genotypes to be laid out on a horizontal two-dimensional surface (standing for what is really a multidimensional mathematical space of possible genotypes). Fitness or unfitness is represented by height; as the genotype varies, the fitness describes a two-dimensional surface, with a great many hills and valleys, in three dimensions. Biologists conventionally represent fitness as increasing with increasing height, so that maxima of fitness correspond to the tops of hills and minima to the bottoms of pits; however, I shall use the reverse convention, which is customary in many other fields, and turn the whole picture upside down. Now fitness increases with depth, and the maxima of fitness are the bottoms of depressions, as shown on page 250.

The landscape is very complicated, with numerous pits ("local maxima of fitness") of widely varying depths. If the effect of evolution were always to move steadily downhill—always to improve fitness—then the genotype would be likely to get stuck at the bottom of a shallow depression and have no opportunity to reach the deep holes nearby that correspond to much greater fitness. At the very least, the genotype must be moving in a more complicated manner than just sliding downhill. If it is also jiggling around in a random way, for example, that will give it a chance to escape from shallow holes and find deeper ones nearby. There must not be too much jiggling, however, or the whole process will cease to work. As we have seen in a variety of connections, a complex adaptive system functions best in a situation intermediate between order and disorder.

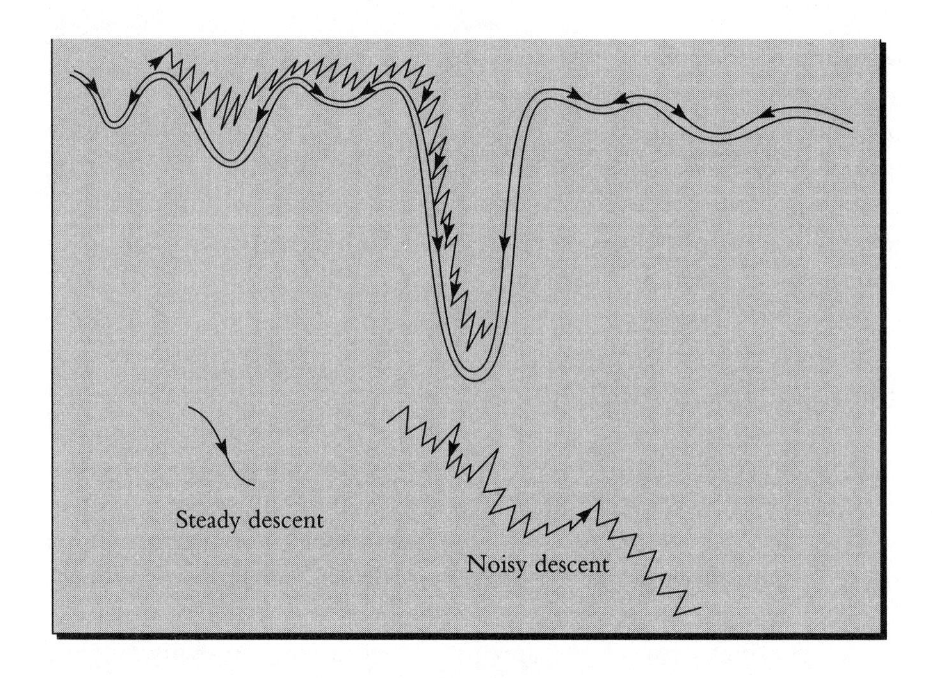

A fitness landscape, with increasing fitness corresponding to greater depth.

Inclusive Fitness

A further complication in utilizing the concept of fitness arises in higher organisms that make use of sexual reproduction. Each such organism conveys only half its genes to a given offspring, while the remaining half derive from the other parent. The offspring are not clones, but merely close relatives. And the organism has other close relatives, the survival of which can also contribute to the propagation of genes similar to its own. Thus biologists have developed the notion of "inclusive fitness," which takes account of the extent to which relatives of a given organism survive to reproduce, weighted according to the closeness of the relationship. (Of course inclusive fitness also takes account of the survival of the organism itself.) Evolution should have a general tendency to favor genotypes exhibiting high inclusive fitness, especially through inherited patterns of behavior that promote the survival of an organism and its close relatives. That tendency is called "kin selection," and it fits in nicely with a picture of evolution in which organisms are merely devices "used" by genes to propagate themselves. That point of view has been popularized under the name of the "selfish gene."

The Selfish Gene and the "Truly Selfish Gene"

An extreme form of the selfish gene phenomenon can occur in what is called "segregation distortion." As described by the sociobiologist Robert Trivers, segregation distortion could result from the operation of a "truly selfish gene." He means a gene that works *directly*, not through the resulting organism, to promote its success in competition with rival genetic patterns. Such a gene present in a male animal may cause the sperm bearing it to outrace or even poison other sperm, and so to win out more easily in the competition to fertilize the eggs of the female. A truly selfish gene need not, however, confer any advantage on the resulting organism, and may even be somewhat harmful.

Apart from such remarkable possible exceptions, selection pressures are exerted indirectly, through the organism produced by the sperm and the egg. That is more in line with the notion of a complex adaptive system, in which the schema (in this case the genome) is tested in the real world (in this case by means of the phenotype) rather than directly.

Individual and Inclusive Fitness

A fascinating case in which ordinary individual fitness and inclusive fitness both seem to be involved is the so-called altruistic behavior of certain species of birds. The Mexican or gray-breasted jay lives in arid habitats in northern Mexico, southeastern Arizona, and southwestern New Mexico. Ornithologists observed years ago that a given nest of that species was often tended by many birds, not just the couple that produced the eggs. What were those other jays doing there? Were they really behaving altruistically? The research of Jerram Brown revealed that the helpers were in most cases themselves offspring of the nesting pair; they were helping to raise their own siblings. That behavior seemed to provide a striking example of evolution of social behavior via inclusive fitness. Evolution had favored a behavior pattern in which young jays, postponing their own reproduction, helped to feed and guard their younger siblings, thus assisting the propagation of genes closely related to their own.

More recently, the picture has become more complicated as a result of the work of John Fitzpatrick and Glen Woolfenden on a related kind of jay, the Florida scrub jay, which is found in the rapidly disappearing arid oak scrub habitat of southern Florida. Until now that bird has usually been considered one of many subspecies of the scrub jay, which is very common in the southwestern United States, but Fitzpatrick and Woolfenden are proposing that it be treated as a separate species, on account of its appearance, vocalizations, behavior, and genetics. The behavior includes helping at the nest, as in the case of the gray-breasted jay. Here again the helpers are apt to be offspring of the nesting pair, but the observations of the Florida researchers indicate that the helpers are serving their own interests as well as their siblings'. Nesting territories in the oak scrub are large (on the order of thirty acres), fiercely defended, and not easy to come by. The helpers are in the best position to inherit all or part of the territory where they are working. In Florida, at least, it looks as if ordinary individual fitness plays a very important role in "altruistic" scrub jay behavior.

I have introduced the scrub jay story not in order to take sides in a controversy among ornithologists, but just to illustrate the subtlety of the whole idea of fitness, whether inclusive or not. Even when fitness is a useful concept, it is still a bit circular. Evolution favors the survival of the fittest, and the fittest are those who survive, or whose close relatives survive.

The Fitness of Sex

The phenomenon of sexual reproduction poses some special challenges to theories of selection pressures and fitness. Like numerous other organisms, higher animals tend to reproduce sexually. Yet in many cases the same animals are not incapable of parthenogenesis, in which females give birth to female offspring with identical genetic material, except for possible mutations. The services of a male are not required. Even the eggs of an animal as complex as a frog can be stimulated, say with a needle prick, to produce tadpoles that grow into adult frogs. In a very few cases, such as the whiptailed lizards of Mexico and the southwestern United States, whole species of vertebrates seem to manage using parthenogenesis alone, with no males at all. Why sex then? What is the enormous advantage that sexual reproduction confers? Why is it usually selected over parthenogenesis? What are males really good for?

Sexual reproduction introduces diversity into the genotypes of offspring. Roughly speaking, the chromosomes (each containing a string of genes) come in corresponding pairs and an individual inherits one of each pair from the father and one from the mother. Which one it inherits from each parent is largely a matter of chance. (In identical twins, these otherwise stochastic choices come out the same for both.) The offspring of organisms with many chromosome pairs generally have different sets of chromosomes from those of either parent.

In addition, sexual reproduction introduces a whole new mechanism, other than ordinary mutation, for change in the chromosomes. In the process called "crossing-over," illustrated on page 254, a pair of corresponding chromosomes from either the father or the mother can get partially switched with each other in the formation of a sperm or an egg, respectively. Say the crossing-over occurs in the case of the egg, contributed by the mother. The egg acquires a mixed chromosome consisting of a part contributed by the mother's father and the rest by the mother's mother, while another egg may receive a chromosome consisting of the remaining parts of the maternal grandfather's and grandmother's chromosomes.

The evolutionary theorist William Hamilton, now the holder of a chair at Oxford, has suggested a simple explanation for the value of sexual reproduction. Roughly speaking, his idea is that enemies of a species, especially harmful parasites, find it harder to adapt to the diverse attributes of a population generated by sexual reproduction than to the comparative uniformity of a population produced by parthenogenesis. The mingling of chromosomes contributed by father and mother and also the process of crossing-over allow all sorts of new combinations to occur among the offspring, forcing the parasites to cope with a wide variety of hosts, presenting different body chemistry, different habits, and so forth. As a result, the enemies have trouble and the hosts are safer.

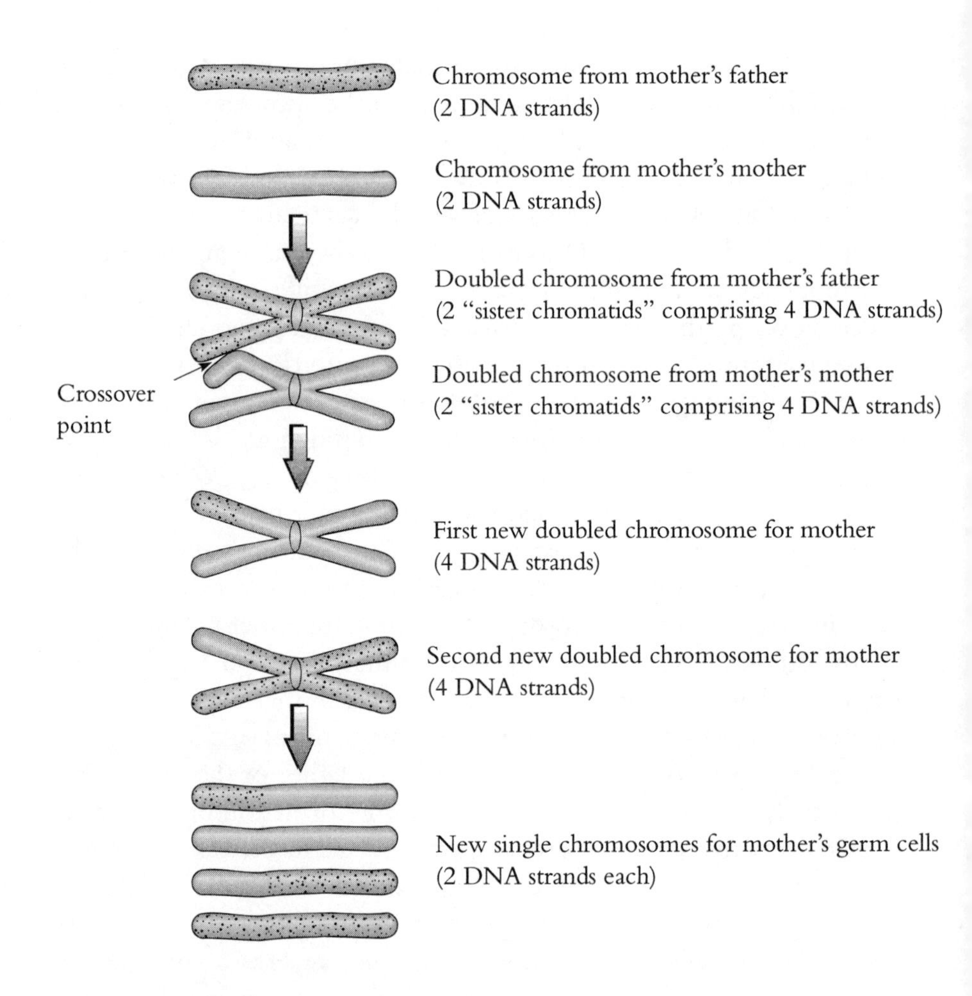

Crossing-over of chromosomes in sexual reproduction.

The theory indicates that species without sexual reproduction should have other mechanisms for coping with parasites, especially in the case of whole groups of lower animals that have been without sex for tens of millions of years. The bdelloid rotifers form such a group. They are wheel animalcules inhabiting places, like beds of moss, that are moist most of the time but become dry every few weeks or months, depending on vagaries of the weather. A student of Hamilton's, Olivia Judson, is studying those rotifers and trying to establish how they deal with parasites. She suggests that their habit of desiccating and blowing away when their surroundings dry up may afford them enough protection from parasites that they can dispense with sex.

In any case, the advantages of sexual reproduction must be considerable to outweigh the obvious disadvantage of breaking up the successful genotypes of parents and grandparents that survived long enough to reproduce. These advantages accrue to the population as a whole, however, while many evolutionary biologists insist that selection pressures are exerted only on individuals. Perhaps that need not be a rigid rule.

At a recent Santa Fe Institute meeting, John Maynard Smith, who teaches at the University of Sussex, was commenting on this issue, when Brian Arthur, chairing the session, recalled the occasion when they first met. Both men have a background in engineering. Maynard Smith became an aircraft designer and then took up evolutionary biology, to which he has made some remarkable contributions. Brian, who grew up in Belfast, went into operations research and then economics, becoming a professor at Stanford and the founding director of the economics program at the Santa Fe Institute. They first encountered each other at a scientific meeting in Sweden, where Maynard Smith remarked in the course of a lecture that while sex had obvious advantages for a population, it was not clear what it did for the individual. Brian called out from the audience, "What a very English view of sex!" Maynard Smith, without missing a beat, replied, "I gather from your accent that you're Irish. Well, in England at least we *have* sex."

Death, Reproduction, and Population in Biology

While sex is by no means universal in biology, death comes pretty close to being so. The death of organisms is one of the more dramatic manifestations of the second law of thermodynamics. As such, it is in a sense common to all complex adaptive systems. However, it is especially significant in biological evolution, where the interplay between death and reproduction is at the forefront of the adaptive process. Competition between clusters of genotypes translates to a considerable degree into competition for population size among the corresponding kinds of organisms. To the extent that fitness is well defined in biological evolution, it is connected with population size.

Comparison of various types of complex adaptive systems reveals systems in which death, reproduction, and population are less important than in biology. For example, take an individual human being deep in thought, solving a problem. The schemata in that case are ideas instead of genotypes. The analogue of death is forgetting. No one can deny the prevalence or the significance of forgetting, but it hardly plays the same role as death in biology. If there were no need for forgetting, no need to "erase the tape," the character of thinking would not change dramatically. Recording an idea is useful, and it counteracts the effect of forgetting, but the number of identical or nearly identical memoranda does not characterize the fitness of an idea to the same extent that population tends to correlate with fitness in biology.

As ideas spread through a society (even the scientific community), the number of people sharing a given idea does have importance. In democratic elections, to the extent that they are concerned with ideas, the majority opinion prevails. Still, it is notorious that even overwhelming numbers of adherents do not necessarily render an idea correct, and may not even guarantee its survival in the long run.

For a case more like that of biological evolution, we can turn to the competition among human societies in the past. To a great extent, fitness was measured by population. In Southeast Asia, for instance, some ethnic groups practiced irrigated rice agriculture while others raised dry rice, often by slashing and burning the forest. The irrigated-rice peoples, such as the Central Thai, the Lao, or the Vietnamese, were able to put many more individuals on the ground per unit area than their neighbors. Denser population helped them to dominate the dry-rice peoples, and in many cases to drive them back into remote hilly terrain. Looking toward the future, we may well ask whether it is desirable for density or total numbers to continue to determine winners and losers in the same way.

The Filling of Niches

Biological evolution, with its emphasis on death and population, is fairly efficient in the long term at filling ecological niches as they arise. When there is an opportunity to make a living in a certain way, it is probable that some organism will evolve to exploit it, even if that way of living may seem rather bizarre to a human observer.

The analogy between an ecological community and a market economy is relevant in this connection. As opportunities for making a profit arise in such an economy, individuals or firms are likely (although by no means certain) to come along to take advantage of them. The analogue of death in this case is going broke, and wealth instead of population is a crude measure of the fitness of a firm.

In both economics and ecology, the advent of a new business or a new organism (or of a new type of behavior in an existing firm or organism) will alter the fitness landscape for the other members of the community. From the point of view of a business or a species, that landscape is constantly changing (besides not being altogether well defined in the first place).

Both of these cases illustrate how a complex adaptive system, once established, can fill niches, create new ones in the process, fill those in turn, and so on, spawning new complex adaptive systems along the way. (As indicated in the diagram on page 20, biological evolution has given rise to mammalian immune systems, to learning and thinking, and, through human beings, to societies that learn and adapt, and recently to computers that function as complex adaptive systems.)

Always exploring, seeking out opportunities, experimenting with novelty, the complex adaptive system tries out increases in complexity and occasionally discovers gateway events that open up the possibility of whole new structures, including new kinds of complex adaptive systems. Given enough time, the likelihood of the evolution of intelligence would seem to be high.

Astronomers and planetary scientists are not aware of any reason why planetary systems should be especially rare in our galaxy or in other similar galaxies elsewhere in the universe. Nor have theorists of the origin of life come up with anything so remarkable about the conditions on our own planet some four billion years ago that the origin of life (or something like life) on a planet would be a particularly improbable event. It is likely that complex adaptive systems abound in the universe and that a great many of them have evolved or will evolve intelligence. As mentioned earlier, the principal statistical unknowns in the Search for Extraterrestrial Intelligence (SETI) are the number of planets per unit volume of space on which intelligent beings have arisen and the length of time that their period of technical civilization, with emission of wireless signals, typically lasts. Given the immense amount we can learn from the diversity of natural communities on Earth, to say nothing of the diversity of human societies, it is mind-boggling to imagine (as science fiction writers sometimes do) the lessons that contact with extraterrestrials could teach us about the variety of circumstances that complex adaptive systems can exploit.

Deception Among Birds

For amusing examples of the exploitation of opportunities by species interacting with other species, we can turn to lying as practiced by animals other than humans. Deception by mimicry is well known; the viceroy butterfly, for instance, resembles the monarch and thus profits by the bad taste of the latter. The cuckoo (in the Old World) and the cowbird (in the New World) practice another kind of deception by laying their eggs in the nests of other birds; the intrusive chicks then do away with the eggs or chicks that belong in the nest and monopolize the attention of the foster parents. But actual lying?

We are accustomed to hearing people lie, but it is somehow more surprising in other organisms. When the Argentine Navy spots a mysterious periscope in the estuary of the Río de la Plata just before the budgets of the armed forces are to be considered by the legislature, we suspect that deception is being practiced so as to capture additional resources, and we are not particularly astonished. But the analogous behavior among birds is more unexpected.

One such case was discovered recently by my friend Charles Munn, an ornithologist studying mixed feeding flocks in the lowland tropical forest of Manu National Park in Peru. Some species forage together in the understory or lower canopy of the forest and others in the middle canopy, where they are sometimes joined by colorful fruit-eating tanagers from the upper canopy. (Among the species found in those flocks in winter are a few North American migrants. Further north in South and Central America there are many more. We residents of North America know them as nesting species in the summer and are intrigued to find them leading a very different life in a distant land. If they are to return year after year to nest, their habitats in the southern countries must be protected. Likewise, their return to those countries will be jeopardized if North American forests are chopped up into still smaller parcels than the ones now remaining. For one thing, thinning out the forests permits further inroads by parasitic cowbirds.)

In each mixed feeding flock, there are one or two sentinel species, which move about in such a way that they are usually near the center of the flock or just below. The sentinels warn the others by a special call of approaching birds that might turn out to be raptors. Charlie noticed that the sentinels for the understory flocks sometimes gave the warning signals even when no danger was apparent. Looking more closely, he found that the fake alarm often permitted the sentinel to grab a succulent morsel that another member of the flock might otherwise have eaten. Careful observation revealed that the sentinels were practicing deception about 15 percent of the time and often profiting by it. Wondering if the phenomenon might be more general, Charlie examined the behavior of the middle canopy flocks and found the sentinels there doing the same thing. For the two species of sentinels, the percentage of false signals was about the same. Presumably, if the percentage were much higher, the signals would not be accepted by the rest of the flock (recall the story *The Boy Who Cried "Wolf"*), and if it were much lower, the opportunity for the sentinel to obtain extra food by lying would be partially or wholly wasted. I am intrigued by the challenge of deriving by some kind of mathematical reasoning the figure of about 15 percent; in a plausible model, might it come out one divided by two pi?

When I asked that question of Charles Bennett, he was reminded of something his father had told him about Royal Canadian Air Force units based in England during the Second World War. They found it useful, when sending out a fighter and a bomber together, to attempt occasionally to deceive the Luftwaffe by positioning the fighter below the bomber rather than above. After a good deal of trial and error, they ended up following that practice at random *one time in seven.*

Small Steps and Large Changes

In our discussion of gateway events, we listed some examples of developments in biological evolution that look like enormous jumps, but we also pointed out that those are rare occurrences at one end of a whole spectrum of changes of various magnitudes, the small changes near the other end of the spectrum being much more common. Whatever the magnitude of the event, biological evolution typically proceeds by working with what is available. Existing organs are adapted to new uses. Human arms, for example, are just slightly modified forelegs. Structures are not suddenly discarded in a revolutionary redesign of the whole organism. The mechanisms of mutation and natural selection do not favor such discontinuities, which would typically be lethal. Yet revolutions do occur.

We discussed how, in the phenomenon of "punctuated equilibrium," the comparatively sudden changes can have several different origins. One is a change in the physicochemical environment that alters selection pressures significantly. Another is the result of "drift," in which neutral mutations, ones that do not disturb the viability of the phenotype (and sometimes do not much alter it) gradually lead to a kind of instability of the genotype. In this situation, one mutation or just a few can make a significant difference to the organism and prepare the way for a cascade of changes in a variety of other species as well. Sometimes small changes set off gateway events, often biochemical in character, that open up whole new realms of life forms. In some cases, these revolutionary changes stem from aggregation of organisms into composite structures. But in every case the basic unit of change is a mutation (or recombination, with or without crossing-over) that operates on what is already present. Nothing is invented out of whole cloth.

How general is that principle for complex adaptive systems? In human thinking, for instance, is it necessary to proceed by small steps? Does the process of invention entail just making strings of minor changes in what already exists? Why shouldn't a human being be able to invent a device that is totally new, totally different from anything known? In science, why not conceive a completely new theory, which bears no resemblance to previous ideas?

Research (as well as everyday experience) seems to indicate that in fact human thought usually does proceed by association and in steps, at each of which specific alterations are made in what has already been thought. Yet remarkable new structures do sometimes emerge, in invention, in science, in art, and in many other fields of human endeavor. Such breakthroughs remind us of gateway events in biological evolution. How do they come about? Does human creative thinking follow different patterns in these different areas of activity? Or are there some general principles involved?

CHAPTER

17

FROM LEARNING TO CREATIVE THINKING

Let us begin with some observations on creative achievement in theoretical science and then explore its relationship to certain kinds of creative achievement in other fields.

A successful new theoretical idea typically alters and extends the existing body of theory to allow for observational facts that could not previously be understood or incorporated. It also makes possible new predictions that can some day be tested.

Almost always, the novel idea includes a negative insight, the recognition that some previously accepted principle is wrong and must be discarded. (Often an earlier correct idea was accompanied, for historical reasons, by unnecessary intellectual baggage that it is now essential to jettison.) In any event, it is only by breaking away from the excessively restrictive received idea that progress can be made.

Sometimes a correct idea, when first proposed and accepted, is given too narrow an interpretation. In a sense, its possible implications are not taken seriously enough. Then either the original proponent of that idea or some other theorist has to return to it, taking it more seriously than when it was originally put forward, so that its full significance can be appreciated.

Both the rejection of a wrong received idea and the return to a correct idea not applied broadly enough are illustrated by Einstein's first paper on special relativity, which he published in 1905, at the age of 26. He had to break away from the accepted but erroneous idea of absolute space and time. Then he could take seriously as a general principle the set of symmetries of Maxwell's equations for electromagnetism—the symmetries that correspond to special relativity. Hitherto they had been viewed narrowly as applying to electromagnetism but not, for example, to the dynamics of particles.

An Example from Personal Experience

It has been my pleasure and good fortune to come up with a few useful ideas in elementary particle theory, not of course in a class with Einstein's, but interesting enough to give me some personal experience of the act of creation as it applies to theoretical science.

One example, from very early in my career, will suffice as an illustration. In 1952, when I joined the faculty of the University of Chicago, I tried to explain the behavior of the new "strange particles," so called because they were copiously produced as though strongly interacting and yet they decayed slowly as though weakly interacting. (Here "slowly" means a half-life of something like a ten billionth of a second; a normal rate of decay of a strongly interacting particle state would correspond to a half-life more like a trillionth of a trillionth of a second, roughly the time it takes light to cross such a particle.)

I surmised correctly that the strong interaction, responsible for the copious production of the strange particles, was prevented by some law from inducing the decay, which was then forced to proceed slowly by means of the weak interaction. But what was the law? Physicists had long speculated about conservation by the strong interaction of a quantity called isotopic spin *I*, which can have values 0, 1/2, 1, 3/2, 2, 5/2, and so on. Experimental evidence in favor of the idea was being gathered at that time by a group of physicists down the hall, led by Enrico Fermi, and I decided to see if the conservation of isotopic spin could be the law in question.

The conventional wisdom was that nuclear (strongly interacting) particle states that are fermions like the neutron and proton would have to have values of I equal to 1/2 or 3/2 or 5/2, and so on, following the example of the neutron and proton, which have $I = 1/2$. (The idea was reinforced by the fact that fermions must have spin angular momentum equal to 1/2 or 3/2 or 5/2, and so on.) Likewise it was believed that the bosonic strongly interacting particles, the mesons, would have to have $I = 0$ or 1 or 2, and so forth, like the known meson, the pion, which has $I = 1$. (Again the parallel with spin angular momentum, which must be a whole number for a boson, strengthened belief in the received idea.)

One set of strange particles (now called sigma and lambda particles) consists of strongly interacting fermions decaying slowly into pion ($I = 1$) plus neutron or proton ($I = 1/2$). I thought of assigning these strange particles isotopic spin $I = 5/2$, which would keep the strong interaction from inducing the decay. But that notion failed to work because electromagnetic effects such as the emission of a photon could change I by one unit at a time and thus evade the law that would otherwise forbid rapid decay. I was invited to talk at the Institute for Advanced Study in Princeton on my idea and why it didn't succeed. In discussing the sigma and lambda particles , I was going to say "Suppose they have $I = 5/2$, so that the strong interaction cannot induce their decay" and then show how electromagnetism would wreck the argument by changing $I = 5/2$ into $I = 3/2$, a value that would permit the decay in question to proceed rapidly by means of the strong interaction.

By a slip of the tongue I said "$I = 1$" instead of "$I = 5/2$." Immediately I stopped dead, realizing that $I = 1$ would do the job. Electromagnetism could not change $I = 1$ into $I = 3/2$ or 1/2, and so the behavior of the strange particles could now be explained after all by means of conservation of I.

But what about the alleged rule that fermionic strongly interacting particle states had to have values of I like 1/2 or 3/2 or 5/2? I realized instantly that the the rule was merely a superstition; there was no real need for it. It was unnecessary intellectual baggage that had come along with the useful concept of isotopic spin I, and the time had come to get rid of it. Then isotopic spin could have wider applications than before.

The explanation of strange particle decay that arose through that slip of the tongue proved to be correct. Today, we have a deeper understanding of the explanation and a correspondingly simpler way of stating it: the strange particle states differ from more familiar ones such as neutron or proton or pions by having at least one s or "strange" quark in place of a *u* or *d* quark. Only the weak interaction can convert one flavor of quark into another, and that process happens slowly.

Shared Experiences of Conceiving Creative Ideas

Around 1970 I was one of a small group of physicists, biologists, painters, and poets assembled in Aspen, Colorado to discuss the experience of getting creative ideas. We each described an incident in our own work. My example was the one involving the slip of the tongue during the lecture in Princeton.

The accounts all agreed to a remarkable extent. We had each found a contradiction between the established way of doing things and something we needed to accomplish: in art, the expression of a feeling, a thought, an insight; in theoretical science, the explanation of some experimental facts in the face of an accepted "paradigm" that did not permit such an explanation.

First, we had worked, for days or weeks or months, filling our minds with the difficulties of the problem in question and trying to overcome them. Second, there had come a time when further conscious thought was useless, even though we continued to carry the problem around with us. Third, suddenly, while we were cycling or shaving or cooking (or by a slip of the tongue, as in the example I described) the crucial idea had come. We had shaken loose from the rut we were in.

We were all impressed with the congruence of our stories. Later on I learned that this insight about the act of creation was in fact rather old. Hermann von Helmholtz, the great physiologist and physicist of the late nineteenth century, described the three stages of conceiving an idea as saturation, incubation, and illumination, in perfect agreement with what the members of our group in Aspen discussed a century later.

Now what goes on during the second stage, that of incubation? For the psychoanalytically oriented, among others, an interpretation that comes immediately to mind is that mental activity continues during the incubation period, but in the "preconscious mind," just outside of awareness. My own experience with the emergence of the right answer in a slip of the tongue could hardly fit better with that interpretation. But some academic psychologists, skeptical of such an approach, offer an alternative suggestion, that nothing really happens during incubation except perhaps a weakening of one's belief in the false principle that is obstructing the search for a solution. The real creative thinking takes place, in their view, just before the moment of illumination. In any case, an appreciable time interval typically elapses between saturation and illumination, and we can think of that interval as a period of incubation whether we picture intense thought out of awareness or just allowing some prejudice gradually to lose its capacity for hindering a solution.

In 1908, Henri Poincaré added a fourth stage, important although rather obvious—verification. He described his own experience in developing a theory of a certain kind of mathematical function. He worked on the problem steadily for two weeks without success. One night, sleepless, it seemed to him that "ideas rose in crowds; I felt them collide until pairs interlocked, so to speak, making a stable combination." Still he did not have the solution. But, a day or so later, he was boarding a bus that was to take him and some colleagues on a geological field trip. "The idea came to me, without anything in my thoughts seeming to have paved the way for it, that the transformations I had used to define these functions were identical with those of non-Euclidean geometry. I did not verify the idea, and on taking my seat, went on with a conversation already begun, but I felt a perfect certainty. On my return to Caen, for conscience's sake, I verified the result."

The psychologist Graham Wallas formally described the four-stage process in 1926, and it has been standard ever since in the relevant branch of psychology, though I think none of us at the Aspen meeting had ever heard of it. I first came across it in a popular book by Morton Hunt entitled *The Universe Within*, from which the above translated quotations are drawn.

Can Incubation Be Hastened or Circumvented?

Now, is it necessary to go through that process? Can the stage of incubation be hastened or circumvented so that we do not have to wait so long for the requisite new idea to come? Can we find a quicker way to escape from an intellectual rut in which we are trapped?

A number of people who offer special programs to teach thinking skills believe that one of the skills they can enhance is creative thinking.

Some of their suggestions for helping to get the thinking process out of a rut fit in quite well with a discussion of that process in terms of complex adaptive systems. Learning and thinking in general exemplify complex adaptive systems at work, and perhaps the highest expression on Earth of that kind of skill is human creative thinking.

A Crude Analysis in Terms of a Fitness Landscape

As in other analyses of complex adaptive systems, it is instructive to introduce the notions of fitness and fitness landscape although, even more than in the case of biological evolution, those concepts are oversimplified idealizations. It is unlikely that a set of selection pressures occurring in the thinking process can be expressed in terms of a well-defined fitness.

That is especially true of an artist's search for creative ideas. In science, the concept is probably more nearly applicable. The fitness of a theoretical idea in science would be a measure of the extent to which it improves existing theory, say by explaining new observations while maintaining or increasing the coherence and explanatory power of that existing theory. In any case, let us imagine that we have a fitness landscape for creative ideas. We shall continue to associate decreasing height with increasing fitness (compare the diagram on page 250).

As we saw in the case of biological evolution, it is too simplistic to suppose that a complex adaptive system merely slides downward on the landscape. When entering a depression, the system would move steadily downhill until it reaches the bottom, the local maximum of fitness. The region from which downhill motion leads to that spot is called the basin of attraction. If the system did nothing but slide downward, it would be overwhelmingly likely to get stuck at the bottom of a shallow basin. On a larger scale there are many basins, and a number of them may be deeper (and therefore more fit, more "desirable") than the one the system has found, as shown on page 250. How does the system get to explore those other basins?

One method of getting out of a basin of attraction, as discussed earlier for biological evolution, involves noise, that is to say, chance motion superimposed on the tendency to descend. Noise gives the system a chance to escape from a shallow depression and seek out a deeper one nearby, and to perform that operation over and over, until the bottom of a really deep basin is reached. The noise must be such that the amplitudes of the chance excursions are not too great, however. Otherwise there would be too much interference with the process of descent, and the system would not remain in a deep basin even after finding one.

Another possibility is to have pauses in the steady downhill crawling process that allow for freer exploration of the vicinity. Those might permit the discovery of deeper depressions nearby. To some extent such pauses correspond to the incubation process in creative thinking, in which the methodical search for the needed idea is suspended and exploration may continue outside of awareness.

Some Prescriptions for How to Escape into a Deeper Basin

Some of the suggestions for speeding up the process of conceiving a creative idea fit in well with the picture of using a controlled level of noise to avoid getting stuck in too shallow a basin of attraction. One can try to escape from the original basin by means of a random perturbation—for example Edward DeBono recommends trying to apply to a problem, whatever it is, the last noun on the front page of today's newspaper.

Another method is akin to brainstorming, which has been used throughout the postwar era. Here several people try to find solutions to a problem by meeting for a group discussion in which one is encouraged to build on someone else's suggestion but not to attack it, no matter how bizarre it is. A crazy or self-contradictory proposal can represent an unstable state of thinking leading to a solution. DeBono likes to cite as an example a discussion of river pollution control, in which someone might say, "What we really need is to make sure that factories are downstream from themselves." That is a manifestly impossible suggestion, but someone else might then come up with a more serious proposal, saying "You can do something like that if you require the intake of water at each factory to be downstream from the effluent." The crazy idea can be regarded as a rise on a fitness landscape that can lead to a much deeper basin than the one from which the discussion started.

Transfer of Thinking Skills?

Edward and many others have prepared teaching materials for special courses in thinking skills for schools, as well as for companies and even for groups of neighbors. Some of those skills relate to getting creative ideas. A number of such courses have been tried out in various parts of the world. For example, a recent president of Venezuela created a Ministry of Intelligence to encourage the teaching of thinking skills in the schools of that country. Under the auspices of the new ministry, a great many students have been exposed to various courses in thinking.

Frequently the materials for such courses emphasize thinking skills in particular contexts. For example, many of Edward's exercises have to do with what I would call policy analysis or policy studies. They refer to choices among courses of action, at the level of the individual, family, organization, village or city, state or province, nation, or transnational body. (Such an exercise may begin, for instance, with the hypothesis that a particular new law has been passed; possible consequences of the law are then discussed.) The materials typically relate to finding and analyzing arguments for and against various known options and also to the discovery of new ones.

One question that naturally arises is to what extent thinking skills learned in one connection are transferable to others. Does exercising one's mind by thinking up new policy options (or ways to weigh the relative merits of old ones) help one to discover good new ideas in a field of science or to create great works of art? Does exercising the mind in such a fashion help one to learn in school about science or mathematics or history or languages? Some day it may be possible to give answers to such questions. Meanwhile, only very preliminary information is becoming available.

Testing Whether Various Proposed Methods Actually Work

When someone takes a course in thinking skills, it is an especially difficult challenge to determine whether any improvement has in fact taken place in the student's capacity for creative thinking. Ideally, a test would be more or less standardized, so that interested parties such as parents, school and government officials, and legislators would be impressed by the results. But how can a standardized test begin to measure creative thinking? One partial answer is provided by design problems. For example, I am told that in Venezuela, at one point, students of thinking were asked to design a table for a small apartment. It is conceivable that the answers to such problems, if graded carefully and imaginatively, might give some indication of whether the students were absorbing creative thinking skills.

David Perkins, of the Harvard Graduate School of Education, who was involved in assigning the table problem, is especially interested in infusing the whole school curriculum with the teaching of thinking skills rather than using special courses. He emphasizes that the need for creative ideas does not arise only in the stratospheric realms of science and art, but also in everyday life. He cites the example of a friend who saves the day, at a picnic where no one has thought to bring a knife, by slicing the cheese neatly with a credit card.

David points out that research has identified a number of characteristics of people who can repeatedly succeed, in the domain of ideas, in escaping from a basin of attraction into another, deeper one. Those characteristics include a dedication to the task, an awareness of being trapped in an unsuitable basin, a degree of comfort with teetering on the edge between basins, and a capacity for formulating as well as solving problems. It seems unlikely that in order to possess those traits one has to be born with them. It may well be possible to inculcate them, but it is far from clear that today's schools do so to any important degree. For example, as David notes, schools are just about the only places where one typically finds problems already formulated.

Problem Formulation and the True Boundaries of a Problem

Problem formulation involves finding the real boundaries of the problem. To illustrate what I mean by that, I shall take some cases that my friend, former neighbor, and Yale classmate Paul MacCready likes to use in his public lectures as examples of novel solutions to problems. (Paul is the inventor of the bicycle-powered aircraft, the solar-powered aircraft, the artificial flapping pterodactyl, and other devices on what he modestly calls the "backwards frontier of aerodynamics.") Although I use the same examples as he does, the lesson that I draw from them is somewhat different.

Consider the famous problem illustrated on the facing page: "Connect all the nine dots by drawing the smallest possible number of straight lines without taking the pencil off the paper." Many people assume that they have to keep the lines within the square formed by the outer dots, although that constraint is not part of the problem as stated. They will then require five lines to solve it. If they allow themselves to extend the lines beyond the square, then they can get away with four, as illustrated. If this were a problem in the real world, a crucial step in formulating it would be to find out whether there is any reason to confine the lines inside the square. That is part of what I call determining the boundaries of the problem.

If the problem allows the lines to be extended beyond the square, perhaps it allows some other kinds of latitude as well. What about crumpling up the paper so that the dots are all in a row and driving a pencil straight through them all in a single stroke? Several such ideas are taken up by James L. Adams in his book *Conceptual Blockbusters.* The best one is contained in a letter he received from a little girl, reproduced on page 272. The main point is in her last sentence: "It doesn't say you mustn't use a thick line." Is a thick line forbidden or not? What are the rules in the real world?

As always, determining the boundaries of the problem is a principal issue in problem formulation. That point is made even more strongly in "The Barometer Story,"* written by a physics professor, Dr. Alexander Calandra of Washington University in St. Louis.

> Some time ago, I received a call from a colleague who asked if I would be the referee on the grading of an examination question. It seemed that he was about to give a student a zero for his answer to a physics question, while the student claimed he should receive a perfect score and would do so if the system were not

---

* From Teacher's Edition of *Current Science,* Vol. 49, No. 14, January 6–10, 1964. Courtesy of Robert L. Semans.

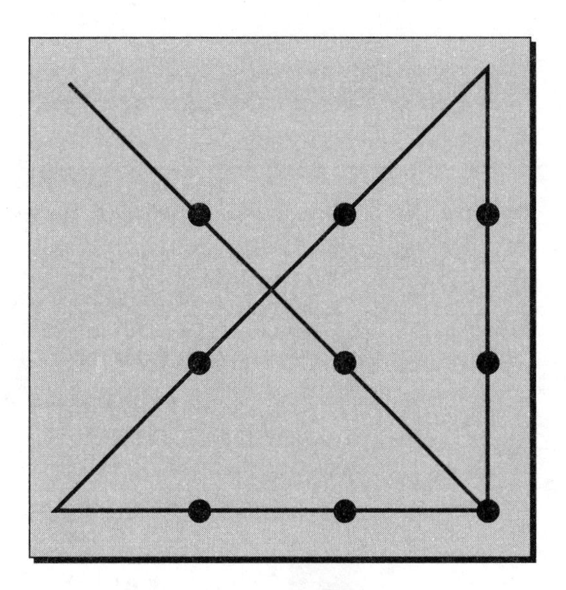

Connecting nine dots in a square with four straight lines without lifting the pencil from the paper.

set up against the student. The instructor and the student agreed to submit this to an impartial arbiter, and I was selected. . . .

I went to my colleague's office and read the examination question, which was, "Show how it is possible to determine the height of a tall building with the aid of a barometer."

The student's answer was, "Take the barometer to the top of the building, attach a long rope to it, lower the barometer to the street, and then bring it up, measuring the length of the rope. The length of the rope is the height of the building."

Now this is a very interesting answer, but should the student get credit for it? I pointed out that the student really had a strong case for full credit, since he had answered the question completely and correctly. On the other hand, if full credit were given, it could well contribute to a high grade for the student in his physics course. A high grade is supposed to certify that the student knows some physics, but the answer to the question did not confirm this. With this in mind, I suggested that the student have another try at answering the question. I was not surprised that my colleague agreed to this, but I was surprised that the student did.

May 30, 1974
5 FDR Navasa
Roosevelt Rds. Na
Ceiba, P.R. 00635

Dear Prof. James L. Adams,
My dad and I were doing Puzzles from "Conceptual BLockbusting". We were mostly working on the dot ones, like My dad said a man found a way to do it with one line. I tried and did it. Not with folding, but I used a fat line. I doesn't say you can't use a fat line.
Like this

Sincerely,
Becky Buechel
age:10

P.S.
acctually you need a very fat writing apparatice

Letter to Professor Adams from a ten-year-old girl. (From *Conceptual Blockbusting—A Guide to Better Ideas,* Third Edition by James L. Adams. Reading, Mass.: Addison-Wesley, p. 31. Copyright 1974, 1976, 1979, 1986 by James L. Adams. Used by permission.)

Acting in the terms of the agreement, I gave the student six minutes to answer the question, with the warning that the answer should show some knowledge of physics. At the end of five minutes, he had not written anything. I asked if he wished to give up, since I had another class to take care of, but he said no, he was not giving up, he had many answers to this problem, he was just thinking of the best one. I excused myself for interrupting him, and asked him to please go on. In the next minute, he dashed off his answer, which was: "Take the barometer to the top of the building, and lean over the edge of the roof. Drop the barometer, timing its fall with a stopwatch. Then, using the formula $s = 1/2at^2$ [distance fallen equals one-half the acceleration of gravity times the square of the time elapsed], calculate the height of the building."

At this point, I asked my colleague if he would give up. He conceded and I gave the student almost full credit. In leaving my colleague's office, I recalled that the student had said that he had other answers to the problem, so I asked him what they were. "Oh, yes," said the student. "There are many ways of getting the height of a tall building with the aid of a barometer. For example, you could take the barometer out on a sunny day and measure the height of the barometer, the length of its shadow, and the length of the shadow of the building, and by the use of simple proportion, determine the height of the building."

"Fine," I said. "And the others?"

"Yes," said the student. "There is a very basic measurement that you will like. In this method, you take the barometer and begin to walk up the stairs. As you climb the stairs, you mark off the length and this will give you the height of the building in barometer units. A very direct method."

"Of course, if you want a more sophisticated method, you can tie the barometer to the end of a string, swing it as a pendulum, and determine the value of g [the acceleration of gravity] at the street level and at the top of the building. From the difference between the two values of g, the height of the building can, in principle, be calculated."

Finally, he concluded, "If you don't limit me to physics solutions to this problem, there are many other answers, such as taking the barometer to the basement and knocking on the superintendent's door. When the superintendent answers, you speak to him as follows:

"Dear Mr. Superintendent, here I have a very fine barometer. If you will tell me the height of this building, I will give you this barometer. . . ."

CHAPTER

18

SUPERSTITION AND SKEPTICISM

In contrast to the distinctive selection pressures that characterize the scientific enterprise (at least science at its best), very different kinds of selection have also affected the evolution of theoretical ideas on the same subjects that now form the province of science. An example is provided by the appeal to authority, independent of comparison with nature. In medieval and early modern Europe, appeals to authority (for instance Aristotle, to say nothing of the Roman Catholic Church) were standard in fields where the scientific method was later to be widely applied. When the Royal Society of London was founded in 1661, the motto chosen was *Nullius in verba*. I interpret that phrase as meaning "Don't believe in anyone's words" and as being a rejection of the appeal to authority in favor of the appeal to nature that distinguished the relatively new discipline of "experimental philosophy," which is now called natural science.

We referred earlier to systems of belief, such as sympathetic magic, that respond predominantly to selection pressures quite different from the comparison of predictions with observation. Over the last few centuries the scientific enterprise has prospered and carved out a realm where authority and magical thinking have largely given way to a partnership of observation and theory. But outside that realm older ways of thinking are widespread and superstitious beliefs flourish. Is the prevalence of superstition alongside science a phenomenon peculiar to human beings or should intelligent complex adaptive systems elsewhere in the universe be expected to have similar proclivities?

Mistakes in the Identification of Regularities

Complex adaptive systems identify regularities in the data streams they receive and compress those regularities into schemata. Since it is easy to make two types of error—mistaking randomness for regularity and vice versa—it is reasonable to suppose that complex adaptive systems would tend to evolve toward a roughly balanced situation in which correct identification of some regularities would be accompanied by both kinds of mistakes.

Contemplating patterns of human thought, we can, in a crude fashion, identify superstition with one kind of error and denial with the other. Superstitions typically involve seeing order where in fact there is none, and denial amounts to rejecting evidence of regularities, sometimes even ones that are staring us in the face. Through introspection and also by observation of other human beings, each of us can detect an association of both sorts of error with fear.

In the one case, people are scared by the unpredictability and especially the uncontrollability of much that we see around us. Some of that unpredictability stems ultimately from the fundamental indeterminacies of quantum mechanics and the further limitations of prediction imposed by chaos. A huge amount of additional coarse graining, with consequent unpredictability, comes from the restricted range and capacity of our senses and instruments: we can pick up only a minuscule amount of the information about the universe that is available in principle. Finally, we are handicapped by our inadequate understanding and by our limited ability to calculate.

The resulting scarcity of rhyme and reason frightens us and so we impose on the world around us, even on random facts and chance phenomena, artificial order based on false principles of causation. In that way, we comfort ourselves with an illusion of predictability and even of mastery. We fantasize that we can manipulate the world around us by appealing to the imaginary forces we have invented.

In the case of denial, we are able to detect genuine patterns but they scare us so much that we blind ourselves to their existence. Evidently the most threatening regularity in our lives is the certainty of death. Numerous beliefs, including some of the most tenaciously held, serve to alleviate anxiety over death. When specific beliefs of that kind are widely shared in a culture, their soothing effect on the individual is multiplied.

But such beliefs typically include invented regularities, so that denial is accompanied by superstition. Moreover, taking another look at superstitions such as those of sympathetic magic, we see that belief in them can be maintained only by denying their manifest defects, especially their frequent failure to work. The denial of real regularities and the imposition of false ones are thus seen to be two sides of the same coin. Not only are human beings prone to both, but the two tend to accompany and support each other.

If this sort of analysis is justified, then we can conclude that intelligent complex adaptive systems on planets scattered throughout the universe should tend to err in both directions in identifying regularities in their input data. In more anthropomorphic terms, we can expect intelligent complex adaptive systems everywhere to be liable to a mixture of superstition and denial. Whether it makes sense, apart from human experience, to describe that mixture in terms of the alleviation of fear is another matter.

A slightly different way of looking at superstition in a complex adaptive system suggests that perhaps superstition might be somewhat more prevalent than denial. The system can be regarded as having evolved in great part to discover patterns, so that a pattern becomes in a sense its own reward, even if it confers no particular advantage in the real world. A pattern of that kind can be regarded as a "selfish scheme," somewhat analogous to the selfish gene or even the truly selfish gene.

Examples from human experience are not difficult to find. A few years ago I was invited to meet with a group of distinguished academics from out of town who had come to discuss a fascinating new discovery. It turned out that they were excited about some features in recent NASA photographs of the surface of Mars that had a vague resemblance to a human face. I cannot imagine what advantage this foray into improbability could have conferred on those otherwise bright people, other than the sheer joy of finding a mysterious regularity.

The Mythical in Art and Society

For human beings, especially at the social level, numerous selection pressures in the real world, besides the alleviation of fear, favor distortions in the process of identifying regularities. Superstitious beliefs may serve to reinforce the power of shamans or priests. An organized belief system, complete with myths, may motivate compliance with codes of conduct and cement the bonds uniting the members of a society.

Over the ages, belief systems have served to organize mankind into groups that are not only internally cohesive, but sometimes intensely competitive with one another, often to the point of conflict or persecution, sometimes accompanied by massive violence. Examples are unfortunately not difficult to find in today's world.

But competing beliefs are just one basis on which people divide themselves into groups that fail to get along with one another. Any label will do. (A label, to quote from the comic strip *B.C.*, is "something you put on [people] so you can hate them without having to get to know them first.") Many large-scale atrocities (and individual cruelties) have been perpetrated along ethnic or other lines, often with no particular connection to beliefs.

Alongside the devastating effects of systems of belief, their positive achievements stand out sharply as well, especially the glorious music, architecture, literature, sculpture, painting, and dance that have been inspired by particular mythologies. Just the example of archaic Greek black figure vases would suffice to bear witness to the creative energies released by myth.

In the face of the overwhelming greatness of so much of the art related to mythology, we need to re-examine the significance of false regularities. In addition to exerting a powerful influence on human intellect and emotions and leading to the creation of magnificent art, mythical beliefs clearly have a further significance that transcends their literal falsity and their connection with superstition. They encapsulate experience gained through centuries and millennia of interaction with nature and with human culture. They contain not only lessons but also, at least by implication, prescriptions for behavior. They are vital parts of the cultural schemata of societies functioning as complex adaptive systems.

The Search for Patterns in the Arts

Belief in myths is only one of many sources of inspiration for the arts (just as it is only one of many sources of hatreds and atrocities). It is not only in connection with the mythical that the arts are nurtured by patterns of association and regularity unrecognized by science. All the arts thrive on the identification and exploitation of such patterns. Most similes and metaphors are patterns that science might ignore, and where would literature, and especially poetry, be without metaphor? In the visual arts, a great work often leads the viewer to new ways of seeing. The recognition and creation of patterns are essential activities in every kind of art. The resulting schemata are subject to selection pressures that are often (though not always) far removed from those operating in science, with wonderful consequences.

We can look upon myth and magic, then, in at least three different and complementary ways:

1. as attractive but unscientific theories, comforting but false regularities imposed on nature;
2. as cultural schemata that help to give identity to societies, for better or for worse; and
3. as part of the grand search for pattern, for creative association, that includes artistic work and that enriches human life.

A Moral Equivalent of Belief?

The question naturally arises whether there is any way to capture the splendid consequences of mythical beliefs without the associated self-delusion and without the intolerance that often accompanies it. Around a century ago, the concept of "the moral equivalent of war" was widely discussed. As I understand it, the point is that war inspires loyalty to comrades, self-sacrifice, courage, and even heroism, and it provides an outlet for the love of adventure, but war is also cruel and destructive to an extraordinary degree. Hence the human race is challenged to find activities with the positive characteristics of war and without the negative ones. Some organizations try to accomplish that goal by introducing to the challenges of outdoor adventure young people who might not otherwise have the opportunity to live the outdoor life. It is hoped that such activity can provide a substitute not only for war but also for delinquency and crime.

One can ask, in connection with superstition instead of war and crime, whether a moral equivalent of belief can be found. Can humans derive the spiritual satisfaction, comfort, the social cohesion, and the brilliant artistic creations that accompany mythical beliefs from something less than acceptance of the myths as literally true?

Part of the answer might lie in the power of ritual. The Greek word *muthos*, from which myth is derived, is said to have referred in ancient times to the spoken words that accompanied a ceremony. The acts were central, in some sense, and what was said about them was secondary. Often, in fact, the original significance of the ritual had been at least partially forgotten, and the surviving myth represented an attempt at explanation by trying to interpret icons from the past and by putting together fragments of old traditions referring to a stage of culture that was long gone. The myths were subject to change, then, while continuity of the ritual was what helped hold the society together. As long as rituals persist, could literal belief in mythology wither away without causing too much disruption?

Another part of the answer might relate to how fiction and drama are perceived. The characters in great literature seem to have a life of their own, and their experiences are regularly cited, much like those of mythical characters, as sources of wisdom and inspiration. Yet no one claims that works of fiction are literally true. Is there some chance, then, that many of the social and cultural benefits of belief can be preserved while the aspect of self-delusion gradually fades away?

Still another partial answer might be provided by mystical experiences. Is it possible that some of the spiritual benefits often derived from superstitious beliefs can be gained instead, at least by some people, through learning techniques that facilitate such experiences?

Unfortunately, in many places in the contemporary world, literal belief in mythology, far from dying out, is on the increase, as fundamentalist movements gather strength and threaten modern societies with the imposition of old-fashioned limitations on behavior and on freedom of expression. (Moreover, even where the strength of mythical beliefs is declining, no great improvement in the relation between different groups of people need occur as a result, since slight differences of almost any kind can be sufficient to sustain hostility between them.)

For a thoughtful discussion of the entire subject of superstitious belief, I recommend *Wings of Illusion* by John F. Schumaker, even though he tends to despair of our being able to dispense, as a species, with our assemblage of comforting and often inspiring illusions.

The Skeptics' Movement

During the last couple of decades, the prevalence of old-fashioned superstitions has been accompanied, at least in Western countries, by a wave of popularity for so-called New Age beliefs, many of which are just contemporary and pseudoscientific superstitions, or sometimes even old superstitions with new names, like "channeling" in place of "spiritualism." Unfortunately these are often portrayed, in the news media and in popular books, as if they were factual or very probable, and a movement has originated to counter such claims, the skeptics' movement. Local groups of skeptics have been formed in communities around the world. (Three places where I have spent a good deal of time are ones that can use a healthy dose of skepticism: Aspen, Santa Fe, and Southern California.)

The local skeptics' organizations are loosely connected to a committee that is based in the United States but includes members from other parts of the world; it is called CSICOP, the Committee for Scientific Investigation of Claims of the Paranormal. CSICOP, which publishes a journal called the *Skeptical Inquirer*, is not a membership organization open to the public, but a body to which one must be elected. Despite some reservations about the organization and its journal, I accepted election some years ago because I like much of its work.

Claims of the so-called paranormal surround us on all sides. Some of the most ridiculous are found in headlines from tabloid publications sold at supermarket counters: "Cat eats parrot . . . now it talks . . . Kitty wants a cracker." "Hundreds back from dead describe heaven and hell." "Incredible fish-man can breathe under water." "Siamese twins meet their two-headed brother." "Space alien made me pregnant." CSICOP doesn't bother to take on such manifest nonsense. But its members do get upset when mainstream newspapers, magazines, or radio and television networks treat as routine and unchallenged, as established or very probable, things that are not at all established: alleged phenomena such as hypnotic regression to previous lives, valuable assistance provided to the police by psychics, or psychokinesis (in which the mind is supposed to cause external objects to move). These claims challenge the accepted laws of science on the basis of evidence that careful investigation reveals to be very poor or entirely lacking. Keeping after the media not to present such things as real or probable is a useful activity of CSICOP.

Claims of the Paranormal? What Paranormal?

Some questions arise, nevertheless, if we look carefully at the implications of the name of the organization. What is meant by claims of the paranormal? Of course, what most of us working in science (and in fact most reasonable people) want to know first about any alleged phenomenon is whether it really happens. We are curious about the extent to which the claims are true. But if a phenomenon is genuine, how can it be paranormal? Scientists, and many nonscientists as well, are convinced that nature obeys regular laws. In a sense, therefore, there can be no such thing as the paranormal. Whatever actually happens in nature can be described within the framework of science. Of course, we may not always be in the mood for a scientific account of certain phenomena, preferring, for example, a poetic description. Sometimes the phenomenon may be too complicated for a detailed scientific description to be practical. In principle, though, any genuine phenomenon has to be compatible with science.

If something new is discovered (and reliably confirmed) that does not fit in with existing scientific laws, we do not throw up our hands in despair. Instead, we enlarge or otherwise modify the laws of science to accommodate the new phenomenon. This puts someone in a strange logical position who is engaged in the scientific investigation of claims of the paranormal, because in the end nothing that actually happens can be paranormal. Perhaps this situation is related to a vague sense of disappointment that I sometimes feel upon reading that otherwise excellent journal the *Skeptical Inquirer*. I experience a lack of suspense. Seeing the title of the article usually gives away the content, namely that whatever is in the title isn't true. Nearly everything that is discussed in the journal ends up debunked. Moreover, many of the authors seem to feel that they have to explain away every last case, even though in the real world an investigation of anything complex usually leaves a few matters somewhat cloudy. I am, it is true, delighted to see such things as psychic surgery and levitation through meditation debunked. But I do think a slight redefinition of the mission would help the organization and the journal to be more lively and interesting, as well as more soundly based. I believe the real mission of the organization is to encourage the skeptical and scientific examination of reports of mysterious phenomena, especially ones that seem to challenge the laws of science, but without making use of the label "paranormal," with its implication that debunking is most likely required. Many of these phenomena will turn out to be phoney, or to have very prosaic explanations, but a few may turn out to be basically genuine and interesting as well. The concept of the paranormal does not seem to me to be a helpful one; and the debunking spirit, while it is entirely appropriate for most of the subjects involved, is not always a perfectly satisfactory approach.

Often we are faced with situations where conscious fraud is involved, credulous people are cheated of their money, seriously ill patients are diverted to worthless fake cures (like psychic surgery) from legitimate treatments that might work, and so forth. In such cases, debunking is a service to humanity. Even then, however, we should spare some thought for the emotional needs of the victims that are being met by the quackery and how those might be satisfied without self-delusion.

I would recommend that skeptics devote even more effort than they do now to understanding the reasons why so many people want or need to believe. If people were less receptive, the news media would not find it profitable to emphasize the so-called paranormal. In fact, it is not just misapprehension about how much good evidence there is for a phenomenon that underlies the tendency to believe in it. In my discussions with people who believe six impossible things before breakfast every day, like the White Queen in *Through the Looking Glass,* I have found that their main characteristic is the dissociation of belief from evidence. Many of those people, in fact, freely confess that they believe what it makes them feel good to believe. Evidence doesn't play much of a role. They are alleviating their fear of randomness by identifying regularities that are not there.

Mental Aberration and Suggestibility

Two subjects that must be included in any discussion of strange beliefs are suggestibility and mental aberration. Polls now reveal, for example, that an astonishing percentage of respondents not only believe in the existence of "aliens from flying saucers" but also claim to have been abducted by them and closely examined or even sexually molested. One seems to be dealing here with people who for some reason have difficulty distinguishing reality from fantasy. It is natural to ask whether some of them are afflicted with serious mental illness.

One might also speculate that a number of believers in such weird events may simply have an unusually high degree of susceptibility to trance, so that they pass in and out of states of suggestibility with the greatest ease. Such subjects can have beliefs imposed upon them when placed in a trance by a hypnotist; perhaps a related process can occur more or less spontaneously. High trance susceptibility may be a potential liability, but it can also be advantageous, because it can facilitate hypnosis, self-hypnosis, or deep meditation, permitting a person to achieve useful forms of self-control that are difficult (although not impossible) to achieve in other ways.

In many traditional societies, people endowed with very high susceptibility to trance may find roles as shamans and prophets. So may others who suffer from certain kinds and degrees of mental disturbance. Both categories of people are thought to be more likely than others to undergo mystical experiences. In modern societies, both are said to be found among the most creative artists. (Of course, all these presumed correlations need to be carefully checked.)

Some research is taking place on the mental characteristics of people who believe in outrageously improbable phenomena, especially individuals who claim to have had personal involvement with them. So far, surprisingly little evidence has turned up either for serious mental illness or for high trance susceptibility. Rather, it appears that in many cases strong belief serves to influence the interpretation of ordinary experiences with physical phenomena or with sleep- or drug-induced mental states. The subject is clearly in its infancy, however. It seems to me desirable to intensify greatly the study of such beliefs and belief systems among human beings and of the underlying causes, since in the long run the subject plays such a crucial role in our common future.

Skepticism and Science

Suppose it is agreed that the skeptics' movement, apart from studying the subject of belief and engaging in such activities as exposing fraud and trying to keep the media honest, is engaged in the skeptical and scientific examination of reports of mysterious phenomena that seem to defy the laws of science. Then the degree of skepticism that is applied should be appropriate to the challenge that the alleged phenomenon presents to the accepted laws. Here one has to be very careful. For example, in complicated fields such as meteorology or planetary science (including geology), bizarre natural phenomena may be alleged to occur that challenge certain accepted principles in those fields but don't appear to violate fundamental laws of nature like conservation of energy. The empirical or phenomenological laws in such fields are sometimes quite difficult to relate to the laws of more basic sciences, and new observational discoveries are being made all the time that require revision of the empirical laws. An alleged phenomenon that violates those laws is not so suspect as one that violates conservation of energy.

Only thirty years ago most geologists, including almost all the distinguished geology faculty at Caltech, were still contemptuously rejecting the idea of continental drift. I remember because I often argued with them about it at the time. They disbelieved in continental drift despite mounting evidence in its favor. They had been taught that it was nonsense mainly because the geological community hadn't thought of a plausible mechanism for it. But a phenomenon may perfectly well be genuine even though no plausible explanation has yet turned up. Particularly in that kind of subject, it is unwise to dismiss an alleged phenomenon out of hand just because experts can't think right away of what might make it happen. Planetary scientists a couple of centuries ago committed the notorious mistake of debunking meteorites. "How can rocks fall from the sky," they objected, "when there are no rocks in the sky?"

Today there is a strong tendency among my friends in the skeptics' movement, as well as my colleagues in physics, to dismiss rather quickly claims of elevated incidence of rare malignancies in people who are exposed more than others to comparatively weak electromagnetic fields from 60 cycle alternating current devices and power lines. The skeptics may well be right in thinking that the claims are spurious, but it is not so obvious as some of them say. Although the fields are too weak to produce gross effects such as substantial increases in temperature, they might still be capable of producing much more subtle effects on certain highly specialized body cells that are unusually sensitive to magnetism because they contain appreciable quantities of magnetite. Joseph Kirschvink of Caltech (who has an unusual set of interests for a Caltech professor) is investigating that possibility experimentally and has found some preliminary indications that such a connection of magnetism with those rare malignancies might be more than a fantasy.

The Kugelblitz—Ball Lightning

A number of phenomena alleged to take place in the atmosphere remain in a sort of limbo to this day. One of them is the "kugelblitz"—ball lightning. Certain observers claim to have seen, in stormy weather, a bright sphere, suggesting lightning in the form of a ball. It may pass between widely spaced slats of a fence; or it may enter a room through a window, roll around inside, and then disappear, leaving slight burns. All sorts of anecdotal reports abound, but there is no incontrovertible evidence, nor is there any really satisfactory theory. One physicist, Luis Alvarez, suggested that the kugelblitz was just a phenomenon of the observer's eyeball. That explanation, however, doesn't agree very well with the anecdotal evidence, for instance as compiled by a scientist from interviews with employees of a national laboratory. Some serious theorists have done research on the phenomenon. While the great Russian physicist Pyotr L. Kapitsa was under house arrest for refusing to work on thermonuclear weapons under the direction of Stalin's secret police chief Lavrenti P. Beria, he and one of his sons wrote a theoretical paper on a hypothetical mechanism for ball lightning. Others have tried to reproduce the phenomenon in the laboratory. But I would say that the results are still inconclusive. In brief, nobody knows what to make of it.

Around 1951, the mention of ball lightning disrupted a seminar at the Institute for Advanced Study in Princeton at which Harold W. ("Hal") Lewis, who is now a professor of physics at the University of California at Santa Barbara, presented a piece of theoretical work on which he and Robert Oppenheimer had collaborated. I think it was Robert's last research effort in physics before he became Director of the Institute, and he was very anxious that people listen carefully to Hal's account of the work, contained in the paper of Oppenheimer, Lewis, and Wouthuysen on meson production in proton-proton collisions. In the course of the discussion after the lecture, someone mentioned that Enrico Fermi had proposed a model in which the two protons stick together for a long time, for unknown reasons, and emit mesons in a statistical manner. Many of us joined in with suggestions about what could cause that kind of behavior. The learned Swiss theoretical physicist, Markus Fierz, interjected the remark that it is not always clear why things stick together. "For example," he said, "take the kugelblitz, ball lightning." (Oppenheimer started to turn purple with fury. Here was his last paper in physics being presented and Fierz was diverting the discussion to ball lightning.) Fierz went on to say that a friend of his was employed by the Swiss government and given a special railroad car so that he could travel around the country and follow up on anecdotal reports of the kugelblitz. Finally, Robert couldn't stand it anymore and stalked out muttering, "Fire balls, fire balls!" I don't think our understanding of the phenomenon has improved much since then (even though Hal Lewis himself has written an interesting article about it).

Fish Falls

One of my favorite examples of mysterious phenomena has to do with fish and frogs falling from the sky. Many of the accounts are quite circumstantial and given by credible observers. Here is one, by A. B. Bajkov, describing a fish fall in Marksville, Louisiana, on October 23, 1947:

> I was conducting biological investigations for the Department of Wildlife and Fisheries. In the morning of that day between 7:00 and 8:00 fish ranging from two to nine inches in length fell on the streets and in yards, mystifying the citizens of that southern town. I was in the restaurant with my wife having breakfast when the waitress informed us that fish were falling from the sky. We went immediately to collect some of the fish. The people in town were excited. The Director of the Marksville Bank, J. M. Barham, said he had discovered upon arising from bed that fish had fallen by hundreds in his yard and in the adjacent yard of Mrs. J. W. Joffrion. The cashier of the same bank, J. E. Gremillion, and two merchants, E. A. Blanchard and J. M.

Brouillette, were struck by falling fish as they walked toward their places of business about 7:45 a.m. . . . . .

(quoted by William R. Corliss from *Science*, 109,402, April 22, 1949).

All the meteorologists that I have consulted assure me that their science does not provide any conclusive objection to the possibility of such creatures being raised up, transported considerable distances, and then dropped as a result of meteorological disturbances. Although one may only speculate about specific mechanisms, such as waterspouts, it is perfectly possible that the phenomenon actually occurs. Conceivably, transport by flocks of birds might provide another explanation.

Moreover, if fish or at least their spawn come down alive, that could make a serious difference to zoogeography, the study of the distribution of animal species. Ernst Mayr actually mentioned in one of his papers on zoogeography that there are many puzzles about the distribution of freshwater fishes that might be resolved if those creatures could be transported by unconventional means, such as fish falls from the sky.

The foregoing discussion makes clear that if fish really do tumble from the heavens, the process doesn't do any damage to the accepted laws of science; in fact it probably helps. Likewise, if one of those "cryptozoological" creatures, like the supposed giant ground sloth in the Amazonian forest, should turn out to be real, it might not harm the laws of science either, any more than the coelacanth that was discovered in the waters off southern Africa fifty years ago although it was thought to be long extinct. But what about alleged phenomena that seem to challenge the fundamental laws of science as we know them?

Alleged Phenomena That Challenge the Known Laws of Science

Although such phenomena are not ipso facto nonexistent, a very high standard of skepticism must be applied to them. Nevertheless, if any of them ever turns out to be genuine, the scientific laws will have to be modified to accommodate it.

Consider the alleged phenomenon (in which, by the way, I don't believe) of telepathy between two people who are very close personally and also closely related, say mother and child or identical twins. Almost everyone has heard anecdotes about such pairs of people, according to which, in moments of extreme stress for one of them, the other becomes alarmed, even if they are very far apart. Most likely these reports are occasioned by a combination of coincidence, selective memory (forgetting false alarms, for example), distorted recollection of the circumstances (including exaggeration of simultaneity), and so on. Besides, it is very difficult to investigate such phenomena scientifically, although in principle not impossible. For instance, one can imagine an experiment, cruel and therefore forbidden by ethical considerations, but not otherwise impractical, in which one hired many pairs of identical twins, separated them by long distances, and then subjected one of each pair to severe stress to see if the other twin would react. (There are some gullible people, including a number of my New Age acquaintances in Aspen, who believe that such an experiment was actually performed with animals while the submarine *Nautilus* was under the polar ice. They think that a mother rabbit was monitored in the submarine and showed signs of anguish when some of her little baby rabbits were being tortured in Holland!)

In any case, suppose for a moment that, contrary to my expectations, such a telepathic phenomenon turned out to be genuine, say for human identical twins. Fundamental scientific theory would have to be profoundly altered, but eventually, no doubt, an explanation could be found. For example, theorists might end up postulating some kind of cord, of a nature not now understood, probably involving important modifications of the laws of physics as presently formulated. Such a cord connecting the twins would carry a signal between them when one of them was in serious trouble. That way the effect could be largely independent of distance, as many of the anecdotes suggest. Let me emphasize again that I am quoting this example not because I believe in telepathy but only to illustrate how scientific theory might be modified to accommodate even very bizarre phenomena in the unlikely event that they turned out to be genuine.

A Genuine Ability—Reading Record Grooves

Occasionally CSICOP finds that a seemingly crazy claim is really justified. Such cases are duly reported in the *Skeptical Inquirer* and discussed at meetings, but in my opinion they should be given more attention than they have received. Then it would be much clearer that the point is to attempt to distinguish genuine claims from false ones and not simply to engage in debunking.

Scientists have, on the whole, a rather poor record of success in investigating suspected fakers. All too often, even well-known savants have been taken in, sometimes becoming promoters of charlatans they should have exposed. CSICOP relies primarily on a magician, James Randi, to devise tests for people who claim extraordinary powers. Randi knows how to put things over on an audience, and he is equally good at figuring out how someone is trying to put things over on him. He gets a thrill out of unmasking fakers and demonstrating how they have obtained their effects.

When it came to the attention of *Discover* magazine that a man was claiming to be able to glean information from the grooves on phonograph records, the obvious move was to dispatch Randi to investigate. The man in question, Dr. Arthur Lintgen of Pennsylvania, said he could look at a record of fully orchestrated post-Mozart classical music and identify the composer, often the piece, and sometimes even the performing artist. Randi subjected him to his usual rigorous tests and discovered that he was telling the exact truth. The physician correctly identified two different recordings of Stravinsky's *Rite of Spring,* as well as Ravel's *Bolero,* Holst's *The Planets,* and Beethoven's *Sixth Symphony.* Naturally Randi showed him some other records as controls. One, labeled "gibberish" by Dr. Lintgen, was by Alice Cooper. On seeing another control, he said, "This is not instrumental music at all. I'd guess that it's a vocal solo of some kind." In fact it was a recording of a man speaking, titled *So You Want to Be a Magician.*

This odd claim that turned out to be genuine violated no important principle. The necessary information was present in the grooves; the question was whether someone had really been able to abstract that information by inspection. Randi confirmed that indeed someone had.

CHAPTER

19

ADAPTIVE AND MALADAPTIVE SCHEMATA

One story that skeptical inquiry has successfully debunked is that of the "hundredth monkey." The first part of the account is true. One member of a monkey colony on an island in Japan learned to wash sand off her food, first in a stream leading down to the sea and then in the sea itself. She communicated the skill to other members of the colony. So far, so good. A New Age legend, however, picks up on those facts and claims further that when a hundred monkeys had picked up the trick, then suddenly, by some mysterious means, members of the species everywhere knew about it and began to practice it. For that, there is no credible evidence whatever.

The true part of the story is quite interesting in itself, as an example of cultural transmission of learned behavior in animals other than humans. Another example is provided by the behavior of some great tits (birds related to the chickadees of North America) in certain English towns several decades ago. Those little birds learned to open milk bottles. More and more great tits acquired the behavior, and a few members of other tit species as well. The physical activity required was already in the birds' repertory; all they needed was the knowledge that the milk bottle contained a suitable reward. There are many other known cases of novel animal behavior transmitted in this way.

Cultural DNA

Human cultural transmission can, of course, be considerably more sophisticated. The explanation presumably lies not only in superior intelligence, but also in the character of human languages, every one of which permits arbitrarily complex utterances. Using those languages, human societies exhibit group learning (or group adaptation or cultural evolution) to a much greater degree than troops of other primates, packs of wild dogs, or flocks of birds. Such collective behavior can be analyzed to some extent by reducing it to the level of individuals acting as complex adaptive systems. However, as usual, such reduction sacrifices the valuable insights that can be gained by studying a phenomenon at its own level. In particular, simple-minded reduction to psychology may not sufficiently stress the fact that, besides the general characteristics of individual human beings, additional information is present in the system, including the specific traditions, customs, laws, and myths of the group. To use the picturesque phrase of Hazel Henderson, all of these can be regarded as "cultural DNA." They encapsulate the shared experience of many generations and comprise the schemata for the society, which itself functions as a complex adaptive system. In fact, the English biologist Richard Dawkins coined the term "meme" to signify a unit of culturally transmitted information analogous to a gene in biological evolution.

Adaptation actually takes place on at least three different levels, and that sometimes causes confusion in the use of the term. First of all, some direct adaptation (as in a thermostat or cybernetic device) takes place as a result of the operation of a schema that is dominant at a particular time. When the climate turns warmer and drier, a society may have the custom of moving to new villages high up in the mountains. Alternatively, it may resort to religious ceremonies for bringing rain, under the supervision of a priesthood. When its territory is invaded by an enemy force, the society may react automatically by retreating to a well-fortified town, already stocked with provisions, and sustaining a siege. When the people are frightened by an eclipse, there may be shamans ready with some appropriate hocus-pocus. None of this behavior requires any change in the prevailing schema.

The next level involves changes in the schema, competition among various schemata, and promotion or demotion depending on the action of selection pressures in the real world. If rain dances fail to bring relief from a drought, the relevant priesthood may fall into disgrace and a new religion may take over. Where the traditional response to climate change has been movement to a higher elevation, poor results from that schema may lead to the adoption of other practices, such as new methods of irrigation or the planting of new crops. If the strategy of retreat to a fortress fails to deal adequately with a series of enemy attacks, the next invasion may provoke the sending of an expeditionary force to the enemy's heartland.

The third level of adaptation is the Darwinian survival of the fittest. A society may simply cease to exist as a consequence of the failure of its schemata to cope with events. The people need not all die, and the remaining individuals may join other societies, but the society itself disappears, carrying its schemata into extinction with it. A form of natural selection has taken place at the societal level.

Examples of schemata leading to extinction are not difficult to find. Some communities (such as the Essenes in ancient Palestine and the Shakers in the nineteenth century United States) are said to have practiced sexual abstinence. All the members of the community, not just a few monks and nuns, were supposed to refrain from sexual activity. Given such a schema, the survival of the community would require that conversions outnumber deaths. That does not seem to have happened. The Essenes disappeared and the Shakers are now few in number. In any case, the prohibition of sexual intercourse was a cultural trait that contributed in an obvious way to the extinction or near-extinction of the community.

The collapse of the Classic Maya civilization in the tropical forests of Mesoamerica during the tenth century is a striking example of the extinction of an advanced culture. As indicated near the beginning of this book, the causes of the collapse are a matter of dispute today; archaeologists are uncertain which schemata failed—those related to the class structure of society, agriculture in the jungle, warfare among the cities, or other facets of the civilization. In any case, it is thought that many individuals survived the collapse and that some of the people speaking Mayan languages in the area today are their descendants. But the construction of stone buildings in the forest cities came to an end, as did the erection of stelae to commemorate the passage of key dates in the Maya calendar. The subsequent societies were very much less complex than those of the Classic Period.

The three levels of adaptation take place, generally speaking, on different time scales. An existing dominant schema can be translated into action right away, within days or months. A revolution in the hierarchy of schemata is generally associated with a longer time scale, although the culminating events may come swiftly. Extinctions of societies usually take place at still longer intervals of time.

In theoretical discussions in the social sciences, for instance in the archaeological literature, the distinctions among the different levels of adaptation are not always clearly maintained, and a good deal of confusion frequently results.

The Evolution of Human Languages

In the case of languages, as well as that of societies, evolution or learning or adaptation takes place in various ways on different time scales. As we discussed earlier, a child's acquisition of language represents a complex adaptive system in operation. On a much longer time scale, the evolution of human languages over centuries and millennia can be regarded as another complex adaptive system. On a time scale of hundreds of thousands or millions of years, biological evolution produced the capacity of human beings (*Homo sapiens sapiens*) to communicate by means of languages of the modern type. (All of those languages have certain common properties such as sentences of arbitrary length, elaborate grammatical structure, and such universal grammatical features as pronouns, genitive constructions of various kinds, and so forth.)

When considering the evolution of grammar, it is important to take the various levels of adaptation into account. Since the pioneering work of Joe Greenberg, a considerable amount of information has been accumulated on the grammatical features that are common to all known human languages ("grammatical universals") and those that apply in almost all cases ("grammatical near-universals"). In accounting for these general features, one must obviously pay attention to the biologically evolved, neurologically pre-programmed constraints emphasized by

Chomsky and his followers. However, one must also consider the results of linguistic evolution over centuries and millennia, which must reflect to some extent selection pressures favoring grammatical features adaptive for communication. Finally, there may be frozen accidents, "founder effects," stemming from arbitrary choices of grammatical features in languages (or even a single language) ancestral to all modern tongues, choices with consequences that persist everywhere to this day. (Recall that in biology the asymmetry between right- and left-handed molecules may be such a frozen accident.) In discussions of linguistics at the Santa Fe Institute, emphasis is laid on the necessity of including all these contributions together in trying to explain grammatical universals and near-universals.

In studying evolution of any complex adaptive system, it is essential to try to pick apart these three strands: the basic rules, frozen accidents, and the selection of what is adaptive. And of course the basic rules may themselves look like frozen accidents when viewed on a cosmic scale of space and time.

Adaptation Versus What Is or Appears Adaptive

Distinguishing different levels and different time scales of adaptation still leaves the set of puzzles associated with why adaptive complex systems like societies seem so often to be stuck with maladaptive schemata. Why haven't they just developed better and better schemata and progressed to higher and higher fitness? Some of the reasons have already been encountered in earlier chapters.

Societies, like other complex adaptive systems, are often subject to selection pressures that are not accurately described by any fitness function. And fitness, as we have seen, is not something that simply increases with time even when it is well-defined. Also, there is not a simple correspondence between features that are adaptive and features that have arisen through the various forms of adaptation. None of these issues is restricted to societies. They are widespread in biology and they are sometimes particularly acute in the experience of individual human beings. What are some of the mechanisms that permit maladaptive schemata to survive?

Maladaptive Schemata—External Selection Pressures

One very general mechanism for the persistence of apparently maladaptive behavior has already been discussed at some length, especially in connection with superstition versus science. The selection pressures affecting the promotion and demotion of theories in science relate mainly to the success of those theories in explaining coherently and predicting correctly the results of observation, at least when the scientific enterprise is working properly. When it is not working well, it is because other selection pressures, many of them stemming from the human weaknesses of scientists, are strong.

In the case of superstitious theorizing, nonscientific kinds of selection pressures play dominant roles. Let us recapitulate the pressures we have mentioned. They include the reinforcement of the authority of powerful individuals and also the maintenance of social cohesion, which confers an advantage in societal evolution. In addition the imposition of a structure of false order and regularity on mostly random events or disconnected facts can provide a degree of comfort; the illusion of understanding and especially mastery relieves fears of the uncontrollable. Related to these selection pressures is the very general kind that we have associated with the catch phrase "the selfish scheme": any complex adaptive system has evolved to discover patterns and so a pattern is in a sense its own reward.

The common element in all these selection pressures is that they are largely external to what is considered adaptive in science, namely success in describing nature. Likewise they are mostly external to what is considered adaptive in engineering, namely controlling nature for some human purpose. Nevertheless, such selection pressures play critical roles in the evolution of cultural DNA.

Clearly there is a general lesson to be learned here. The system being discussed is often defined in such a way that it is not closed. Selection pressures of great importance are exerted from outside. A simple example is provided by one of the processes that take place in the evolution of human languages. Suppose certain tribes or nations speaking different languages come into contact and, after a few generations, some of the languages survive, with certain modifications, while others become extinct. Which ones die out depends less on the relative merits of the various tongues as means of communication than on quite different considerations, such as the relative military strength or cultural attainments of the different tribes or nations. Those are selection pressures exerted outside the linguistic realm.

In the domain of biological evolution, where selection normally takes place at the phenotypic level, there may be, as we discussed earlier, exceptional cases where it acts directly on the germ cells: a "truly selfish gene"promoting, for sperm carrying it, the successful fertilization of an egg, even though that gene may not be helpful, and could even be harmful, to the resulting organism.

What all of these examples suggest is that the apparent persistence of maladaptive schemata in complex adaptive systems may often arise simply from too narrow a choice of criteria for what is considered adaptive, given all the selection pressures that are operating.

Pressures Exerted by Influential Individuals

In studying the evolution of human organizations, it is not always advantageous to consider the individual members of the organization merely as simplified generic agents. Often the particular decisions made by specific individual human beings make a great deal of difference to future history. While it may be that in the long run many such effects prove to be temporary aberrations that are "healed" through the operation of long-term trends, still it is impossible to ignore the fact that individuals do matter. Thus the element of design enters into the picture. The constitution of a state or federation is written by individuals. Even though many of the conflicts that arise in the course of its drafting represent the competition of large-scale interests, still the specific compromises that are worked out are forged by particular statesmen. Likewise a business firm is directed by individuals, and the character and ideas of the boss or bosses (and sometimes of other individuals as well) are critical to the success or failure of the enterprise.

At the same time, an organization does behave in many ways as a complex adaptive system, with schemata and selection pressures. A business firm operates according to a certain set of practices and procedures, sets goals for its various departments or divisions, makes plans for the future, and generates mental models for the functioning of the whole enterprise. The models, together with the goals, plans, practices, and procedures, constitute schemata, subject to direct pressures exerted by managers at various levels, from the boss to the foremen or office managers. The actual selection pressures on the firm in the real world, however, have to do with profits, with survival in the marketplace. It matters whether customers are attracted and then satisfied. In general, when organizations are regarded both as complex adaptive systems and as theatres for the exercise of the management skills of individuals, the question arises as to the relationship between the ultimate selection pressures that govern the survival of the organization and the internal selection pressures exerted by the individual managers.

W. Edwards Deming, the American statistician (with a Ph.D. in physics) who advised the Japanese on the reconstruction of their industries after the Second World War, became something of a hero in Japan as a result of his wise recommendations. For a decade or more before his recent death at the age of 93, he was finally honored in his native country, where his ideas are now widely disseminated and accepted by many industrial firms. Perhaps best known is his emphasis on "total quality management" or TQM. Of the many facets of TQM, it is perhaps most useful to cite here his strictures on the internal pressures exerted by managers, including middle managers. Those are the people who dispense rewards and sanctions. By creating incentives for employees to act in particular ways, they directly affect some of the principal schemata for the organization. But are those direct effects consonant with the selection pressures exerted in the real world? Are the employees being rewarded for activity that actually leads to satisfied customers? Or is it just the whim of some manager that they are satisfying? Are managers apt to behave like the truly selfish gene, acting directly on the survival of the schema in a way that may not promote the survival of the organism?

Adaptive Systems with Humans in the Loop

The case of managers in a business firm exemplifies the more general situation of adaptive systems with one or more human beings in the loop—systems subject to what is sometimes called directed evolution, in which selection pressures are exerted by individual human beings.

The simplest situations involve direct adaptation, with no variant schemata. Consider an eye examination by an optometrist. You look with one eye at a chart covered with rows of letters as well as vertical and horizontal lines. The optometrist presents you with a sequence of binary choices. For each pair of images, you are asked if the left- or the right-hand one is clearer. Before long you have converged on the right prescription to take care of whatever combination of astigmatism, myopia, and presbyopia the eye in question presents. The single eye-chart schema has adjusted itself to your eye.

A less cut-and-dried example of a human in the loop is provided by the work of Karl Sims, now at Thinking Machines, a company that designs and manufactures parallel processing computers. Sims utilizes a computer screen consisting of 256 by 256 pixels, in each of which the color can vary over the whole spectrum. Patterns result from specifying the color of each pixel. Using Sims's program, the computer starts with a particular pattern and then generates a set of variant patterns, using a particular algorithm. The person "in the loop" chooses the variation that looks best to him or her. The computer then offers another set of choices, and so on. After not very long, the system has converged on a picture that appeals to the human being involved. I am told that the results are often quite remarkable and also that participation in the process is addictive.

One can imagine elaborations of this method in which chance would play a role in the algorithm for computing the choices that are offered at each stage. Or, in what amounts to nearly the same thing, the computer could employ a pseudo-random process as part of the algorithm.

At a Santa Fe Institute Science Board Symposium, Chris Langton gave a brief description of Sims's work. Bob Adams, the archaeologist who was then the Secretary of the Smithsonian Institution, raised the point that the algorithm governing the way the computer keeps offering sets of choices could itself be subject to variation. If so, it would become a kind of schema, each variant of which could be regarded as a different search process for hunting through the enormously long list of possible patterns. The particular search process adopted (which may or may not include an element of chance), together with the results of the choices made by the human subject, would determine the pattern on the computer screen.

The patterns could then be transferred to a permanent medium and subjected to selection pressures, for example sale in the marketplace or comments by critics. The computer programs that led most often (through the human subjects) to pictures commanding relatively high prices or relatively favorable comments might be promoted, and the others demoted. Likewise the tastes (conscious or unconscious) of the human subject might change under the influence of the prices or comments. The program, the computer, the human subject, and the marketplace or critic would then constitute a complex adaptive system, with human beings in the loop. In fact, such a system may serve as a kind of crude caricature of how the creative process of real artists sometimes functions.

We are all familiar with another complex adaptive system that operates in this way, namely the breeding of animals or plants for human use. Plant and animal breeding played an important role in the history of modern biology. Darwin repeatedly referred to them in his *Origin of Species* under the rubric of "artificial selection," with which natural selection was compared and contrasted. The Mendelian laws of genetics were discovered by the monk Gregor Mendel in the course of his breeding of peas. In addition, at the turn of the century, around the time when Mendel's work was rediscovered and disseminated to the world, the Dutchman de Vries discovered mutations while breeding tulips.

The breeder exerts selection pressures by choosing for further breeding only some of the organisms produced. Of course, natural selection is at work as well, and many of the animals or plants fail to survive or procreate for reasons that have nothing to do with the breeder's decision. The genome is a schema, as usual in biological evolution, but here the evolution is in part directed and the breeder's principles also form a schema, albeit of a different kind.

When a horse breeder puts a horse up for sale or enters it in a race (or both), his methods, analogous to Karl Sims's computer program plus the choices made by his human subject, are being exposed to the selection pressures of the marketplace and the race track. Thus a complex adaptive system with a component of directed evolution can become part of a higher order complex adaptive system in which the character of the human direction can itself evolve.

However, suppose a wealthy amateur breeder doesn't care how his horses perform at the race track or whether anyone wants to buy them.

In that case, in the context of the higher order complex adaptive system, the results of the horse breeding methods will probably appear to be maladaptive. Like managers who offer incentives for behavior unlikely to attract or keep customers, such a dilettante horse breeder may be pleasing himself but he is not acting like a businessman. From a purely business point of view, the breeding is a failure, yet it can go on.

Persistence of Maladaptive Schemata: Windows of Maturation

Maladaptive schemata sometimes persist because the relevant kind of adaptation has come to a halt, or nearly so. Young children form relationships with important people in their lives: parents, stepparents, siblings, nannies, mothers' boyfriends, and so on. According to Dr. Mardi Horowitz, the attitudes and behavior of a child in such a relationship are governed by a "person schema" relating to the child's perceptions of the individual in question. At first such a schema is subject to alteration, but later in childhood it becomes very resistant to change. As the child grows up, those person schemata may profoundly affect the way he or she relates to other people. We are all familiar, for example, with adults who keep reenacting with various surrogates a childhood relationship with a parent. Often person schemata appear quite maladaptive, and living in accordance with them amounts to what is often called neurotic behavior, notoriously difficult to cure.

One useful way to look at such situations is in terms of "windows of maturation" versus "plasticity." An extreme example of a window of maturation is the phenomenon of imprinting, which was made famous by Konrad Lorenz in his book *King Solomon's Ring*. A newborn greylag goose regards the first suitable animal that it sees as its parent and follows it around. If that animal is Lorenz or another human being, then the goose comes to regard itself in some sense as a human and its ability to live as a normal goose is permanently compromised. The window of maturation is the very short period after birth during which the gosling identifies its "mother" and after which the identification is fixed for good. The typical gosling sees its real mother early on, and the genetic program of imprinting is then phenotypically successful. For the rare gosling that adopts an ethologist like Lorenz as its mother, the program is evidently a failure. In such a case the learning schema supplied by imprinting is maladaptive for the particular individual involved. Because that learning schema works out fine for most individuals, however, the genetic schemata that lead to imprinting have not been eliminated in biological evolution. Still, genetic schemata providing for such windows of maturation must also confer some general advantage in order to have survived. Presumably that advantage comes from the possibility, when the window closes down, of turning off the machinery for acquiring certain new information.

Windows of maturation are known in human beings, too. For instance, some babies are born with visual problems that must be corrected early if recovery is to be possible (at least without some new, so far undiscovered form of intervention). In other cases, the windows are not absolute, so to speak. The consequences of various forms of neglect during crucial periods of infancy and early childhood may be serious if nothing is done to reverse the damage, but under suitable conditions there may be significant possibilities of recovery. Those possibilities are discussed under the rubric of plasticity, the capacity of the nervous system to reorganize itself so that patterns that might otherwise persist indefinitely can in fact be changed.

A major public policy issue, particularly in the United States, involves the extent to which deficits in learning ability acquired before the age of two and a half can be remedied by programs like Head Start, which give children special help during the following two and a half years or so. Some investigators claim that an early window of maturation plays a crucial role here and that remedial programs at the later ages are not nearly so effective in the long run as improving the learning environment of babies. Others claim to have shown that there is enough plasticity in this case to permit substantial and long-lasting reversal of learning deficits by means of interventions such as Head Start, provided they are carried out (as often they are not) with sufficient intensity and duration.

Whatever the merits of the arguments about general learning deficits in young children, it is known that for the acquisition of a first language the early years of life are critical. The few known cases of children raised with little or no human language contacts indicate that the innate machinery for mastering the grammar of a language ceases to be effective. Apparently a true window of maturation is involved.

Persistence of Maladaptive Schemata: Time Scales

One of the most common reasons, and perhaps the simplest, for the existence of maladaptive schemata is that they were once adaptive but under conditions that no longer prevail. The environment of the complex adaptive system has changed at a faster rate than the evolutionary process can accommodate. (Windows of maturation are in a sense an extreme example of such a mismatch of time scales.)

In the realm of human thought, it often happens that we are confronted with a rapidly changing situation that overtaxes our ability to alter our thought patterns. Gerald Durrell, the founder of the zoo on the island of Jersey, who has written so many charming books on his expeditions to bring back rare animals, recounts what happened once when he was holding a certain West African snake in his hands. He was not taking any special precautions because he "knew" it to belong to a harmless blind species (like the blindworm of Europe). Suddenly the serpent opened its eyes, but Durrell did not react quickly enough to the new information that the snake belonged to an unknown and possibly dangerous species. In fact it was poisonous, and Durrell was bitten and nearly killed.

Rather than change our way of thinking, we tend to cling tenaciously to our schemata and even twist new information to conform to them. Many years ago, two physicists associated with the Aspen Center for Physics were climbing in the Maroon Bells Wilderness. While descending, they lost their bearings and came down on the south side of the mountains, instead of the north side near Aspen. They looked below them and saw what they identified as Crater Lake, which they would have spotted from the trail leading home. One of them remarked, however, that there was a dock on the lake, which Crater Lake does not possess. The other physicist replied, "They must have built it since we left this morning." Needless to say, that desperate defense of a failed schema turned out to be incorrect. The physicists were looking at Avalanche Lake on the wrong side of the mountains, and it took them a couple of days to get home.

Realizing that the snake, since it was not actually blind, could be poisonous fits the description that we gave of having a creative idea in everyday life, escaping from one basin of attraction into another. So does reflecting that the lake, with its dock, was unlikely to be Crater Lake and was therefore elsewhere in the mountains. The present discussion emphasizes that the process of getting such ideas may, in many cases, not keep pace with the need for them.

It is notorious that business firms often have trouble adjusting their practices rapidly enough to changing market conditions. Right now, in the United States, the reduced appropriations for military preparedness mean that industries hitherto devoted mainly to defense have to find civilian customers in a hurry. Often those industries have formed their ideas of marketing through decades of dealing with the armed services and related government agencies. The prevailing schema for selling a product may be to go to lunch with an admiral, not necessarily a winning move in the competition for civilian business. Moreover, the mechanisms for varying such schemata and responding to selection pressures often take many years to operate, whereas the demand for defense-related systems may be drastically reduced over a year or two. If the managers (or new managers replacing them) do not introduce new mechanisms with a faster response time, the prospects for their company are not bright.

The challenge of circumstances that change more rapidly than a given evolutionary process can accommodate is one that profoundly affects the prospects for the biosphere and for the human race as a whole. Human cultural evolution, especially through advances in the technological sphere, has made possible in a brief span of time an extraordinary expansion of the human population and of the capacity of each person to affect adversely other people and the environment. Biological evolution, in humans and in the other organisms, has no chance of keeping up. Our own genetic schemata reflect in great part the world of fifty thousand years ago and cannot, through the normal mechanisms of biological evolution, undergo important changes in just a few centuries. Likewise, other organisms and whole ecological communities cannot evolve quickly enough to cope with the changes wrought by human culture.

The implication is that cultural change itself is the only hope for dealing with the consequences of a gigantic human population armed with powerful technologies. Both cooperation (in addition to healthy competition) and foresight are required to an unprecedented degree if human capabilities are to be managed wisely. The need for cooperation and foresight will be even greater if reliance is placed, for dealing with some of the most urgent concerns, on artificial transformations of human beings and other organisms, utilizing future developments in genetic and other engineering.

Given the immense complexity of the numerous interlocking issues facing humanity, foresight demands the ability to identify and gather great quantities of relevant information; the ability to catch glimpses, using that information, of the choices offered by the branching alternative histories of the future, and the wisdom to select simplifications and approximations that do not sacrifice the representation of critical qualitative issues, especially issues of values. Powerful computers are essential for assistance in looking into the future, but we must not allow their use to bias the formulation of problems toward the quantifiable and analyzable at the expense of the important.

It is appropriate at this point to take a brief look at the kinds of simplified models of complex problems that computers can provide. Computers acting as complex adaptive systems can serve us both by learning or adapting themselves and by modeling or simulating systems in the real world that learn or adapt or evolve.

CHAPTER

20

MACHINES THAT LEARN OR SIMULATE LEARNING

Computers can function as complex adaptive systems. Either the hardware can be designed so that they do, or else computers with ordinary hardware can be programmed to learn or adapt or evolve. So far, most such designs or programs have depended on imitating a simplified picture of how some living complex adaptive system works.

Neural Net Computation

One well known type of computer complex adaptive system is the neural net, which can be implemented either with software or hardware. Here the analogy is with a crude model of how the brain of a mammal (especially a human being) might operate. One starts with a set of many nodes or units (often called neurons, although it is far from clear to what extent they really correspond to individual neurons in a brain). Each unit is characterized at every instant of time by a bit (0 or 1) that is supposed to indicate whether the "neuron" is firing or not. Each unit is connected to some or all of the others, and the strength of the influence of any unit on any other is some positive or negative number depending on the two units. That number is positive if the first unit excites the second and negative if it inhibits the second. As the learning progresses, those strengths keep changing.

Neural net computation can be carried out on conventional computers; in that case the software is responsible for the units and their excitatory or inhibitory effects on one another. The units then exist only as elements of the computation. It is also possible to employ special computer hardware to realize the network, which is then composed of separate computing units arranged in a parallel processing format.

Of the many problems to which neural net computation has been applied, one example is learning to read an English text aloud with correct pronunciation. Since English spelling is notoriously far from phonetic, that is not a trivial exercise. The computer has to discover a huge number of general rules together with their exceptions, which can be regarded as special additional rules. Given enough text, not only the general rules but the special ones as well will crop up enough times to function as regularities. For the neural net computer to learn to read English aloud, it has to function as a complex adaptive system. It must try out various tentative identifications of sets of regularities in a batch of text, compress the information about them into schemata, and apply those schemata to more text, letting them compete with one another to come closest to the correct pronunciation, which is supplied by a "teacher." This kind of learning is called supervised learning, as opposed to the kind where, for example, schemata for pronunciation would be tried out on English speakers to see if they understood, but no teacher would be available to supply right answers. Supervision allows fitness to be defined, in terms of the amount of difference between the correct pronunciation of a text and the pronunciation resulting from the schema.

In NETtalk, as developed by Terry Sejnowski and C. R. Rosenberg in 1987, the input data consisted of seven consecutive characters (each of which could be one of the 26 letters or else a space, a comma, or a period) from some written English text, presented in a moving window that gradually scanned the whole passage. The output was a pronunciation code for the middle character of the seven; that code was fed into a speech generator.

The inputs were identified with the bits attached to a set of 7 times 29 (= 203) units, and the outputs with bits associated with 26 other units. There were 80 additional units to help with the learning. The schemata were represented by sets of interaction strengths, where the interactions were restricted to effects of input units on helper units and effects of helper units on output units. All other interaction strengths were fixed at zero, just to keep the process manageable.

The network was trained using the characters constituting a text of 1024 words, accompanied by the "teacher," which was the sequence of pronunciation codes for all those characters, giving correct English pronunciation of the text.

In the first training run through the text, the strengths would start out at some arbitrary values and then change (by a kind of learning) as the center of the seven-letter window moved letter by letter through the text. At each step, the inputs and the strengths would be fed into a simple formula yielding candidate output values indicating attempted pronunciations. The discrepancy between the correct and the candidate outputs was then reduced by modifying the strengths, using a related simple formula. The strengths at the end of the first training run were used as the initial strengths for a second training run with the same 1024 word text. And so on.

Success meant that the strengths, instead of fluctuating wildly from run to run, would zoom in, with only minor deviations, toward values giving a fair approximation to correct pronunciation. In the event, ten training runs on the 1024 words were enough to give intelligible speech, while after fifty training runs the pronunciation accuracy was estimated at 95 percent. After training, the resulting strengths were used to pronounce another batch of English text, with no further teaching. Intelligible speech, at an accuracy level of 78 percent, was achieved.

There are many other versions of neural nets and a plethora of problems to which they have been applied, often with considerable success. The schema is always represented by a set of interaction strengths, each representing the effect of one unit on another. My Caltech colleague John Hopfield pointed out in 1982 a condition that, if artificially imposed on those strengths, would permit fitness not only to be well defined but to increase steadily during the learning process. The strength of the effect of any unit A on another unit B is required to be the same as that of B on A. This condition is almost certainly unrealistic for real brains and is also violated by many successful neural nets, including NETtalk. It is instructive, however, that there are situations in which fitness is both well-defined and steadily increasing and others in which it is not.

As usual when fitness is well defined, the learning process consists of exploring valleys on a fitness landscape. If fitness is also steadily increasing, so that height keeps decreasing, then the problem of getting stuck in a shallow depression when there are deep pits nearby crops up as always and can be ameliorated by introducing noise. That could be accomplished, for example, by altering the strengths slightly, in a random way, from time to time. Such random changes in the schema resemble those proposed to jog the mind out of a rut when one is seeking a creative idea. As usual, there is an optimum level of noise.

Genetic Algorithms as a Complex Adaptive System

Since neural nets are based on a crude analogy with learning in higher animals, one might ask whether there are also computer-based complex adaptive systems suggested by an analogy with biological evolution. Indeed there are. They were pioneered by John Holland of the University of Michigan, a mainstay of the Santa Fe Institute, and they make use of "genetic algorithm" software, a special "classifier system," and conventional computer hardware. So far, these systems have been used mostly for problems in which fitness is well defined, such as devising strategies for winning at checkers or methods of installing wiring systems that minimize costs. However, there is no reason why the systems cannot be applied to other kinds of problems as well.

A highly simplified description of genetic algorithm software would go something like this. Each schema is a computer program for a candidate strategy or method. Every such program is composed of a number of computer instructions. The variation in schemata is accomplished by changing those instructions, say by having two of them undergo a crossing-over process (as shown on the facing page) like the one (illustrated on page 254) that occurs in the sexual reproduction of living organisms. The two instructions are both divided at a certain point into a beginning and an end. Crossing-over causes two new instructions to emerge. One of those is composed of the beginning of the first old instruction and the end of the second. The other new one consists of the beginning of the second old one and the end of the first.

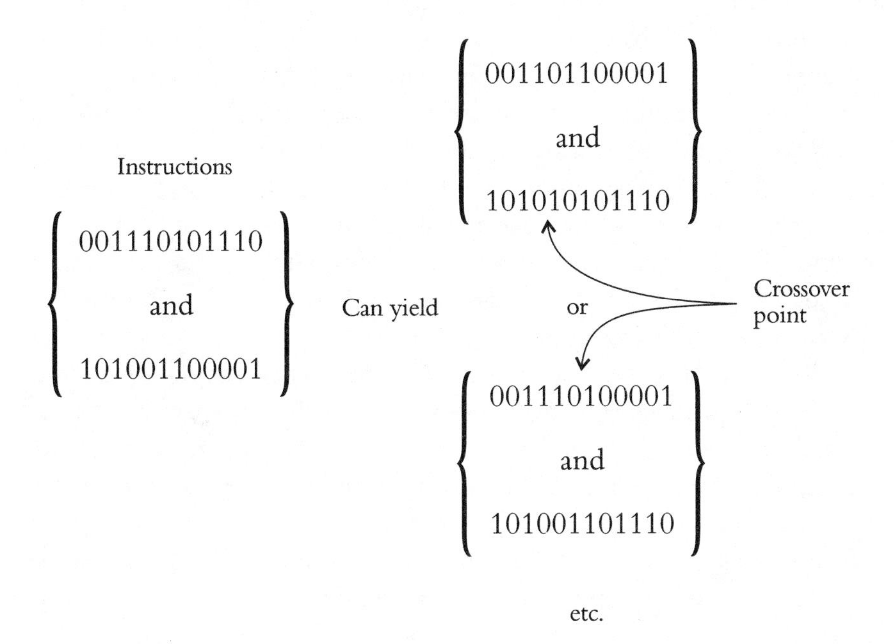

Crossing-over for computer instructions.

Modifying a computer program by replacing one or more instructions with new ones generated in this manner sometimes improves the fitness of program and sometimes degrades it. Only by trying out the different programs on the problem to be solved can the computer judge the worth of each modification. (Making that kind of difficult judgment is known as "credit assignment.") John Holland's classifier system provides a kind of marketplace in which competing instructions are bought and sold. Those that have a record of improving the performance of programs command higher prices than those that lead to no improvement or to worse performance. In that way an ordered list of instructions is established. New instructions are continually entered, and the ones at the bottom of the list are deleted to make room for them. The instructions at the top of the list are the ones used in the modified programs that constitute the mutated schemata.

This is only the crudest sketch of what is a quite sophisticated procedure. Even so, it should be clear that programs evolve as a result, and that fitness tends to increase in the course of the evolution. The feedback from the performance of a program to the promotion and demotion of the instructions that make it up is not a rigid rule, however, but a general tendency affected by market conditions. Hence there is enough noise in the system to permit escape from minor basins of fitness so that depths nearby can be plumbed. Typically, the system is exploring a huge space of possible methods or strategies and not reaching a steady state—an absolute optimum. The optimum strategy for playing chess, for example, has not been found. If the game in question were tic-tac-toe, however, the machine would soon find the best way to play and the search would be over.

Although the genetic algorithm method has been applied mainly to search and optimization problems where fitness (or "payoff") is well defined, it can also be used in other cases, just as neural nets can be employed in both kinds of situation. Both neural nets and genetic algorithms yield computer-based complex adaptive systems that can evolve strategies no human being ever devised. It is natural to ask whether there is anything special about these two classes of techniques, suggested by vague analogies with the functioning of brains and of biological evolution respectively. Can another class be invented that is based on an analogy with mammalian immune systems? Is there in fact a huge but well-defined set of computer-based complex adaptive systems that includes those that are known or hypothesized and many others besides? Can such an overarching category be described in practical terms so that a potential user could search through the different possible computer-based systems to find one appropriate for his or her problem?

Such questions are among the ones that students of computer-based complex adaptive systems are trying hard to answer.

Simulation of Complex Adaptive Systems

The use of computers in connection with complex adaptive systems is by no means restricted to developing hardware or software for computer-based complex adaptive systems used to solve problems. Another vast area of computer applications is the simulation of the behavior of complex adaptive systems.

The most striking feature of those simulations is the emergence of complex behavior from simple rules. Those rules imply general regularities, but the working out of an individual case exhibits special regularities in addition. This situation is similar to that of the whole universe, governed by simple laws allowing for an infinity of scenarios, each of which exhibits its own regularities, especially for a given region of space and epoch of time, so that more and more complex forms can emerge as time goes on.

The trick in designing a manageable simulation is to prune the rules so as to make them even simpler, but in such ways that the most interesting kinds of emergent behavior remain. The designer of a simulation must then know a good deal about the effects of changes in the rules on behavior in many different scenarios. Some designers, such as Robert Axelrod, a political scientist at the University of Michigan, have developed a keen intuition that helps them guess how to simplify without throwing the baby out with the bath water. Naturally, that intuition is based partly on *a priori* reasoning and partly on experience of fiddling with the rules and then watching what happens under the modified rules in particular computer runs. Still, the design of simple simulations rich in interesting consequences remains more of an art than a science.

Can the study of sets of rules and their consequences be made more scientific? Additional experience is needed, together with the formulation of inspired empirical guesses about what kinds of rules lead to what kind of behavior. Then rigorous theorems may be conjectured, and finally some of those theorems may be proved, presumably by mathematicians.

In that way, a kind of science of rules and consequences may emerge, with the computer runs functioning as experiments and the conjectured and proved theorems constituting the body of theory. In fact, with the advent of rapid and powerful computers, more and more simple simulations are being run, and on more and more subjects. The raw material for the future science is already accumulating.

In the end, though, what really matters is the relevance of the simulations to the real-world situations that they imitate. Do the simulations supply valuable intuition about real situations? Do they suggest conjectures about real situations that could be tested by observation? Do they reveal possible behaviors that had not been thought about before? Do they indicate new possible explanations of known phenomena?

In most fields simulations are still too primitive for these questions to be answered in the affirmative. Nevertheless, it is astonishing how, in certain cases, a very simple set of rules can give insight into the operation of a complex adaptive system in the real world.

A Simulation of Biological Evolution

A splendid and by now quite celebrated example is the TIERRA program written by Thomas Ray of the University of Delaware and the Santa Fe Institute. He was an ecologist working in the lowland rain forest of Costa Rica at the biological research station called La Selva. Ecological research attracted him because he wanted to study evolution. Unfortunately, not much biological evolution takes place during a human lifetime and so he began to find his field work frustrating. He therefore decided to simulate evolution on a computer.

He was planning to develop a suitable program in stages, starting with a highly oversimplified one and then gradually building in more features, such as punctuated equilibrium or the existence of parasitism. He taught himself painfully to write a program in "machine language" and managed to get a single very simple one written and debugged. That initial program was TIERRA, and it has turned out to be extraordinarily rich. Running it over and over and understanding the lessons of all the different runs has occupied him ever since. Moreover, a number of features he was planning to build in later, including both punctuated equilibrium and the prevalence of parasitism, emerged from TIERRA itself. The program even turned up something very much like sex.

TIERRA uses "digital organisms," which are sequences of machine instructions that compete for space in the memory of the computer and for time on the central processing unit, which they use for self-replication. The community of complex adaptive systems provided by TIERRA is degenerate in a sense because the genotype and the phenotype of each digital organism are both represented by the same object, namely the sequence of instructions. That sequence is what undergoes mutation and it is also what is acted upon by selection pressures in the real world. Still, it is a good idea (as emphasized by Walter Fontana) to keep the two functions separate when thinking about the system, even though both are performed by the same entity. (According to some theories of the origin of life on earth, an early stage of that process had the same degenerate character, with RNA playing the roles of both genotype and phenotype.)

Mutations are introduced in two ways. First, from time to time bits are flipped (from 0 to 1 or vice versa) at random anywhere in the whole set of organisms (much as real organisms are affected by cosmic rays). The rate used is around one bit flipped for every ten thousand instructions executed. Second, in the course of replication of digital organisms, bits are flipped at random in the copies. Here the rate is set somewhat higher, about one bit flipped for every couple of thousand instructions copied. These are average rates; the errors are irregularly timed to avoid periodic effects.

The importance of death in biology was not neglected in the design of TIERRA. Memory space is severely limited, and in the absence of death self-replicating creatures would soon fill it up, leaving no room for further replication. Hence the "reaper," which kills off organisms on a regular basis according to a rule that depends on the age of the organism and on errors it made in executing certain instructions.

Tom Ray designed a self-replicating sequence of eighty instructions, which is always used as the ancestor—the initial digital organism—in any run of TIERRA. When he ran the system the first time, he expected a long period of trouble and trouble-shooting to ensue. Instead, interesting results started to emerge right away, many of them suggestive of real biological phenomena, and that situation has prevailed ever since.

One intriguing development was the appearance, after a long period of evolution, of a refined version of the ancestor. It has only thirty-six instructions instead of eighty and yet manages to pack into them a more complex algorithm. When Tom showed this trick of compression to a computer scientist, he was told that it was an example of a known technique called "unrolling the loop." In TIERRA, evolution had figured out how to unroll the loop. Tom writes, "The optimization technique is a very clever one invented by humans. Yet it is implemented in a mixed-up but functional style that no human would use (unless perhaps very intoxicated)."

How do such organisms with other than eighty instructions arise? Mutations cannot produce them directly. Initially, the system contains only the ancestor and its eighty-instruction descendants. (They multiply until the memory is nearly full; that is when the reaper starts its work. The changing population of organisms then continues to occupy most of the memory.) Eventually, mutations appear that alter the genotype of an eighty-instruction organism in a special way: when the organism examines itself to determine its size, so that it can pass on that size to its descendants, the answer comes out wrong and a new size is passed on instead. In that way, the population comes to contain organisms of many different sizes.

If so many insights have emerged from the first attempt to model biological evolution in this way, there must be a huge territory still waiting to be explored. New ways of simulating how evolution, operating over enormous stretches of time, has generated the information now stored in organisms and natural communities throughout the world may help not only to improve our understanding of existing diversity, but also to create a climate of ideas in which that diversity can be better protected.

A Tool for Teaching About Evolution

TIERRA, together with related computer simulations of biological evolution to be developed in the future, will be especially valuable for conveying to nonscientists a feeling for how evolution works. Most people find it easy to appreciate, even without computer simulations, how comparatively minor variations combined with a few generations of selection can produce changes in a population. Personal experience with the breeding of dogs, budgerigars, horses, or roses can easily convince almost anyone of the reality of evolution on a small scale. But evolution on a longer time scale, with the emergence of new species, genera, families, and still higher taxa is a different matter. Even the comparatively close relationship of the elephant to the rock hyrax is hard for most people to grasp. It is still more difficult to visualize the interrelationship of all forms of life, including the immense changes that can be wrought over billions of years.

What is especially hard for many people to accept is that *chance plus selection pressures* can lead from a simple initial condition to highly complex forms and to complex ecological communities comprising such forms. They cannot really bring themselves to believe that such evolution can take place without some kind of guiding hand, some kind of design. (Others balk especially at the evolution of consciousness, the self-awareness of which we humans are so proud; they feel somehow that consciousness cannot arise without antecedent consciousness.) Never having entertained any of those doubts, I can only view them from the outside. But it seems clear to me that one way to relieve them is to let people experience the remarkable transformations effected by millions of generations of largely random processes combined with natural selection. That can be done only by simulation, as in TIERRA, which can run through a huge number of generations in a manageable period of time, and in more sophisticated and realistic simulations that will be available in the future.

In describing biological evolution in terms of chance and selection, we are treating the various mutation processes as purely stochastic. A few investigators have claimed, however, to find deviations from chance behavior. They have interpreted certain observations as indicating that sometimes mutations arise in nonrandom ways, even in ways that seem biased in favor of increasing fitness in response to changing selection pressures. In at least one case, the evidence adduced has been satisfactorily explained without such an interpretation. Perhaps all the alleged cases can be similarly explained. But even if certain organisms should turn out to have evolved mechanisms permitting occasional exceptions to chance behavior of mutations, our description of biological evolution would, as far as we know, be largely unchanged.

Before Tom Ray had developed TIERRA, I convened a small group of thoughtful people at the Santa Fe Institute to discuss whether we could invent a computer game that might become popular and would convince the players of the immense power of the evolutionary process extended over very many generations. One excellent result of the meeting was that when John Holland went home he invented ECHO, a rich computer simulation of an ecology of simple organisms. However, the game that was to be a teaching aid was not forthcoming. Then, shortly afterward and quite independently, came Tom Ray's TIERRA, which, although not really a game, may ultimately produce the same effect.

Some participants in the meeting pointed out that a pocket in the cover of the first paperback edition of Dawkins's book *The Blind Watchmaker* holds software for a computer game illustrating evolution. That kind of game is not, however, exactly what I had in mind. The point is that in real biological evolution there is no designer in the loop. But Dawkins, whose book is devoted to making that very point in an elegant manner, has invented a game in which the player keeps supplying the selection pressures as evolution proceeds, much like the user of Karl Sims's software for producing pictures. (The game does come with a "drift" option, in which the player can let the organisms alone, but they are still not subject to selection pressures from an ecological community to which they belong.) Using the (only partially justified) language of fitness, one can say that in Dawkins's game, the fitness is *exogenous*, supplied from the outside, whereas in nature (as his book explains) the fitnesses are *endogenous*, ultimately determined, without external interference, by the character of the earth and the sun and by chance events, including the evolution of huge numbers of particular species. Can a game be designed in which the players, like Tom Ray using TIERRA, supply only an initial situation and a set of rules for biological evolution while chance and natural selection do the rest?

Simulation of Collectivities of Adaptive Agents

Any serious simulation of evolution must include the interaction of populations belonging to numerous species; the environment of each of those species comprises all the other organisms as well as the physicochemical surroundings. But what if we are trying to understand what happens to such an ecological community over a comparatively short period of time, during which not much biological evolution is taking place? We are then attempting a simulation of ecological processes.

A number of theorists associated with the Santa Fe Institute have used computer models to learn about the properties of those complex adaptive systems that are collectivities of co-adapting adaptive agents, which construct schemata to describe and predict one another's behavior. Those researchers have come up with a body of lore about such systems, consisting of plausible conjectures together with results demonstrated for particular models. The picture that emerges is one in which the region of intermediate algorithmic information content, between order and disorder, may contain a régime resembling that of self-organized criticality, exemplified by sand piles. In that régime, key quantities may be distributed according to power laws. Most important of all, there may be a tendency for the whole system to evolve toward the condition in which those power laws apply.

Stuart Kauffman has done a good deal of theoretical research on these ideas, as has Per Bak. Stuart is among those who describe them by using the term "adaptation toward (or to or at) the edge of chaos," where "edge of chaos" is used somewhat metaphorically to indicate a critical condition between order and disorder. The entire expression, which is now widespread in popular literature, was first employed by Norman Packard (using the preposition "toward") as the title of a paper on the approach to such a critical condition by a very simple computer-based learning system. Related research was carried out independently around the same time by Chris Langton.

In the ecological and economic domains where some of the obvious applications lie, such power laws are well known from observation, particularly ones governing the distribution of resources. The famous empirical law of wage distribution in a market economy, discovered in the nineteenth century by the Italian economist Vilfredo Pareto, approximates a power law for the higher incomes. Pareto also discovered a rough power law for individual wealth, again applicable to the high end of the spectrum.

Ecologists often look at the share of resources utilized by all the individuals of a given species taken together, considered as a function of the various species in a natural community. They too find empirical power laws. For example, along the rocky part of the shore of the Sea of Cortés near its nothern extremity just south of the U.S. border, the intertidal zone contains a number of different organisms, such as barnacles and mussels, occupying various proportions of the surface area of the rocks. The total areas occupied by the different species obey a power law to a fairly good approximation. Preying on these rock dwellers are other creatures, higher in the food chain. Among them, at or near the top of the chain, is a 22-armed starfish, *Heliaster kubiniji*. What would happen if the starfish were removed from the picture? That actually occurred, through some catastrophe, over a certain stretch of the coastline, and ecologists were able to observe the consequences. The result was that the system consisting of the remaining organisms readjusted itself, with new values for the total rock areas covered by the various species clinging to them. However, the approximate power law once again held. There may thus be some empirical support for the idea that systems of coadapting agents are attracted to a kind of transition régime characterized by power laws for resource distribution.

Rule- and Agent-Based Mathematics

In much of today's research on complex adaptive systems, mathematics plays a very significant role, but in most cases it is not the kind of mathematics that has traditionally predominated in scientific theory. Suppose that the problem is one in which a system is evolving in time so that at each moment its state changes according to some rule. Many of the striking successes of scientific theory have been achieved with the aid of continuum mathematics, in which the time variable is continuous and so are the variables describing the state of the system. That state changes from moment to moment according to a rule that is expressed in terms of the continuous variables that characterize the system. In technical language, the time development of the system is said to be described by a differential equation or a set of such equations. Much of the progress in fundamental physics over the last several centuries has taken place with the aid of such laws, including Maxwell's equations for electromagnetism, Einstein's equations for general-relativistic gravitation, and Schrödinger's equation for quantum mechanics.

When such equations are solved with the aid of a digital computer, it is usual to approximate the continuous time variable by a so-called discrete variable, which takes on values separated by finite intervals instead of all possible values between the initial and final instants that bound the time period under study. Moreover, the continuous variables characterizing the state of the system are also approximated by discrete ones. The differential equation is replaced by a difference equation. As the intervals between the nearby values of the discrete variables, including time, become smaller and smaller, the difference equation looks more and more like the differential equation it is replacing, and the digital computer comes closer and closer to solving the original problem.

The kind of mathematics that is often used in the simulation of complex adaptive systems resembles the discrete mathematics used on a digital computer to approximate continuous differential equations, but now the discrete mathematics is used for its own sake and not just as an approximation. Furthermore, the variables describing the state of the system may take on just a few values, with the different values representing alternative events. (For instance, an organism may or may not eat another organism; or two organisms may or may not engage in combat, and, if they do, one or the other will win; or an investor can buy, hold, or sell shares of a stock.)

Even the time variable may run over only a few thousands of values, representing, for instance, generations of a species or financial transactions, depending on the kind of problem. In addition, the changes in the system at each of those discrete moments are, for many problems, determined by a rule that depends not only on the state of the system at the time, but also on the result of a chance process.

Discrete mathematics of the kind we have been discussing is often called rule-based. It is a natural kind of mathematics for digital computers, and it is often applied to the simulation of complex adaptive systems composed of many individual adaptive agents, each of which is itself a complex adaptive system. Typically, the agents—such as organisms in an ecological community or individuals and businesses in an economy—are evolving schemata describing the behavior of other agents and how to react to it. In such cases, rule-based mathematics becomes agent-based mathematics, as used, for example, in TIERRA.

Making Economics Less Dismal

Exercises using agent-based mathematics are among the tools that have been employed recently to guide economics toward a more evolutionary approach to its subject matter. A preoccupation with a kind of ideal equilibrium, based on perfect markets, perfect information, and perfect rationality of agents, has characterized a great deal of economic theory during the past few decades. That is true despite the efforts of some of the best economists to incorporate imperfections of all three kinds into the post–World War II neoclassical synthesis.

In a story that has long circulated among economists, a neoclassical theorist and his well-behaved little granddaughter are walking along the street in a large American city. The girl spots a twenty-dollar bill on the pavement and, being very polite, asks her grandfather if it is all right to pick it up. "No, Dear," he replies, "if it were real someone would already have picked it up."

For several years, a number of scholars, including the members of an interdisciplinary network assembled by the Santa Fe Institute, have directed their efforts toward studying economies as evolving complex adaptive systems composed of adaptive economic agents endowed only with bounded rationality, possessing imperfect information, and acting on the basis of chance as well as perceived economic self-interest. The fairly successful predictions of equilibrium theory then appear as approximations, while the newer approach admits departures from those predictions, and especially fluctuations around them, in better agreement with reality.

In one exceedingly simple model, developed by Brian Arthur, John Holland, and Richard Palmer (a physicist at Duke University and the Santa Fe Institute), investors in a single security (say a stock) are represented by adaptive agents dealing with one another through a central clearinghouse. A share of stock pays a yearly dividend, which may vary with time in some arbitrary way. The going annual rate of interest is a constant, and the ratio of the dividend to that interest rate determines, more or less, the fundamental value of the share. The actual price of the share may, however, deviate greatly from the fundamental value. Each agent keeps constructing elementary schemata, based on the history of the stock price, that tell him or her when to buy or hold or sell. At any time, different agents may be using different schemata. Moreover, a given agent may have a list of several schemata and switch from one to another depending on performance. In this way price fluctuations are generated, often wild ones involving speculative booms and busts, with the slowly changing fundamental value supplying a kind of rough lower bound for the jagged curve of price versus time. Such fluctuations, reminiscent of what goes on in real markets, turn up here in an evolutionary model, with agents that are far from perfect but try to learn.

A number of the participants in the movement to reform economics have shown that perfect rationality is not only in obvious contradiction with the facts of human affairs, but is actually inconsistent with any situation in which market fluctuations occur. I personally have always been astonished by the tendency of so many academic psychologists, economists, and even anthropologists to treat human beings as entirely rational or nearly so. My own experience, whether engaging in introspection or observing others, has always been that rationality is only one of many factors governing human behavior and by no means always the dominant factor. Assuming that humans are rational often makes it easier to construct a theory of how they act, but such a theory is often not very realistic. There, alas, is the chief flaw in much of today's social and behavioral science. When it comes to theories of complex phenomena, making them more analyzable may be convenient but does not necessarily make them better at describing the phenomena—and may easily make them much worse.

The great contribution of economic theory to understanding human affairs is, in my opinion, simply the repeated emphasis on incentives. In any situation, what are the incentives for different courses of action? When the first Dead Sea Scrolls were discovered and archaeologists wanted more scraps of the scrolls to be found and turned in by wandering Arab shepherds, the misguided scholars offered a fixed reward *per scrap*, thereby making it likely that the fragments would be broken up into tiny pieces before being delivered. Economists study, often in sophisticated ways, how incentives operate throughout society and they point out the flaws in scheme after scheme, in government or business, analogous to the flaws in the reward system for the Dead Sea Scrolls. Incentives provide selection pressures in an economy. Even when the responses to them are not fully rational, and even if there are other pressures at work, economic incentives still help to determine which schemata for economic behavior will prevail. Human ingenuity will often find some way to profit by the incentives that exist, just as biological evolution will frequently manage eventually to fill some vacant ecological niche. Approaching economics in an evolutionary way and recognizing the bounded rationality of human beings can only improve economists' insights into the ways in which incentives operate.

The economics program has been one of the most successful activities of the Santa Fe Institute, in terms of stimulating new theoretical and modeling activities of high quality. Ultimately, of course, success must be measured, as in all theoretical science, by explanations of existing data and by correct predictions of the results of future observations. The Institute is still too young and the problems it studies too difficult for very much success of that kind to have been achieved so far. The next few years will be critical for judging the results of the Institute's work, and economic modeling is likely to be one of the efforts that result in verified predictions.

However, there are other reforms that are much needed in economic theory. Attempts to deal with some of those were contemplated in parts of the original plan for the Institute's economics program that have not yet been implemented. One vital problem has to do with taking proper account of values difficult to quantify.

Economists have sometimes been lampooned as people who would measure the value of love by the price of prostitution. The value of some things is easy to assess in terms of money, and the temptation is strong to count only such things in cost-benefit calculations and to ignore everything else. If the construction of a dam is proposed, old-fashioned cost-benefit analyses take into account benefits like electrical power and flood control. In addition, the resulting reservoir may be assigned a recreational value measured by the cost of the marinas and docks that will be built for power boats. The cost of the buildings in the valley that will be drowned by the filling of the reservoir may be counted against the dam, but not the value of the plants and animals in that same valley, nor the historical associations that the valley may have had, nor the community ties that are destroyed. It is difficult to assign a monetary value to such things.

The apparently hard-headed practice of ignoring values difficult to quantify is often advertised as being value-free. On the contrary, it represents the imposition on any analysis of a rigid system of values, favoring those that are easily quantifiable over others that are more fragile and may be more important. All our lives are impoverished by decisions based on that kind of thinking.

Many economists and political scientists have recommended leaving fragile values to the political process. But if that is done, all the quantitative studies, with their careful calculations of what happens to easily quantified values, have to be weighed by decision makers against qualitative arguments that are not similarly bolstered by impressive numbers. Nowadays the idea is gaining ground of actually polling people to see what kind of value they would assign to such things as a given improvement in air quality or the preservation of a park or neighborhood. In economic theory, people's preferences are often treated as well defined, fixed, and given. That is a point of view in harmony with democratic ideals. But is the fate of the planet just a matter of untutored opinion? Doesn't science have some insights to offer?

Natural science would seem to be particularly relevant when changes are contemplated that are irreversible or nearly so. Does economics as presently formulated pay sufficient attention to irreversibility? In physics, the first law of thermodynamics is the conservation of total energy, and keeping track of energy in physics somewhat resembles the process of keeping track of money in economics. But where is the analogue in economics of the second law of thermodynamics, the tendency of entropy to increase (or remain the same) in a closed system? Entropy helps to define irreversibility in physics, and many thinkers have tried to define a corresponding notion in economics, so far without conspicuous success. Perhaps the quest is not hopeless, however. And perhaps it is worth pursuing, since it might lead to an improvement on the widespread notion that whatever is nearly used up can be replaced by some substitute, such as plastic trees.

Meanwhile, leading economic thinkers have developed concepts that address some of the concerns about following only things that are easily monetized. The notion of "psychic pay" takes account of the fact that people gain satisfaction from, and can be paid in, coin that is intangible, such as pride in helping others. The "cost of information" addresses the fact that people may not know how to make reasonable free market decisions (for instance about purchases) if they don't have the necessary facts or insights. The "social rate of discount" is supposed to deal with the debt between the generations—how steeply a given generation discounts the future is related to how much it is planning to leave to future generations.

However, working economists in business, government, and international agencies may not find it easy to include such advanced concepts in their reports and recommendations. Furthermore, it may be very difficult to quantify some of those concepts even though they have been introduced into the theory.

In both theory and practice, then, there seems to be some room for improvement in the way economics addresses questions of fragile values, especially in cases where those values are in danger of disappearing irreversibly. Any improvements that are made can be particularly valuable in connection with the preservation of biological and cultural diversity.

PART IV

DIVERSITY AND SUSTAINABILITY

CHAPTER

21

DIVERSITIES UNDER THREAT

We have examined how simple rules, including an orderly initial condition, together with the operation of chance, have produced the wonderful complexities of the universe. We have seen how, when complex adaptive systems establish themselves, they operate through the cycle of variable schemata, accidental circumstances, phenotypic consequences, and feedback of selection pressures to the competition among schemata. They tend to explore a huge space of possibilities, with openings to higher levels of complexity and to the generation of new types of complex adaptive system. Over long periods of time, they distill out of their experience remarkable amounts of information, characterized by both complexity and depth.

The information stored in such a system at any one time includes contributions from its entire history. That is true of biological evolution, which has been going on for four billion years or so, and also of the cultural evolution of *Homo sapiens sapiens*, for which the time span is more like a hundred thousand years. In this chapter we take up some of the problems and dilemmas encountered in trying to preserve, at least in great part, the diversity that those two kinds of evolution have produced.

In contrast to the previous chapters, the emphasis here will be more on actions and policies than on knowledge and understanding for their own sake. Likewise, the voice will be as much that of the advocate as of the scholar. In the next chapter we move on to the broad context within which a sustainable and desirable future could be sought, and how that context might be studied.

While much of our discussion will focus on science and scholarship and the role of experts, we must bear in mind that in the long run attempts to impose solutions on human societies from above often have destructive consequences. Only through education, participation, a measure of consensus, and the widespread perception by individual people that they have a personal stake in the outcome can lasting and satisfying change be accomplished.

The Conservation of Biological Diversity

We have mentioned the importance of conveying to everyone (for instance by means of computer simulations) a feeling for how a single ancestor could give rise, through transmission errors and genetic recombination accompanied by natural selection, to the effective complexity represented by the astonishing diversity of life forms in existence today. Those life forms contain an extraordinary amount of information, accumulated over geologic time, about ways to live on the planet Earth and ways for different life forms to relate to one another. How little of that information has been gathered so far by human beings!

Yet humans, through procreation combined with a high environmental impact per person (especially per rich person), have started to produce an episode of extinction that might eventually compare in destructiveness with some of the great extinctions of the past. Does it make any sense to destroy in the course of a few decades a significant fraction of the complexity that evolution has built up over such a long period?

Are we humans going to behave like some other animals, filling up every available nook and cranny in response to a biological imperative, until our population is limited by famine, disease, and conflict? Or are we going to make use of the intelligence that, we like to boast, distinguishes our species from the others?

The conservation of biological diversity is one of the most important tasks facing humanity as the twentieth century nears its end. The enterprise involves people in many walks of life and in various parts of the world, using diverse methods to decide what needs to be done and especially what needs to be done first. Although the choice of how to assign priorities will vary from place to place, there are some principles and practices that may be widely applicable.

The Importance of the Tropics

It seems that the greatest need for conservation efforts (especially on land) is in the tropics, where there is the greatest species diversity and also the greatest pressure to utilize natural resources to meet the needs of a poor and rapidly growing human population. This conjunction—more to lose and more danger of loss—makes biological conservation in the tropics especially urgent.

The tropics are different from the temperate world not only in the number of species now threatened, but also in how much is known about them. In temperate latitudes it is generally possible to define conservation needs by looking at individual species (at least for the "higher" plants and animals) and determining which ones are in difficulty on a local, national, or world level. When biomes (ecological communities) are considered, as they should be, they can be defined as associations of known species.

In the tropics, numerous species are still unknown to science and some whole biomes remain underexplored. Under those conditions, it is impractical, as a general rule, to state the aims of conservation in terms of species. Instead, one usually has to concentrate on saving representative systems in which the individual species are represented, and the definition of those systems is not always easy.

The Role of Science

Science plays a crucial role in tropical conservation. That is especially clear when we remember that the aim of science is not just to accumulate facts but to promote understanding by finding structure (that is, regularities) in the information and also, where possible, mechanisms (dynamical explanations) for phenomena.

A whole spectrum of approaches is available for gathering, organizing, and interpreting data about the status of natural communities throughout the tropics. Systematic biologists (those who study the classification and distribution of plants and animals) tend to favor long-term research, which may take place over many decades and produce knowledge that will be important for a long time to come. At the other end of the scale are techniques such as satellite imagery and aerial photography, which yield immediately some crude indications of differences in ground cover. To understand what those differences mean, one needs to establish "ground truth," which can be more or less detailed, but typically involves expeditions and a good deal of taxonomic work. Such efforts are situated in the middle of the spectrum, between long-term studies on the ground and rapid surveys from air or space.

There is no longer any serious disagreement that a large-scale, man-made episode of extinction has begun in the tropics. To some, it is self-evident that we should not wantonly destroy the product of billions of years of evolution. Others need additional reasons to protect what is in danger of being lost. Those reasons include the potential utility to human beings of species that we are exterminating before we even know they exist, to say nothing of the value to future generations of understanding the operation of complex ecosystems in a comparatively undegraded condition. One of the important tasks for scientists is to explain those arguments in detail. Science can provide not only guidance for setting priorities in conservation but also an understandable rationale for those priorities.

In other words, preservation of biological diversity requires more scientific knowledge so that conservationists have a good idea how to proceed and also so that they can demonstrate that what they are doing makes sense. Accurate, well-marshaled information is a powerful tool that can help mobilize the broader social will needed to protect viable examples of the various ecological communities. In this endeavor, I would venture to guess, it is important to use and develop the discipline of biogeography.

Biogeography is the study of the distributions of plants and animals and how those distributions evolved, taking into account the influence of geology and topography. It is concerned with processes of variation, dispersal, survival, and extinction, including developments over past time as well as ongoing processes that determine the limits of distributions for the various organisms today. Biogeography, in close association with both systematics and ecology, can provide a body of theory that helps to organize the data on occurrences of plant and animal species. It may assist in providing a classification of biomes and it can be of great utility in planning the configuration of a viable system of protected areas and in identifying gaps in existing systems.

Rapid Assessment

From the point of view of science, it is essential to maintain the long-term research that does not provide quick results but can give lasting ones. Obviously, however, conservation action cannot always wait for those results. By the time field biologists have completed a careful, thorough study of flora and fauna in a particular area of the tropics, it may be too late to recommend the preservation of natural communities in all or part of that area, because those communities may no longer exist.

Pursuing the whole spectrum of scientific activities essential for conservation requires taking creative advantage of all potential resources. In particular, a few individual field biologists (botanists, ornithologists, and herpetologists, for example) have learned from their personal training, field experience, and scientific knowledge how to take a quick, rough census of the species in their fields of study present in an area of a given tropical region. They have acquired an idea of the composition of various biomes and they have also developed rapid methods of determining the degree of degradation of an environment. Their knowledge and their wisdom can and should be utilized in conservation work. By estimating the biological diversity of a particular area, as well as the state of preservation of its natural communities, and by helping to determine which biomes are restricted to small regions and which ones are gravely threatened, they can give immensely valuable advice to those who set priorities for protection. The same field biologists can also contribute greatly to the success of short-term expeditions that provide ground truth for aerial and satellite photography, as well as to the success of long-term studies in systematic biology and biogeography. It is particularly important to train more scientists like them, especially among nationals of the tropical countries themselves.

Through the John D. and Catherine T. MacArthur Foundation, of which I am a director, I helped set up the Rapid Assessment Program under the auspices of Conservation International. A core group was assembled, consisting of an ornithologist, a mammalogist, and two botanists. In association with other field biologists, they formed teams to explore particular places (so far mostly in the Americas). The teams have by now examined areas of many different kinds, including dry forest, montane cloud forest, and lowland rain forest, initially identified by aerial survey, in order to find out whether they possessed enough biological diversity and were sufficiently undisturbed to warrant protection.

In 1989 I participated in one of those aerial surveys along with Spencer Beebe, then an official of Conservation International, and Ted Parker, the program's ornithologist. We found a remarkably large and well-preserved area of forest in Bolivia, the Alto Madidi, and identified it as an early target for the program. The terrain stretches from lowland Amazonian rain forest (drained by tributaries of the Amazon, although many hundreds of miles distant from the great river itself) to high mountain forest of several kinds. Later on, the team visited the region and studied it on the ground, finding it even more striking in diversity and quality than we had guessed while looking at it from the air. Now the Bolivian Academy of Sciences and the Bolivian government are considering the possibility of extending protection to the Alto Madidi.

Walking through South American forests with Ted Parker, I found myself agreeing with the superlative opinions I had heard expressed about him. Of all the highly skilled field ornithologists I have accompanied, he was the most impressive. He knew by heart and could recognize the songs and call notes of more than three thousand New World bird species. For days on end, he would identify every forest sound as made by a frog, an insect, or a particular species of bird. When we recorded the birds and called them in by playing their songs back to them, his identifications would always be proved correct. But then, one day, he might exclaim, on hearing a faint "Psst" from the underbrush, "I don't know what that is!" Sure enough, it would be a new bird for the area or the country, or even, very occasionally, a species new to science.

Listening at dawn, he could estimate, from the calls and songs he heard, both the ornithological diversity and the quality of the habitat. His colleagues in mammalogy (Louise Emmons) and botany (Alwyn Gentry and Robin Foster) could perform comparable feats in their specialties.

Recently, tragedy struck this outstanding team. Ted and Alwyn were killed, along with an Ecuadorian colleague, Eduardo Aspiazu, when their plane crashed during an aerial survey. The pilot was killed too. The biologists, as usual, were urging the pilot to fly lower so that they could inspect the forest carefully from the air. (They were looking for a small stretch of remaining dry forest near Guayaquil that might be protected before it was all gone.) Suddenly the aircraft entered a cloud, visibility was lost, and they collided with a nearby mountain.

While mourning the loss of our friends, who seemed almost indispensable, those of us involved in tropical conservation hope that the work of the Rapid Assessment Program will somehow continue. We hope that their places will be taken by other specialized field biologists, nearly as skilled, and that new ones will be trained, especially citizens of tropical countries.

In general, the future of the preservation of ecological diversity in the tropics depends to a great extent on the activities of the growing body of scientists and conservationists from the tropical countries themselves. By and large, major conservation decisions will be made on the national level, and an increasing number of citizens' organizations in the various countries are providing leadership in the protection of biological diversity. Internationally known scientists from temperate countries can sometimes exert a useful influence, but conservation will not happen without local and national support.

Participation of Local People

In fact, conservation needs the support both of influential individuals, many of them in large cities, to get projects started and of local rural populations to maintain nature reserves over time. Long-term protection of large areas cannot succeed unless it is regarded with favor by the local people. This means emphasis on the contributions of conservation to aspects of rural development. For instance, agriculture often depends on the protection of watersheds, and the long-term availability of forest products for use and sale often requires the maintenance of nearby protected forest. Local people must have an economic stake in conservation, and they need to understand that stake. Often they can be directly involved with protected areas, for example through nature tourism or opportunities to serve as guides or rangers in national parks.

It is particularly important to involve local indigenous people, such as the American Indians of the neotropics. In many cases, their cultural continuity and even their physical existence are more threatened than are the plants and animals of the areas where they live. Their knowledge about their environment, accumulated over many centuries, can help to identify human uses for native organisms, as well as methods of earning a living without destroying the ambient ecological communities. In some cases, indigenous peoples have taken the lead in conservation efforts, for example, the Kuna of Panama, who have made a park out of a large fraction of their mainland territory. (Many of the Kuna live on the San Blas Islands, where they are well known as the makers of the colorful *molas*, often used to decorate dresses and handbags.)

The struggle for survival of organisms in tropical forests leads to chemical arms races and other processes that generate chemical substances with potent biological effects, many of them useful to human beings, especially in medicine. Such chemicals are being sought by two different means. One method, ethnobotany, exploits the knowledge of indigenous peoples, obtained by trial and error over hundreds or thousands of years, and thus makes use of cultural evolution as well as the biological evolution that produced the chemicals in the first place. The other method is direct chemical prospecting, in which specimens of plants and animals (insects, for example) are brought from the forest to the laboratory, where new chemicals are isolated using modern methods of extraction. Here, the results of biological evolution are exploited without the helpful intervention of indigenous cultures. Both methods aim to find at least a few chemicals that will finally be utilized, say by drug manufacturers, often in developed countries. Even when such chemicals are used in modified or synthetic form, ways must be found for a significant fraction of the profits to be returned to the people of the forest or the surrounding areas. Only then can the process of exploration and utilization give those local people an additional stake in the preservation of the forest. The same is true of the many schemes for marketing other nontimber forest products, such as nuts and succulent tropical fruits. As usual, incentives create selection pressures on the schemata for human behavior.

A Spectrum of Conservation Practices

The collection of certain nontimber forest products (such as those that require hunting) can be carried out, like the harvesting of timber itself, only in areas that are at best partially protected. One scheme that has been widely adopted and endorsed by the United Nations is the creation of biosphere reserves. A typical biosphere reserve has a core area, often a wild watershed, that is fully protected, and a surrounding region in which some harvesting practices are permitted but with careful attention to conservation. Still further out, but still within the reserve, there may be areas in which agriculture and other forms of normal economic activity are allowed, but with some restrictions.

Clearly, the establishment of a system of fully protected natural areas, including some in biosphere reserves, is only part of what needs to be done. A wide variety of conservation practices is required outside of those areas. These include reforestation (with native species wherever that is practical); implementing wise energy and water policies; coping with the environmental effects of agriculture, mining, and manufacturing; and attending to the all-important matter of population growth. It is highly desirable, moreover, to develop integrated national and regional conservation strategies.

Many aspects of conservation in this broad sense require financial expenditures that the poorer tropical countries cannot afford by themselves. For the developed nations of the temperate zone to assume a large part of the burden is in their long-term, enlightened self-interest. All of us on the surface of this planet will be much worse off if the biological riches of the tropics continue to be wasted. Whenever resources are transferred from the developed countries, whether through gifts, loans, or partial forgiveness of debt, a sizable fraction should be earmarked for conservation in the broad sense. An agreement to practice conservation in exchange for aid is part of what has sometimes been called the "planetary bargain." In recent years, a number of "debt swaps" have been carried out, in which debts owed by a tropical country, deeply discounted on the world financial market, are bought up by conservation organizations and then recognized at face value by the government of the country concerned for use in buying up land for protected areas. (The same principle can be applied to other desirable objectives, such as economic development of a less developed country or higher education abroad for its nationals.) Debt swaps are excellent examples of the planetary bargain in operation.

If one were to stand back and estimate the prospects for a successful, comprehensive program of conservation of biological diversity in the tropics, the results might not be encouraging. However, history shows clearly that humanity is moved forward not by people who stop every little while to try to gauge the ultimate success or failure of their ventures, but by those who think deeply about what is right and then put all their energy into doing it.

The Preservation of Cultural Diversity

Just as it is crazy to squander in a few decades much of the rich biological diversity that has evolved over billions of years, so is it equally crazy to permit the disappearance of much of human cultural diversity, which has evolved in a somewhat analogous way over many tens of thousands of years. Yet human unity (as well as solidarity with the other life forms with which we share the biosphere) is now a more important goal than ever before. How can those concerns be reconciled?

I first became aware of the tension between unity and diversity at an early age. When I was a child, I raised with my father the old question of whether humanity could promote universal peace by using only a single world language. He described to me, in reply, how two hundred years ago, in the era of the Enlightenment and the French Revolution, the German thinker Herder, a pioneer of the Romantic Movement as well as a figure of the Enlightenment, wrote about the need to preserve linguistic diversity by saving the endangered Latvian and Lithuanian languages—so archaic, so close to the ancestral Indo-European. With the aid of native writers of that time, such as the Lithuanian poet Donelaitis, the work of conserving those chunks of cultural DNA was accomplished. Now Latvia and Lithuania are once again independent countries, and those tongues saved from extinction two centuries ago are their national languages.

The most challenging problems of cultural conservation involve indigenous peoples, especially those who are sometimes called primitive, largely because of the state of their technology. In many cases, these indigenous peoples are being either physically exterminated by disease and violence or else displaced or dispersed and culturally annihilated. A century ago, in some parts of the western United States, a few people were still shooting "wild Indians" on weekends. That is how Ishi, the last Yahi, lost his family and friends, as recounted by Alfred and Theodora Kroeber. Today, North Americans deplore similar atrocities being committed in other countries. Let us hope that the present desperate situation can be quickly ameliorated so that those peoples have better opportunities to survive and to choose, either to be left more or less alone for the time being or else to undergo an organic kind of modernization, with a degree of cultural continuity and memory of the past.

The rich lore, as well as institutions and ways of life, of indigenous peoples around the world constitute a treasure house of information about the possibilities of human organization and modes of thought. Many of them also possess precious knowledge of how to live as part of a tropical ecological community. (Others, it should be noted, have been destructive of nature, particularly peoples who have lived on previously uninhabited islands, large or small, for less than a millennium or two. In some cases, the notion of indigenous peoples living in harmony with nature turns out to be wishful thinking.)

Imagine, though, the knowledge of the properties of plants in the minds of certain tribal shamans. Many of those witch doctors are now dying without replacement. The great Harvard ethnobotanist, Richard Schultes, who spent many years studying medicinal plants in the Amazon Basin, says that every time such a shaman dies, it is as if a library had burned down. Schultes has trained many younger ethnobotanists, who are engaged in salvaging as many secrets as possible from those libraries before they disappear altogether. One of them, Mark Plotkin, recently published a delightful account of his adventures, under the title *Tales of a Shaman's Apprentice*.

Human beings have distilled, over hundreds or thousands of years of learning by trial and error, a remarkable amount of information about the uses of organisms for food, medicine, and clothing. Sometimes the process of learning must have been quite dramatic, as in the case of the bitter manioc of the Amazonian rain forest. Not many plants grow on the forest floor because so much of the sunlight is captured by the trees of the upper, middle, and lower canopies. Under those conditions, the bitter manioc (the tuber from which tapioca is made) is a valuable resource, edible and nutritious. However, the raw tuber contains a good deal of prussic (hydrocyanic) acid and is therefore highly poisonous. Only when heat is used to break up and drive off the acid is the flesh of the tuber edible. A number of people must have lost their lives as hungry members of the various Amazonian bands and tribes learned how to make use of the bitter manioc.

It is not only in such less developed regions that discovery by trial and error has revealed useful properties of plants and plant preparations. Folk medicine has made a great deal of difference to people's lives all over the planet. Naturally, not all the claims of folk medicine are justified, but modern science has confirmed some of them. An experience of my own father provides an example. When he was a boy, the son of a forester, living in the beech woods of what was then eastern Austria, near the Russian border, he accidentally chopped off the last joint of one of his fingers with an axe. He retrieved the fallen joint, rinsed it off, and replaced it on the finger, which he then wrapped with a poultice made of bread. He bore the circular scar for the rest of his life, but the joint stayed on. It was to be many years before modern science recognized the bacteriostatic properties of the bread mold *Penicillium notatum*, but undoubtedly those same properties saved my father's finger.

In the adaptive process by which groups of people made such useful discoveries, the selection pressures must have involved some questions fairly similar to those that are asked by science. Does the process actually work? Can people eat this food safely? Do wounds heal when wrapped this way? Does this herb help a woman to begin labor when her child is overdue?

The folk remedies stemming from sympathetic magic present a different picture. Among the purported cures based on similarity is one for jaundice (actually a symptom of liver disease) that involves staring into the golden eye of a stone curlew. If my father had tried that, it would hardly have been very useful, except perhaps for a slight psychosomatic effect. In the development of sympathetic magic, so widespread among the peoples of the earth, the selection pressures, as we have emphasized earlier, were mostly very different from those pertaining to objective success.

Yet those peoples did not necessarily draw any sharp distinction between magic on the one hand and the discovery of real uses of plant and animal products on the other. Witch doctors were still witch doctors, even if they did teach modern people to use materials such as cinchona bark, which yields quinine for use against malaria. Cultural traditions are not always easy to dissect into the parts that fit in easily with modern ideas and those that are in conflict with them.

The Tension Between Enlightenment and Cultural Diversity

The tension continues today between our need for the universality envisioned by the Enlightenment and our need for the preservation of cultural diversity. In discussing the future of the planet, using the results of scientific investigation and attempting to employ rational ways of thinking about the implications of those results, we are hampered by the prevalence of superstition. The persistence of erroneous beliefs exacerbates the widespread anachronistic failure to recognize the urgent problems that face humanity on this planet. We are, of course, severely threatened by philosophical disunity and especially by destructive particularism in all its many forms. Such particularism is still manifested in many places in the ancient form of tribalism, but today it may be related to differences in nationality, language, or religion or to other differences, sometimes so small that an outsider can scarcely detect them, but still sufficient to give rise to deadly rivalry and hatred, especially when exploited by unscrupulous leaders.

Yet at the same time, cultural diversity is itself a valuable heritage that should be preserved: that Babel of languages, that patchwork of religious and ethical systems, that panorama of myths, that potpourri of political and social traditions, accompanied as they are by many forms of irrationality and particularism. One of the principal challenges to the human race is to reconcile universalizing factors such as science, technology, rationality, and freedom of thought with particularizing factors such as local traditions and beliefs, as well as simple differences in temperament, occupation, and geography.

Universal Popular Culture

The erosion of local cultural patterns around the world is not, however, entirely or even principally the result of contact with the universalizing effect of scientific enlightenment. Popular culture is in most cases far more effective at erasing distinctions between one place or society and another. Blue jeans, fast food, rock music, and American television serials have been sweeping the world for years. Moreover, universalizing influences cannot be categorized simply as belonging either to scientific or to popular culture. Instead, they form a continuum, a whole spectrum of different cultural impacts.

Occupying an intermediate position between high and popular culture are institutions like Cable News Network. In some places and on some occasions, CNN broadcasts are a valuable, timely source of memorable images and reasonably accurate information not otherwise obtainable. In other situations, they seem to represent a form of entertainment, part of the universalizing popular culture. In any event, news broadcasts received around the world and news articles that appear in daily and weekly publications in many countries are considered to be part of the worldwide "information explosion," along with an astonishing proliferation of other nonfiction periodicals and of books, to say nothing of the rapidly growing electronic mail network and the coming explosion of interactive multimedia communications.

The Information (or Misinformation?) Explosion

Unfortunately, that information explosion is in great part a misinformation explosion. All of us are exposed to huge amounts of material, consisting of data, ideas, and conclusions—much of it wrong or misunderstood or just plain confused. There is a crying need for more intelligent commentary and review.

We must attach a higher prestige to that very creative act, the writing of serious review articles and books that distinguish the reliable from the unreliable and systematize and encapsulate, in the form of reasonably successful theories and other schemata, what does seem reliable. If an academic publishes a novel research result at the frontier of knowledge in science or scholarship, he or she may reap a reward in the form of a professorship or a promotion, even if the result is later shown to be entirely wrong. However, clarifying the meaning of what has already been done (or picking out what is worth learning from what is not) is much less likely to advance an academic career. Humanity will be much better off when the reward structure is altered so that selection pressures on careers favor the sorting out of information as well as its acquisition.

Tolerating the Intolerant—Is It Possible?

But how do we reconcile the critical examination of ideas, including the identification and labeling of error with tolerance—and even celebration and preservation—of cultural diversity? We have discussed how each specific cultural tradition has ideas and beliefs embedded in it as artistic motifs, defining and unifying social forces, and sources of personal comfort in the face of tragedy. As we have emphasized, many of those ideas and beliefs are ones that science would label erroneous (or at least unjustified by evidence), while others represent precious discoveries about the natural world and about possible forms of human individual and social development (including, perhaps, the exploration of new realms of mystical experience and the formulation of value systems that subordinate the appetite for material goods to more spiritual appetites). The preservation of cultural diversity, however, must somehow transcend that distinction. The patterns or schemata that are elements of cultural DNA cannot readily be divided into those that are worth preserving and those that are not.

Yet the difficulty goes far deeper. Many of the local patterns of thought and behavior are associated not only with harmful error and destructive particularism but specifically with harassment and persecution of those who espouse the universalizing scientific and secular culture, with its emphasis on rationality and the rights of the human individual. And yet it is within that very culture that one often finds people concerned, as a matter of principle, with the preservation of cultural diversity.

Somehow the human race has to find ways to respect and make use of the great variety of cultural traditions and still resist the threats of disunity, oppression, and obscurantism that some of those traditions present from time to time.

CHAPTER 22

TRANSITIONS TO A MORE SUSTAINABLE WORLD

Concern for the preservation of biological diversity is inseparable from concern about the future of the biosphere as a whole, but the fate of the biosphere is in turn closely linked with virtually every aspect of the human future. I intend to describe here a kind of research agenda on the future of the human race and the rest of the biosphere. That agenda does not call, however, for open-ended forecasting. Instead, it calls for people from a great many institutions and a wide variety of disciplines to think together about whether there may be evolutionary scenarios that lead from the present situation toward a more nearly sustainable world during the twenty-first century. Such an approach is more focused than simple speculation about what might happen in the future.

Why should anyone try to think on such a grand scale? Shouldn't one plan a more manageable project that concentrates on a particular aspect of the world situation?

We live in an age of increasing specialization, and for good reason. Humanity keeps learning more about each field of study; and as every specialty grows, it tends to split into subspecialties. That process happens over and over again, and it is necessary and desirable. However, there is also a growing need for specialization to be supplemented by integration. The reason is that no complex, nonlinear system can be adequately described by dividing it up into subsystems or into various aspects, defined beforehand. If those subsystems or those aspects, all in strong interaction with one another, are studied separately, even with great care, the results, when put together, do not give a useful picture of the whole. In that sense, there is profound truth in the old adage, "The whole is more than the sum of its parts."

People must therefore get away from the idea that serious work is restricted to beating to death a well-defined problem in a narrow discipline, while broadly integrative thinking is relegated to cocktail parties. In academic life, in bureaucracies, and elsewhere, the task of integration is insufficiently respected. Yet anyone at the top of an organization, a president or a prime minister or a CEO, has to make decisions *as if* all aspects of a situation, along with the interaction among those aspects, were being taken into account. Is it reasonable for the leader, reaching down into the organization for help, to encounter only specialists and for integrative thinking to take place only when he or she makes the final intuitive judgments?

At the Santa Fe Institute, where scientists, scholars, and other thinkers from all over the world, representing virtually all disciplines, meet to do research on complex systems and on how complexity arises from simple underlying laws, people are found who have the courage to take a *crude look at the whole* in addition to studying the behavior of parts of a system in the traditional way. Perhaps the Institute can help to spark collaborative research, by institutions from around the globe dedicated to the study of particular aspects of the world situation, on potential paths toward a more nearly sustainable world. The aspects in question will have to include political, military, diplomatic, economic, social, ideological, demographic, and environmental issues. A comparatively modest effort has already begun, under the name of Project 2050, under the leadership of the World Resources Institute, the Brookings Institution, and the Santa Fe Institute, with participation of people and institutions from many parts of the world.

Now what is meant here by sustainable? In *Through the Looking Glass*, Humpty Dumpty explains to Alice how he uses words to mean anything he wants, paying them for the privilege each Saturday night (the end of the nineteenth-century work week). These days a great many people must be paying wages to the word "sustainable." For example, if the World Bank finances some old-fashioned massive development project destructive of the environment, that project may well be labeled "sustainable development" in the hope of making it more acceptable.

This practice reminds me of the Monty Python routine in which a man enters an office to get a license for his fish, Eric. Told that there is no such thing as a fish license, he points out that he had received the same reply when he asked about cat licenses, but that he has one anyway. Producing it, he is told, "That's not a cat license. That's a dog license with the word 'dog' crossed out and the word 'cat' written in with a pencil."

Today many people are busy writing in the word "sustainable" in pencil. The definition is not always clear. Thus it is not unreasonable to try to *assign* a meaning here. The literal signification of the word is evidently not adequate. The complete absence of life on Earth might be sustainable for hundreds of millions of years, but that is not what is meant. Universal tyranny might be sustainable for generations, but we do not mean that either. Imagine a very crowded and highly regimented, perhaps extremely violent world with only a few species of plants and animals surviving (those with intimate connections with human society). Even if such conditions could somehow be kept going, they would not correspond to what is meant here by a sustainable world. Clearly, what we are after embraces a modicum of desirability along with sustainability. Remarkably, there is a certain measure of theoretical agreement today on what is desirable, on the aspirations of the human race, as embodied, for example, in declarations of the United Nations.

What kind of future, then, are we envisaging for our planet and our species when we speak of sustainability, tempering our desires with some dose of realism? Surely we do not mean stagnation, with no hope of improvement in the lives of hungry or oppressed human beings. But neither do we mean continued and growing abuse of the environment as population increases, as the poor try to raise their standard of living, and as the wealthy exert an enormous per capita environmental impact. Moreover, sustainability does not refer to environmental and economic concerns alone.

In negative terms, the human race needs to avoid catastrophic war, widespread tyranny, and the continued prevalence of extreme poverty, as well as disastrous degradation of the biosphere and destruction of biological and ecological diversity. The key concept is the achievement of quality of human life and of the state of the biosphere that is not purchased mainly at the expense of the future. It encompasses survival of a measure of human cultural diversity and also of many of the organisms with which we share the planet, as well as the ecological communities that they form.

Some people may be technological optimists, believing that we humans do not need to change course very much in order to avoid a disastrous future, that we can achieve approximate sustainability without special effort, merely through an endless series of technological fixes. Some may not believe in the goal of sustainability at all. Nevertheless, we can all think about it. Even those of us who do not accept sustainability as a goal can still ask whether there are ways to approach it during the next fifty to a hundred years and if so, what those ways might be and what the world might look like as a result. Discussion of the questions does not require sharing the values of those who posed them.

Historians tend to be impatient with people who say, "This is a unique period in history," because that claim has been made about so many eras. Still, our time is special in two well-defined and closely related ways.

First, the human race has attained the technical capability to alter the biosphere through effects of order one. War is old, but the scale on which it can now be fought is entirely new. It is notorious that a full-scale thermonuclear war could wipe out a significant fraction of life on the planet, not to mention the trouble that could be caused by biological or chemical warfare. Moreover, through population growth and certain economic activities, humans are altering the global climate and exterminating significant numbers of plant and animal species. Actually, human beings caused more destruction in the past than is usually admitted. Deforestation by the axe and by goats and sheep, followed by erosion and desiccation, is thousands of years old and was remarked, for example, by Pliny the Elder. Even the tiny numbers of people living in North America ten thousand years ago may have contributed to the extinction of the North American ice-age megafauna, such as mammoths and giant sloths, dire wolves, sabre-toothed cats, and species of camels and horses. (One theory blames some of the extinctions at least partially on the habit of driving whole herds of animals over cliffs in order to use the meat and skins of just a few.)

Nevertheless, today the potential for damage to the entire biosphere is much greater than ever before. Human activity has already created a multiplicity of environmental problems, including climate change, ocean pollution, diminishing quality of fresh water, deforestation, soil erosion, and so on, with strong interactions among them. As with conflict, many of the environmental ills are old ones, but their scale is unprecedented.

Second, the rising curves of world population and natural resource depletion cannot go on rising steeply forever; they must soon pass through inflection points (when the rate of increase starts to decrease). The twenty-first century is a crucial time (in the original sense of a crossroad) for the human race and the planet. For many centuries, total human population as a function of time hewed closely to a simple hyperbolic curve that reaches infinity in about the year 2025. Ours is obviously the generation in which world population must start to peel away from that hyperbola, and it has already begun to do so. But will the population curve flatten out as a result of human foresight and progress toward a sustainable world, or will it turn over and fluctuate as a result of the traditional scourges of war, famine, and pestilence? If the curves of population and resource depletion do flatten out, will they do so at levels that permit a reasonable quality of human life, including a measure of freedom, and the persistence of a large amount of biological diversity, or at levels that correspond to a gray world of scarcity, pollution, and regimentation, with plants and animals restricted to a few species that co-exist easily with mankind?

A similar question can be posed about the progressive development of the means and scale of military competition. Will people allow large-scale, thoroughly destructive wars to break out, or will they use intelligence and foresight to limit and redirect competition, to damp down conflict, and to balance competition with cooperation? Will we learn, or have we perhaps already learned, to manage our differences in ways short of catastrophic war? And what of smaller conflicts arising from political disintegration?

Gus Speth, who was the first president of the World Resources Institute (which I am proud to have played a role in founding), has suggested that the challenge to the human race over the next few decades is to accomplish a set of interlinked transitions. I propose to amplify slightly his conception of those transitions so as to incorporate more political, military, and diplomatic considerations in addition to the social, economic, and environmental ones that he emphasizes. With those modifications, the rest of this chapter is organized around that crude but useful notion of a set of transitions.

The Demographic Transition

We have seen that the coming decades must witness a historic change in the curve of world population versus time. Most authorities estimate that world population will level off during the next century, but at a figure something like twice the present number of 5.5 billion or so. Today, high rates of population growth (associated particularly with improvements in medicine and public health without corresponding declines in fertility) still prevail in many parts of the world. That is especially true of tropical, less developed regions, including countries, such as Kenya, that can least afford it ecologically or economically. Meanwhile, the developed countries have generally achieved rather stable populations, except for the effects of migration, which will certainly be a major issue in the coming decades.

Scholars have engaged in much discussion of the factors thought to be responsible for the decline in net fertility that has taken place in most of the developed countries. They now suggest measures that may help to produce similar declines in various parts of the tropical world. Those measures include improved provisions for women's health, literacy, further education, and opportunities to participate in the work force, as well as other advancements in the position of women; reduced infant mortality (which initially works in the opposite direction, of course, but may later prevent couples from compensating for expected deaths by producing more children than they really want); and social insurance for the elderly, still a distant goal in many developing countries.

Naturally the availability of safe and effective contraception is crucial, but so is the erosion of traditional incentives for having large families. In some parts of the world the average couple (and especially the average male) still wants to have many children. What kinds of rewards can be offered to one- and two-child families? How can people be persuaded, in culturally appropriate ways, that in the modern world such families are in the common interest, with higher levels of health, education, prosperity, and quality of life than would be possible for families with many children? With swings of fashion having such importance in human affairs, what can be done to help the idea of small families to become popular? These questions are still sadly neglected in many places, even by organizations that claim to be helping to solve the world population problem.

If human population is really going through an inflection point and will level off, globally and in most places, in a few decades, not only is that a historical process of the greatest significance, but its timing and the resulting numbers are likely to be of critical importance as well. The exact character and magnitude of the effect of population growth on environmental quality depend on many variables, such as patterns of land tenure, and are worth careful study in various different areas. Nevertheless it already seems overwhelmingly probable that, on the whole, population growth encourages environmental degradation, whether through the huge consumption rates of the wealthy or through the desperate struggle of the poor to survive at whatever cost to the future.

The environmental consequences are likely to be much more serious if the world simply waits for improved economic conditions among impoverished populations to effect reductions in net fertility, as opposed to trying to encourage such reductions in parallel with economic development. The total environmental impact per person is likely to be considerably greater after economic improvement than before, and the fewer the numbers when relative prosperity is finally achieved, the better for the people and for the rest of the biosphere.

The Technological Transition

Decades ago some of us (particularly Paul Ehrlich and John Holdren) pointed out the fairly obvious fact that environmental impact, say in a given geographical area, can be usefully factored into three numbers multiplied together: population, conventionally measured prosperity per person, and environmental impact per person per unit of conventional prosperity. The last factor is the one that particularly depends on technology. It is technological change that has permitted today's giant human population to exist at all, and while billions of people are desperately poor, quite a few others manage to live in reasonable comfort as a consequence of advances in science and technology, including medicine. The environmental costs have been huge, but nowhere near as great as they may be in the future if the human race does not exercise some foresight.

Technology, if properly harnessed, can work to make the third factor as small as can be practically arranged, given the laws of nature. How much the prosperity factor can be improved, especially for the very poor, depends to a considerable extent on how much is squandered on the first factor, mere numbers of people.

Evidence of the beginning of the technological transition is starting to show up in many places, even though the bulk of it is yet to occur. Even apparently simply technological fixes, however, can end up posing extremely complex problems.

Consider the example of eradicating malaria in human populations. Not too long ago, the draining of swamps was still the principal method of control. But now it is understood that the destruction of wetlands is to be avoided whenever possible. Meanwhile, science has identified the plasmodia responsible for malaria and the mosquito vectors that carry them. Spraying with chemical pesticides like DDT to eliminate the mosquitos seemed to be a step forward, but turned out to have serious environmental consequences. For one thing, birds at the top of the aquatic food chain got very concentrated doses of the metabolic product DDE, which caused thinning of egg shells and reproductive failure in many species, including the American national bird, the bald eagle. Twenty years ago, DDT was phased out in the developed world, and the threatened bird populations started to recover. It is still used elsewhere, although resistant strains of the mosquito vectors are starting to appear.

It then turned out that some of the immediately available replacements for DDT were fairly dangerous to humans. Nowadays, however, much more sophisticated methods are available for reducing the populations of the vectors, including the use of chemicals that specifically target them, as well as the release of sterile mating partners and other "bio-environmental controls." Such measures can be coordinated in what is called "integrated pest management." So far, they are still fairly expensive, if deployed on a large scale. In the future, cheaper and equally gentle techniques may be developed. Insect repellents are also available, of course, but they are expensive too and cause problems of their own.

Meanwhile, a simple, behavioral approach that is effective in many places is to use mosquito netting and stay under it for half an hour at dawn and half an hour at dusk, when the vector mosquitoes are biting. Unfortunately, in many tropical countries, the rural poor are very busy outdoors at those times and cannot stay under netting.

Some day antimalarial vaccines will probably be developed, which may even wipe out the various forms of the disease entirely, but then another difficulty will arise: important wild areas that had been protected by the dangers of malaria will be exposed to unwise development.

I have no doubt spent too long on this apparently simple example, in order to expose some of its complexities. Analogous complexities can be expected to crop up anywhere in the technological transition to lower environmental impact, whether in industrial production, the extraction of minerals, food production, or energy generation.

Like the conversion from defense industries to civilian production, the technological transition requires financial assistance and retraining for workers as opportunities close down in one kind of employment and develop in another. Policy makers may be well advised to consider these different types of conversion as posing related challenges. Thus, ceasing to manufacture chemical warfare agents would be regarded as similar to phasing out logging in the old growth forests of the Pacific Northwest of the United States. Moreover, such policy issues come up again when society tries to reduce the consumption of products injurious to human health, whether legal, like tobacco, or illegal, like crack cocaine.

However, on the demand side the three kinds of conversion present somewhat different problems. In the case of chemical weapons, the principal challenge was to persuade governments not to order them any more and to ferret out and destroy the stocks that exist. In the case of drugs, the issues are matters of angry dispute. In the case of the technological transition to lower environmental impact, the question is what the incentives are to develop gentler technologies and to use them. That brings us to the economic transition.

The Economic Transition

If the air or water is treated as a free good in economic transactions, then polluting it, using up its quality, costs nothing; the associated economic activity is carried on by stealing from the environment and from the future. Authorities have attempted for centuries to deal with such problems by means of prohibitions and fines, but those were often ineffective. Today regulation is being attempted on a massive scale in some places, and some successes have been achieved. However, it seems that the most efficient way for governments to deal with such issues is to charge, more or less, for the cost of restoring quality. That is what economists call internalizing externalities. Regulation, with its fines and other punishments, is itself a form of charging. Regulators, however, usually require specific actions by polluters, whereas internalizing costs encourages restoring quality, or avoiding its degradation in the first place, by whatever means is cheapest. The engineers and accountants of the industry concerned are the ones who prescribe the measures to be taken. Micromanagement by bureaucrats is unnecessary.

Attempting to charge real costs is a principal element of the required economic transition from living in large part on nature's capital to living mainly on nature's income. While charging is usually better than regulation, it is certainly much better than mere exhortation. For one thing, it reduces ambiguities.

Suppose you are engaged in awarding green medallions to products with low environmental impact. Soon you encounter a problem. A particular detergent may be lower in phosphates than another and thus produce less eutrophication (growth of algae) in lakes, but it may require greater energy use because it needs hotter water in the wash. As you go on you find more such tradeoffs. How do you balance one consideration against another? If at least a crude attempt is made to charge producers for eutrophication caused by their detergents and if the cost of the energy needed for a wash is clearly marked on the package, a consumer can just use total expenditure to make decisions, and the market will work out the prices. The green medallion may become unnecessary.

The great difficulty in charging true costs, of course, is estimating them. We discussed earlier how economics has never really succeeded in coming to grips with subtle problems of quality and irreversibility, issues analogous to those that arise in connection with the second law of thermodynamics in natural science. Such problems can, of course, be shoved over into the political arena and treated as matters of public opinion only, but surely in the long run science will have something to say about them too. Meanwhile, the simplest approach is to estimate the cost of restoring whatever is lost. In the case of the irreplaceable, some form of strictly enforced prohibition may be necessary, but otherwise the sustainability of quality is closely tied to the idea of paying to restore it, and the definition of quality will be dealt with by science and public opinion in interaction with each other.

A critical part of any program to charge true costs is the elimination of subsidies for destructive economic activity, much of which would not be economic at all were it not for those subsidies. In the work of the World Commission on Environment and Development (the Brundtland Commission), composed of distinguished statesmen from many parts of the world, it took the brilliant Secretary-General of the Commission, Jim MacNeill of Canada, to point out that in order to see what is happening to the environment, one must look not so much at the activities of the Environment Ministry as at the Ministry of Finance and the budget. It is there that the destructive subsidies can be hunted down and sometimes, albeit with great political difficulties, killed.

Discussion of budgets leads directly to the question of whether national accounting procedures include the depletion of nature's capital. Usually they do not. If the president of a tropical country contracts with a foreign lumber company to have a large chunk of the nation's forests cut down for a low price and a bribe, the national accounts show the price as part of the national income, and maybe even the bribe as well if it is spent at home and not sent to a Swiss bank, but the disappearance of the forest, with all its benefits and potential, does not appear as a corresponding loss. Nor is it only tropical countries that sell their forests too cheaply, as attested by the fate of the temperate rain forests of the U.S. Pacific Northwest, British Columbia, and Alaska. Clearly the reform of national accounting systems is a major need in all countries. Fortunately, efforts to accomplish that reform are already being undertaken in some places. Our example also makes clear that the struggle against major corruption is a key element in achieving the economic transition.

Another indicator of the level of concern over living on nature's capital is the discount rate. I understand that the World Bank, in financing projects with large environmental impacts, still applies a discount rate of 10 percent per year to the future. If that is true, it means that the loss of some great natural asset thirty years in the future is discounted by a factor of 20. The natural heritage of the next generation is valued at 5 percent of its assigned value today, if indeed it is counted at all.

The discount rate, used in this way, is a measure of what is called intergenerational equity, which is crucial to the notion of sustainable quality. Discounting the future too steeply amounts to robbing the future. If the notion of discount rate is generalized somewhat, it can be used to encapsulate much of what is meant by sustainability.

The Social Transition

Some economists make much of possible tradeoffs between intergenerational equity and intragenerational equity, that is, between concern for the future and concern for today's poor, who need to exploit some resources in order to survive. Although some of the degradation of the biosphere today is caused by the very poor scrabbling for a living, much of it can be attributed to the wealthy squandering resources on frills. A great deal of it, however, is connected with massive projects that are supposed to help, for example, the rural poor of a developing country, but often do so, if at all, rather inefficiently and destructively. In contrast, the same people can often be aided very effectively through large numbers of small efforts, applied locally, as for example in the practice known as microlending.

In microlending, a financial institution is established to provide very small loans to local entrepreneurs, many of them women, to start small enterprises that provide a living locally to a number of people. Frequently such businesses provide comparatively nondestructive employment and contribute to intergenerational as well as intragenerational equity. Fortunately, microlending to support sustainable economic activity is becoming more widespread.

It is hard to see how quality of life can be sustainable in the long run if it is very inequitably shared, if there are large numbers of people starving, lacking shelter, or dying young of disease when they can see a more comfortable existence attained by billions of other people. Clearly, large-scale moves in the direction of intragenerational equity are needed for sustainability. As in the case of microlending for sustainable development, there is often more synergy than conflict between intergenerational and intragenerational equity. Policies that really help the rural poor in developing countries are much more compatible with those that preserve nature than is often claimed. Policies that truly benefit the urban poor certainly include provisions for avoiding urban environmental catastrophes. Such policies also include measures to resolve the problems in the countryside that are producing large-scale migrations to the cities, many of which are already swollen to such proportions as to be almost unmanageable. In fact, the social transition clearly must include the alleviation of some of the worst problems of the megacities.

Today, even more than in the past, no nation can deal with problems affecting either urban or rural economic activity without taking account of international issues. The emergence of the global economy is a dominant feature of the contemporary scene, and the desire to participate more actively in that economy is a major force affecting the policies of governments and businesses around the world. Together with rapid transport, global communications, and global environmental effects, the prominence of global economic issues means that a greater degree of worldwide cooperation is essential to deal with the serious and interlocking issues that face the whole human race. That brings us to the institutional or governance transition.

The Institutional Transition

The need for regional and global cooperation is hardly restricted to environmental matters, or even environmental and economic matters. The maintenance of peace, so-called international security, is at least as important.

Recently, with the dissolution of the Soviet Union and the "Soviet bloc" of nations, and with a greater degree of cooperation on the part of China, it has become possible for world institutions, including organs of the United Nations, to function more effectively than in the past. For the U.N. to organize the monitoring of elections or to sponsor negotiations for ending a civil war is now a matter of routine. "Peacekeeping" activities are in progress in many parts of the world. The outcomes are by no means always satisfactory, but at least the processes are becoming established.

Meanwhile, transnational cooperation is taking place in many other ways, and indeed the role of the national state is necessarily weakened in a world where so many important phenomena increasingly transcend national boundaries. In many spheres of human activity, transnational and even universal (or nearly universal) institutions, formal or informal, have been functioning for a long time. Now there are many more. Typically, they channel competition into sustainable patterns and temper it with cooperation. Some are more important or more effective than others, but they are all of some significance. A few diverse examples are the air travel system; the International Postal Union; the Convention on Broadcasting Frequencies; Interpol; migratory bird treaties; CITES (the Convention on International Trade in Endangered Species); the Convention on Chemical Weapons; The International Union of Pure and Applied Physics, The International Council of Scientific Unions, etc.; World Congresses of Mathematics, Astronomy, Anthropology, Psychiatry, etc.; PEN, the international writers' organization; financial institutions such as the World Bank and the International Monetary Fund; multinational corporations, including McDonald's as well as IBM; U.N. Agencies such as WHO, UNEP, UNDP, UNFPA, UNICEF, and UNESCO; and the Red Cross, Red Crescent, Red Shield of David, and Red Sun and Lion. Moreover, the increasing importance of English as an international language should not be ignored.

Gradually, bit by bit, the human race is beginning to come to grips, on a global or highly transnational basis, with some of the problems of managing the biosphere and human activities in it. Here the effect of the changed situation in the former Soviet Union and in Eastern Europe is extremely encouraging. It results in the probability of near-universality for numerous activities for which there was little hope of anything like universality before.

Also, negotiations are going forward on issues of the global commons—those aspects of the environment that are not recognized as belonging to anyone and therefore belong to all, where selfish exploitation without cooperation can only lead to results bad for all parties. Obvious examples are the oceans, space, and Antarctica.

Agreements between more and less developed countries can follow the pattern of the planetary bargain, which we encountered earlier in connection with the conservation of nature. Here it assumes a more general significance: resource transfers from wealthier countries to poorer ones carry an obligation for the poorer ones to take measures that advance sustainability in the broad sense, so that avoiding nuclear proliferation is included along with activities such as protecting wilderness areas. (Another manifestation of the planetary bargain is that electric utilities in temperate countries offset their emissions of carbon dioxide by paying to preserve forests in tropical countries.)

However, the problem of destructive particularism—the sharp and often violent competition among peoples of different language, religion, race, nation, or whatever—has come into even sharper focus than usual in the last few years, especially with the lifting of some of the lids that had been put on these competitions by authoritarian regimes. Dozens of violent ethnic or religious struggles are under way in different parts of the globe. Many different brands of fundamentalism are on the march. The world is experiencing simultaneous trends toward unity and toward fragmentation within that unity.

We have mentioned that seemingly no difference is so small that it cannot be used to divide people into harshly antagonistic groups. Look, for example, at the bitter struggle going on in Somalia. Language difference? No, all speak Somali. Religious difference? Virtually all Muslims. Different sects within Islam? No. Clan differences? Yes, but they are not causing so much trouble. It is mainly *subclans* that are at war with each other, under rival war lords, as legal order has collapsed.

The Ideological Transition

What will happen to these trends? If our long-outdated proclivities toward destructive particularism are excessively indulged, we will have military competitions, breeding competitions, and competitions for resources at levels that will make the sustainability of quality difficult or impossible to achieve. Seemingly a dramatic ideological transition is needed, comprising the transformation of our ways of thinking, our schemata, our paradigms, if we humans are to approach sustainability in our relations with one another, to say nothing of our interactions with the rest of the biosphere.

Scientific research has not yet made clear to what extent human attitudes toward other people who are perceived as different (and toward other organisms) are governed by inherited, hard-wired tendencies developed long ago in the course of biological evolution. It may be that to some degree our propensities to form groups that don't get along with one another and to wreak unnecessary destruction on the environment have such origins. They may be biologically evolved tendencies that were perhaps once adaptive but are so no longer, in a world of interdependence, destructive weapons, and greatly increased capacity to degrade the biosphere. Biological evolution is too slow to keep up with such changes. Still, we know that cultural evolution, which is much more rapid, can modify biological propensities.

Sociobiologists emphasize that we humans, like other animals, inherit a tendency to protect ourselves and our close relatives so that we and they can survive to procreate and pass on some of our genetic patterns. But in human beings that instinct to promote inclusive fitness is profoundly transformed by culture. A sociobiologist, invoking the image of someone jumping into a river to save another person from a crocodile, would argue that such "altruistic" behavior is more likely if the other person is a close relative. A cultural anthropologist might point out that in many tribes certain relatives, including fairly distant relatives, are "classificatory" siblings or parents or offspring, who are treated in many respects as if they really were those close relatives. Perhaps members of such a tribe are just as willing to risk their lives to save their classificatory brothers and sisters as their real ones. In any event, sociobiologists now agree that patterns of altruistic behavior in humans are greatly affected by culture. A certain willingness to risk one's life for another human being can easily extend to all the members of one's tribe.

Such behavior occurs at higher levels of organization as well. On the scale of a nation state, it is known as patriotism. As people have aggregated into larger and larger societies, the concept of "us" has tended to grow in scope. (Unfortunately, stress can reveal lines of weakness in the social fabric that cause it to tear apart again into smaller units. That is what has happened, for example, in the vicinity of Sarajevo, where one resident was quoted as saying: "We have lived next door to those people for forty years, and we have intermarried with them, but now we realize that they are not fully human.") Despite such setbacks, the undeniable trend is toward a more and more inclusive sense of solidarity.

The greatest ideological question is whether, on a short time scale, that sense of solidarity can come to encompass the whole of humanity and also, in some measure, the other organisms of the biosphere and the ecological systems to which we all belong. Can parochial and short-term concerns be accompanied increasingly by concerns that are global and long-term? Can family consciousness undergo a rapid enough cultural evolution to planetary consciousness?

When political unity has been achieved in the past, it has often come about through conquest, sometimes followed by attempts to suppress cultural diversity, because cultural diversity and ethnic competition are two sides of the same coin. To meet the requirement of sustainable quality, however, evolution toward planetary consciousness must accommodate cultural diversity. The human race needs unity in diversity, with the diverse traditions evolving so as to permit cooperation and the accomplishment of the many interlinked transitions to sustainability. Community is essential to human activity, but only communities motivated to work together are likely to be adaptive in the world of the future.

Meanwhile, human cultural diversity has given rise to a multiplicity of ideologies or paradigms, schemata that characterize ways of thinking across the globe. Some of those ways of looking at the world, including particular views of what is the good life, may be especially conducive to sustainable quality. It is desirable that such attitudes become more widespread, even though cultural diversity would suffer through the decline of other attitudes with more destructive consequences. As usual, the preservation of cultural diversity can engender not only paradoxes but conflict with other goals as well.

A few years ago I attended a remarkable lecture given at UCLA by Václav Havel, then president of the soon-to-be-divided Czech and Slovak Federated Republic and now president of the Czech Republic. His topic was the environmental damage to his country during the last decades, with serious effects on human health. He blamed the damage on anthropocentrism, especially the notion that we humans own the planet and have enough wisdom to know what to do with it. He complained that neither greedy capitalists nor dogmatic communists have sufficient respect for the larger system of which we are merely a part. Havel, of course, is a writer and a fighter for human rights as well as a politician. Most ordinary politicians refrain from attacking anthropocentrism, since the voters are all human. But it may indeed be healthy for our species to attribute intrinsic worth to nature and not only perceived utility for a particular kind of primate that calls itself *sapiens.*

The Informational Transition

Coping on local, national, and transnational levels with environmental and demographic issues, social and economic problems, and questions of international security, as well as the strong interactions among all of them, requires a transition in knowledge and understanding and in the dissemination of that knowledge and understanding. We can call it the informational transition. Here natural science, technology, behavioral science, and professions such as law, medicine, teaching, and diplomacy must all contribute, as, of course, must business and government as well. Only if there is a higher degree of comprehension, among ordinary people as well as elite groups, of the complex issues facing humanity is there any hope of achieving sustainable quality.

It is not sufficient for that knowledge and understanding to be specialized. Of course, specialization is necessary today. But so is the integration of specialized understanding to make a coherent whole, as we discussed earlier. It is essential, therefore, that society assign a higher value than heretofore to integrative studies, necessarily crude, that try to encompass at once all the important features of a comprehensive situation, along with their interactions, by a kind of rough modeling or simulation. Some early examples of such attempts to take a crude look at the whole have been discredited, partly because the results were released too soon and because too much was made of them. That should not deter people from trying again, but with appropriately modest claims for what will necessarily be very tentative and approximate results.

An additional defect of those early studies, such as *Limits to Growth*, the first report to the Club of Rome, was that many of the critical assumptions and quantities that determined the outcome were not varied parametrically in such a way that a reader could see the consequences of altered assumptions and altered numbers. Nowadays, with the ready availability of powerful computers, the consequences of varying parameters can be much more easily explored. The sensitivity of the results to different assumptions can be checked, and the structure of the study can thus be made more transparent. Moreover, part of the study can take the form of games, such as *SimCity* or *SimEarth*, which are commercial products developed by the Maxis Corporation under the leadership of Will Wright. Games permit a critic to revamp the assumptions to suit his or her own taste and see what results.

Peter Schwartz, in his book *The Art of the Long View*, relates how the planning team of the Royal Dutch Shell Corporation concluded some years ago that the price of oil would soon decline sharply and recommended that the company act accordingly. The directors were skeptical, and some of them said they were unimpressed with the assumptions made by the planners. Schwartz says that the analysis was then presented in the form of a game and that the directors were handed the controls, so to speak, allowing them to alter, within reason, inputs they thought were misguided. According to his account, the main result kept coming out the same, whereupon the directors gave in and started planning for an era of lower oil prices. Some participants have a different recollection of what happened at Royal Dutch Shell, but in any case the story beautifully illustrates the importance of transparency in the construction of models. As models incorporate more and more features of the real world and become correspondingly more complex, the task of making them transparent, of exhibiting the assumptions and showing how they might be varied, becomes at once more challenging and more critical.

Those of us participating in a study such as Project 2050, aimed at sketching out paths that may lead toward a more sustainable world in the middle of the next century, face difficult questions. How can these transitions toward sustainable quality be accomplished, if at all, during the next fifty to one hundred years? Can we hope to understand, even crudely, the complex interactions among the transitions and especially the issues that arise from their delicate relative and absolute timing? Is there any hope of taking sufficient account of the wide variations in conditions around the world? Are there other transitions, or other ways of looking at the whole set of issues, that are more important? These questions concern the period, around the middle of the twenty-first century, when the various transitions may be partly accomplished or at least well under way. Thinking usefully about that era is difficult, but not necessarily impossible. As Eilert Lövborg said, in Ibsen's *Hedda Gabler*, when surprise was expressed that his history book had a continuation describing the future, "there is a thing or two to be said about it just the same."

As to the more distant future, what kind of global conditions might prevail, after the middle of the next century, that would really approach the sustainability of quality? What are our visions of such a situation? What would we see and hear and feel if we were there? We should really try to envision it, especially a world with growth in quality finally predominating over growth in quantity. We should imagine a world in which, Utopian as it sounds, the *State of the World Report* and the *World Resources Report* do not look worse every year, population is stabilizing in most places, extreme poverty is disappearing, prosperity is more equitably shared, serious attempts are made to charge true costs, global and other transnational institutions (as well as national and local ones) are beginning to cope with the complex interlocking issues of human society and the rest of the biosphere, and ideologies favoring sustainability and planetary consciousness are gaining adherents, while ethnic hatreds and fundamentalisms of all kinds are losing out as divisive forces even though a great deal of cultural diversity remains. We can scarcely hope to attain anything approaching such a world if we cannot even imagine what it would look like or estimate on a quantitative basis how it might function.

Of the three ranges of time, it is naturally hardest to get people to think about the long-term vision of a more sustainable world, but it is vital that we overcome our reluctance to make concrete images of such a world. Only then can our imagination escape from the confines of the practices and attitudes that are now causing or threatening to cause so much trouble, and invent improved ways to manage our relations with one another and with the rest of the biosphere.

As we try to envision a sustainable future, we must also ask what kinds of surprises, technological or psychological or social, could make that fairly distant future totally different from what we might anticipate today. A special team of imaginative challengers is required to keep posing that question.

The same team could also ponder the question of what new serious problems might arise in a world where many of today's worst fears are somewhat allayed. Just a few years ago, most pundits were not predicting that the Cold War era would soon turn into a new age with different problems, but even those few that were predicting it were not speculating seriously on which concerns would replace the familiar ones that were no longer dominant.

What of the short term, the next few decades? What kinds of policies and activities in the immediate future can contribute to the possibility of approaching sustainable quality later on? It is not at all difficult to get discussions going about the near future, and some of the problems we face in the short run are becoming clear to many observers. Perhaps the chief lesson to be learned from contemporary experience is one that we touched on when we mentioned microlending. It is the importance of bottom-up as opposed to top-down initiatives. If local people are deeply involved in a process, if they help to organize it, and if they have a perceived stake, especially an economic stake, in the outcome, then the process often has a better chance of success than if it is imposed by a distant bureaucracy or a powerful exploiter. In helping tropical areas to achieve objectives in the preservation of nature along with at least partially sustainable economic development, conservationists have found that what pays off the most is investment in local groups and local leadership, and particularly in training for local leaders.

Although it is fairly easy to persuade people to discuss the middle range of time—the era during which the interlinked transitions must be largely accomplished if anything like sustainability is to be achieved—the extraordinary complexity of the challenge may be daunting. All those transitions must be considered, each with character and timing to be determined, perhaps different in different parts of the world, and all strongly coupled to one another. Still, that very complexity may lead to a kind of simplicity. Certainly it is true in physical science (which is much less difficult to analyze, to be sure, but may still have some lessons to teach) that in the neighborhood of a transition, say from a gas to a liquid, near a mathematical singularity, there are only a few crucial parameters on which the nature of the transition depends. Those parameters cannot always be characterized in advance, however; they must emerge from a careful study of the whole problem. It is true in general that the behavior of highly complex nonlinear systems may exhibit simplicity, but simplicity that is typically emergent and not obvious at the outset.

Integrated policy studies of possible paths toward a more nearly sustainable world can be exceedingly valuable. But we must be careful to treat all such studies as "prostheses for the imagination," and not to attribute to them more validity than they are likely to possess. Trying to fit human behavior, and especially problems of society, into the Procrustean bed of some necessarily limited mathematical framework has already brought much grief to the world. For instance, the science of economics has often been used in that way with unfortunate consequences. Besides, ideologies destructive of human freedom or welfare have often been justified by arguments loosely based on science, and especially on analogies between sciences. The social Darwinism preached by some political philosophers of the nineteenth century is one of many examples, and by no means the worst.

Nevertheless, taken in the proper spirit, a multiplicity of crude but integrative policy studies, involving not just linear projection but evolution and highly nonlinear simulation and gaming, may provide some modest help in generating a collective foresight function for the human race. An early Project 2050 document puts it this way: We are all in a situation that resembles driving a fast vehicle at night over unknown terrain that is rough, full of gullies, with precipices not far off. Some kind of headlight, even a feeble and flickering one, may help to avoid some of the worst disasters.

If humanity does equip itself somehow with a measure of collective foresight—some degree of understanding of the branching histories of the future—a highly adaptive change will have taken place, but not yet a gateway event. The accomplishment of the interlinked transitions to greater sustainability, however, would be such an event. In particular, the ideological transition implies a major step for humanity toward planetary consciousness, perhaps with the aid of wisely managed technical advances now only dimly foreseeable. After the transitions, humanity as a whole—together with the other organisms inhabiting the planet—would function, much more than now, as a composite, richly diverse complex adaptive system.

CHAPTER

23

AFTERWORD

In this brief chapter, I try to respond to the need for a kind of executive summary, not of every topic in the whole book, but of the central theme of simplicity, complexity, and complex adaptive systems—the theme that connects the quark, the jaguar, and humanity.

The Quark and the Jaguar is not a treatise. It is comparatively non-technical, and it reaches into a large number of areas that it cannot explore thoroughly or in depth. Furthermore, much of the work that *is* described in some detail is work in progress, which means that even if it were treated in full, with equations and more scientific jargon than already employed, it would still leave a great many important questions unanswered. Evidently, the main function of the book is to stimulate thought and discussion.

Running through the entire text is the idea of the interplay between the fundamental laws of nature and the operation of chance. The laws governing the elementary particles (including quarks) are beginning to reveal their simplicity. The unified quantum field theory of all the particles and forces may well be at hand, in the form of superstring theory. That elegant theory is based on a form of the bootstrap principle, which requires that the elementary particles be describable as made up out of one another in a self-consistent manner. The other fundamental law of nature is the simple initial condition of the universe at the time its expansion began. If the proposal by Hartle and Hawking is correct, then that condition can be expressed in terms of the unified theory of the particles, and the two basic laws become one.

Chance necessarily enters the picture because the fundamental laws are quantum-mechanical, and quantum mechanics supplies only probabilities for alternative coarse-grained histories of the universe. The coarse graining must be such as to permit probabilities to be well defined. It also allows for an approximately classical, deterministic description of nature, with frequent small excursions from classicality and occasional large ones. The excursions, especially the larger ones, result in the branching of histories, with probabilities for the different branches. In fact all the alternative coarse-grained histories form a branching tree, or "garden of forking paths," called a "quasiclassical domain." The indeterminacy of quantum mechanics thus goes far beyond the famous uncertainty principle of Heisenberg. Moreover, that indeterminacy can be amplified in nonlinear systems by the phenomenon of chaos, which means that the outcome of a process is arbitrarily sensitive to the initial conditions, as often happens, for example, in meteorology. The world we human beings see around us corresponds to a quasiclassical domain, but we are restricted to a very much coarser-grained version of that domain because of the limited capabilities of our senses and instruments. Since so much is hidden from us, the element of chance is still further enhanced.

On certain branches of history and at certain times and places in the universe the conditions are propitious for the evolution of complex adaptive systems. Those are systems (as illustrated in the diagram on page 25) that take in information—in the form of a data stream—and find perceived regularities in that stream, treating the rest of the material as random. Those regularities are compressed into a schema, which is employed to describe the world, predict its future to some extent, and to prescribe behavior for the complex adaptive system itself. The schema can undergo changes that produce many variants, which compete with one another. How they fare in that competition depends on selection pressures, representing the feedback from the real world. Those pressures may reflect the accuracy of the descriptions and predictions or the extent to which the prescriptions lead to survival of the system. Such relationships of the selection pressures to "successful" outcomes, however, are not rigid correlations but only tendencies. Also, the response to the pressures can be imperfect. Thus the process of adaptation of the schemata leads only approximately to "adaptive" results for the systems. "Maladaptive" schemata can occur as well.

Sometimes maladaptation is only apparent, arising because important selection pressures are overlooked in defining what is adaptive. In other cases, genuinely maladaptive situations occur because adaptation is too slow to keep up with changing selection pressures.

Complex adaptive systems function best in a régime intermediate between order and disorder. They exploit the regularities provided by the approximate determinism of the quasiclassical domain, and at the same time they profit from the indeterminacies (describable as noise, fluctuations, heat, uncertainty, and so on), which actually can be helpful in the search for "better" schemata. The notion of fitness, which could give a meaning to the word "better," is often difficult to pin down, in which case it may be more useful to concentrate on the selection pressures that are operating. Sometimes a fitness quantity is well defined because it is "exogenous," imposed from the outside, as in the case of a computer programmed to search for winning strategies in a game like checkers or chess. When fitness is "endogenous," emerging from the vagaries of an evolutionary process that lacks any external criterion for success, it is in many case quite ill defined. Still, the idea of a fitness landscape is useful, if only as a metaphor. The fitness variable corresponds to height (which I take arbitrarily to be lower when fitness is greater), and all the variables specifying the schema are imagined to be laid out, say on a horizontal line or plane. The search for fitter schemata then corresponds to exploring a wiggly line or two-dimensional surface, hunting for very low places. As illustrated in the figure on page 250, that search would most likely lead to getting stuck in a comparatively shallow depression if it were not for some appropriate amount of noise (or heat, obeying what Seth Lloyd calls the Goldilocks principle—not too hot, not too cold, but just right). The noise or heat can shake the system out of a shallow pit and permit it to discover a much deeper one nearby.

The variety of complex adaptive systems here on Earth is illustrated in the diagram on page 20, which shows how one such system has a tendency to give rise to others. Thus the earthly systems, all of which have some connection with life, range from the prebiotic chemical reactions that first produced living things, through biological evolution and the cultural evolution of humanity, all the way to computers equipped with appropriate hardware or software and to possible future developments treated in science fiction, such as composite human beings formed by wiring people's brains together.

When a complex adaptive system describes another system (or itself), it constructs a schema, abstracting from all the data the perceived regularities and expressing those in concise form. The length of such a concise description of the regularities of a system, for instance by a human observer, is what I call the effective complexity of the system. It corresponds to what we usually mean by complexity, whether in scientific usage or in everyday discourse. Effective complexity is not intrinsic but depends on the coarse graining and on the language or coding employed by the observing system.

Effective complexity, whether internal or not, is insufficient by itself to describe the *potentialities* of a complex system, adaptive or nonadaptive. A system may be comparatively simple, but capable of evolving with high probability, within a given time interval, into something much more complex. That was true, for example, of modern human beings when they first appeared. They were not much more complex than their close relatives the great apes, but because they were likely to develop cultures of enormous complexity, they possessed a great deal of what I call potential complexity. Similarly, when, early in the history of the universe, certain kinds of matter fluctuations occurred that led to the formation of galaxies, the potential complexity of those fluctuations was considerable.

Effective complexity of a system or a data stream should be contrasted with algorithmic information content (AIC), which is related to the length of a concise description of the whole system or stream, not just its regularities but its random features as well. When AIC is either very small or near its maximum, effective complexity is near zero. Effective complexity can be large only in the region of intermediate AIC. Again the régime of interest is that intermediate between order and disorder.

A complex adaptive system discovers regularities in its incoming data stream by noticing that parts of the stream have features in common. The similarities are measured by what is called mutual information between the parts. Regularities in the world arise from a combination of the simple fundamental laws and the operation of chance, which can produce frozen accidents. Those are chance events that turned out a particular way, although they could have turned out differently, and produced a multiplicity of consequences. The common origin of all those consequences in an antecedent chance event can give rise to a great deal of mutual information in a data stream. I used the example of the succession to the English throne of Henry VIII—after the death of his older brother—resulting in the existence of a huge number of references to King Henry on coins and in documents and books. All those regularities stem from a frozen accident.

Most accidents, for example a great many fluctuations at the molecular level, take place without being amplified in such a way as to have significant repercussions, and they do not leave much regularity behind. Those accidents can contribute to the random part of the data stream reaching a complex adaptive system.

As time goes on, more and more frozen accidents, operating in conjunction with the fundamental laws, have produced regularities. Hence, complex systems of higher and higher complexity tend to emerge with the passage of time through self-organization, even in the case of nonadaptive systems like galaxies, stars, and planets. Not everything keeps increasing in complexity, however. Rather, the highest complexity to be found has a tendency to increase. In the case of complex adaptive systems, that tendency may be significantly strengthened by selection pressures that favor complexity.

The second law of thermodynamics tells us that the entropy (measuring disorder) of a closed system has a tendency to increase or stay the same. For example, if a hot body and a cold body are in contact (and not interacting much with the rest of the universe), heat tends to flow from the hot one to the cold one, thus reducing the orderly segregation of temperature in the combined system.

Entropy is a useful concept only when a coarse graining is applied to nature, so that certain kinds of information about the closed system are regarded as important and the rest of the information is treated as unimportant and ignored. The total amount of information stays the same and, if it is initially concentrated in important information, some of it will tend to flow into unimportant information that is not counted. As that happens, entropy, which is like ignorance of important information, tends to increase.

A kind of fundamental coarse graining of nature is supplied by the histories forming a quasiclassical domain. For the universe observed by a complex adaptive system, the effective coarse graining can be taken to be very much coarser, because the system can take in only a comparatively tiny amount of information about the universe.

As time goes on, the universe winds down, and parts of the universe that are somewhat independent of one another tend to wind down too. In every part, the various arrows of time all point forward, not only the arrow corresponding to the increase of entropy, but also those corresponding to the sequence of cause and effect, the outward flow of radiation, and the formation of records (including memories) of the past and not the future.

Sometimes people who for some dogmatic reason reject biological evolution try to argue that the emergence of more and more complex forms of life somehow violates the second law of thermodynamics. Of course it does not, any more than the emergence of more complex structures on a galactic scale. Self-organization can always produce *local* order. Moreover, in biological evolution we can see a kind of "informational" entropy increase as living things come into better adjustment with their surroundings, thus reducing an informational discrepancy reminiscent of the temperature discrepancy between a hot and a cold object. In fact, complex adaptive systems all exhibit this phenomenon—the real world exerts selection pressures on the systems and the schemata tend to respond by adjusting the information they contain in accordance with those pressures. Evolution, adaptation, and learning by complex adaptive systems are all aspects of the winding down of the universe.

We can ask whether the evolving system and the surroundings come into equilibrium, like a hot body and a cold body reaching the same temperature. Occasionally they do. If a computer is programmed to evolve strategies to play a game, it may find the optimal strategy and the search is over. That would certainly be the case if the game is tic-tac-toe. If the game is chess, the computer may discover the optimal strategy some day, but so far that strategy is unknown, and the computer continues to hunt around in a huge abstract space of strategies seeking better ones. That situation is very common.

We can see a few cases where, in the course of biological evolution, a problem of adaptation seems to have been solved once and for all early in the history of life, at least on the phenotypic level. The extremophiles living in a hot, acidic, sulfurous environment deep in the ocean at the boundaries between tectonic plates are probably quite similar, at least metabolically, to organisms that lived in that environment more than three and a half billion years ago. But most problems of biological evolution are not in the least like tic-tac-toe, in fact not even like chess, which will no doubt be a solved problem some day. For one thing, the selection pressures are not at all constant. In most parts of the biosphere, the physicochemical environment keeps changing. Moreover, in natural communities the various species form parts of the environment of other species. The organisms are co-evolving, and there may not be any true equilibrium to be attained.

At various times and places, approximate and temporary equilibria do seem to be reached, even for whole communities, but after a while those are "punctuated," sometimes by physicochemical changes and sometimes by a small number of mutations following a long period of "drift," meaning sequences of genetic changes that affect phenotypes only slightly and in ways that don't matter much to survival. Drift can have the effect of preparing the way for very small alterations in the genotype to cause important phenotypic changes.

From time to time such comparative modest changes in genotypes can lead to gateway events, in which whole new kinds of organisms come into being. An example is the appearance of single-celled eukaryotes, so called because the cell possesses a true nucleus and also other organelles—chloroplasts or mitochondria—which are thought to be descended from originally independent organisms incorporated into the cell. Another example is the origin of multi-celled animals and plants from single-celled organisms, presumably by aggregation, with the aid of a biochemical breakthrough, a new kind of glue-like chemical that held the cells together.

When a complex adaptive system gives rise to a new kind of complex adaptive system, whether by aggregation or otherwise, that can be considered a gateway event. A familiar example is the evolution of the mammalian immune system, the operation of which somewhat resembles biological evolution itself, but on a much more rapid time scale, so that invaders of the body can be identified and attacked in hours or days, as compared with the hundreds of thousands of years often needed to evolve new species.

Many of the same features that are so conspicuous in biological evolution are found, in rather similar form, in other complex adaptive systems, such as human thought, social evolution, and adaptive computing. All these systems keep exploring possibilities, opening up new paths, discovering gateways, and occasionally spawning new types of complex adaptive system. Just as new ecological niches keep turning up in biological evolution, so new ways to make a living continue to be discovered in economies, new kinds of theories are invented in the scientific enterprise, and so on.

Aggregation of complex adaptive systems into a composite complex adaptive system is an effective way to open up a new level of organization. The composite system then consists of adaptive agents constructing schemata to account for and deal with one another's behavior. An economy is an excellent example, and so is an ecological community.

A good deal of research is being carried out on such composite systems. Theories are developed and compared with experience in various fields. Much of that research indicates that such systems tend to settle into a well-defined transition zone between order and disorder, where they are characterized by efficient adaptation and by power law distributions of resources. Sometimes that zone is called, rather metaphorically, the "edge of chaos."

There is no indication that there is anything terribly special about the formation of a planetary system like the solar system or about its including a planet like the Earth. Nor is there any evidence that the chemical reactions that initiated life on this planet were in any way improbable. It is likely, therefore, that complex adaptive systems exist on numerous planets scattered through the universe and that at least some of those complex adaptive systems have many features in common with terrestrial biological evolution and the resulting life forms. It is, however, still a matter of dispute whether the biochemistry of life is, on the one hand, unique or nearly so, or, on the other hand, just one of a great many different possibilities. In other words, it is not yet certain whether it is determined mostly by physics or owes its character in great part to history.

The nearly four billion years of biological evolution on Earth have distilled, by trial and error, a gigantic amount of information about different ways for organisms to live, in the presence of one another, in the biosphere. Similarly, modern humans have, over more than fifty thousand years, developed an extraordinary amount of information about ways for human beings to live, in interaction with one another and with the rest of nature. Both biological and cultural diversity are now severely threatened and working for their preservation is a critical task.

The preservation of cultural diversity presents a number of paradoxes and conflicts with other goals. One challenge is the very difficult one of reconciling diversity with the pressing need for unity among peoples now faced with a variety of common problems on a global scale. Another is presented by the hostility that a number of parochial cultures evince toward the universalizing, scientific, secular culture, which supplies many of the most vigorous advocates of the preservation of cultural diversity.

The conservation of nature, safeguarding as much biological diversity as possible, is urgently required, but that kind of goal seems impossible to achieve in the long run unless it is viewed within the wider context of environmental problems in general, and those in turn must be considered together with the demographic, technological, economic, social, political, military, diplomatic, institutional, informational, and ideological problems facing humanity. In particular, the challenge in all of these fields can be viewed as the need to accomplish a set of interlinked transitions to a more sustainable situation during the course of the coming century. Greater sustainability, if it can be attained, would mean a leveling off of population, globally and in most regions; economic practices that encourage charging true costs, growth in quality rather than quantity, and living on nature's income rather than its capital; technology that has comparatively low environmental impact; wealth somewhat more equitably shared, especially in the sense that extreme poverty would no longer be common; stronger global and transnational institutions to deal with the urgent global problems; a public much better informed about the multiple and interacting challenges of the future; and, perhaps most important and most difficult of all, the prevalence of attitudes that favor unity in diversity—cooperation and nonviolent competition among different cultural traditions and nation states—as well as sustainable coexistence with the organisms that share the biosphere with humankind. Such a situation seems Utopian and perhaps impossible to achieve, but it is worthwhile to try to construct models of the future—not as blueprints but as aids to the imagination—and see if paths can be sketched out that may lead to such a sustainable and desirable world late in the next century, a world in which humanity as a whole and the rest of nature operate as a complex adaptive system to a much greater degree than they do now.

INDEX

Action
  effective action, 209
  $\hbar$ as unit of, 207
  in Newtonian physics, 207
  in superstring theory, 208–209
Action at a distance, 87, 172
Adams, James L., *Conceptual Blockbusters,* 270
Adams, Robert McC., 299
Adaptive schemata, 291–305
  cultural DNA, 292–294, 296
  with humans in the loop, 298–301
  language evolution, 294–295, 296
  levels of, 292–294
  role of influential individual, 297–298
  transmission of animal behavior, 291
AIC. *See* Algorithmic information content
Algorithm, defined, 35, 46
Algorithmic complexity, 35, 50, 223
Algorithmic information content, 35, 48–50, 101–102, 104, 116, 370
  depth and, 103–105
  entropy and, 224
  information vs., 37
  intermediate, between order and disorder, 58–60, 318
  introduction of, 34–36
  and large effective complexity, 58–60
  as measure of randomness, 41
  for random strings, 38, 41
  uncomputability of, 38–41
  of universe, 134
Algorithmic randomness, 35, 41, 49–50
Alternative histories
  in quantum mechanics, 142–143
  at race track, 141–142
  of universe, 138, 140–141, 160, 162, 210
Altruistic behavior, 251–252, 360
Alvarez, Luis, 286
Amazon, 3
Ampère's conjecture, 82
Ampère's law, 82
Anderson, Carl, 179, 190
Anderson, Philip, 193
Anderson-Higgs mechanism, 193–194, 195, 196, 203
Annihilation, 124

"Anthropic principles," 212–213
Antiexclusion principle, 124, 177
Antifamily, 190
Antiparticle, 179–180
Antiquarks, 183, 185
Apollonians, xiii
Arrows of time, 215–231
  and biological evolution, 238
  defined, 130
  entropy and second law, 217–218, 220, 226–227, 372
  and greater complexity, 227–230
  and initial condition of universe, 129–131, 195, 227
  order in the past, 220–222
  past and future, 215–216
  radiation and records, 216
  thermodynamic, 220, 226
Arthur, W. Brian, 255, 322
Artificial selection, 300
Asimov, Isaac, 19
Aspen Center for Physics, 303
Aspen Institute seminars, 96, 264
Aspiazu, Eduardo, 335
Axelrod, Robert, 313

Baby universes, virtually created and destroyed, 211
Bacteria. *See* Drug resistance in bacteria
Bajkov, A. B., 287
Bak, Per, 97, 319
Ball lightning (kugelblitz), 286
Basin of attraction, 266–267
Battery (voltaic pile), invention of, 81
Beebe, Spencer, 334
Belize, 4, 9
Bell, John, 172
Bell's theorem (Bell's inequalities), 171–172
Bennett, Charles, 100–102, 173, 175, 223, 259
Beria, P. Lavrenti, 286
Bertlmann's socks, 172
Big bang, 148, 215
Binding energy, 110
Biochemistry
  effective complexity vs. depth, 114–115
  on other planets, 115–116
Biogeography, 332–333
Biological diversity, conservation of, 330–331
Biological evolution, 17, 19, 115, 220, 228, 231. *See also* Selection in biological evolution
  as complex adaptive system, 61, 69–70, 235, 245
  cultural evolution vs., 304
  information contained in genome, 64, 69
  question of drive toward complexity, 244–246
  simulation of, 314–316
  time span of, 329
Biology
  of brain, 117
  reduction of, 113–114, 115
  role of chance events, 115, 134, 316–317
  terrestrial biology, 113–115
Birdwatching, 5, 7–8, 13
Bit, defined, 34
Bit string, 35–38, 48, 50, 57–59
Blackett, Patrick, 179
Black holes, 221, 230
Bohm, David
  and Einstein, 170
  EPRB experiment, 171
Bohr, Niels, 165, 168
Bootstrap principle, 128–129, 209
Borges, Jorge Luis, 149
Bosons, 124, 177
  higgson (Higgs boson), 193–194
Bottom quarks, 191
Brain and mind, 116–118
  and free will, 157
  neural net as model of brain, 307–310
  study of left and right brain, 117–118
Brassard, Gilles, 173
Brillouin, Léon, 223

Bronze disease, 63–64
Brookings Institution, 346
Brout, Robert, 193
Brown, Jerram, 251
Brown, Robert, 152
Brownian motion, 152
Brueckner, Keith, 45
Brun, Todd, 27
Brundtland Commission, 355

Calandra, Alexander, "The Barometer Story," 270–272
California Institute of Technology, xii
  brain and mind studies at, 117
  as reductionist institution, 116–119
Caltech. *See* California Institute of Technology
CERN accelerators, 125, 189, 193, 204
Chaitin, Gregory, 35–36, 39–40, 102, 223
Chan Chich, 4
Chance processes. *See* Randomness
Chaos, 143, 164, 276
  classical, 25–26
  deterministic, in financial markets, 47–48
  fractals, power laws, and, 97
  and indeterminacy, 24–27
Chao Tang, 97
Charmed quarks, 191
Chartists, 47–48
Chemical prospecting, 336
Chemistry
  derivable from elementary particle physics, 112
  and fundamental physics of electron, 109–111
  at its own level, 111
  QED calculation of chemical processes, 110, 111
Childhood, 12
Chomsky, Noam, 53, 295
Closed system, 217–221, 235–236
Coarse-grained histories, 144–145
  decoherence of, 146–147, 148, 153
  and interference terms, 145–146
  maximally refined, 159–160, 162
  of universe, 160, 162, 211
Coarse graining, 29–30, 34, 153
  entropy and, 225–226
Color charge, 184
Color force, 184–185
Colorful gluons, 182–183, 185
Colors of quarks, 181
Committee for Scientific Investigation of Claims of the Paranormal, 281–282, 289
Competition, 305
  for population size, 255–256
  of schemata, 24, 87, 329
Completeness, challenge to quantum mechanics, 168–169, 171
Complex adaptive systems, 11, 16–21
  and biological evolution, 17, 19, 61, 235, 245
  common processing features of, 17
  computers as, 73, 307–308
  defining "complex," 27–28, 32
  economies as, 321
  effect of directed evolution, 300
  genetic algorithms as, 310–312
  humans as, 17
  illustrations, 17–19
  and language acquisition, 294–295
  and maladaptive schemata, 295–297
  and mythology, 278
  nonadaptive evolution vs., 9, 235, 245
  as observer, 58–60, 155–156
  operation of, 24
  between order and disorder, 249
  organization as, 297–298
  scientific enterprise as, 75, 76
  simulation of, 312–314
Complexity. *See also* Crude complexity; Effective complexity
  algorithmic, 35, 50, 223
  computational, 28, 100–101
  cultural, 70

Complexity (*cont.*)
different kinds of, 28–29
increasing, 227–230, 241–242, 244–246
and length of description, 30–32
of pattern of connection, 30–32
potential, 70, 230
problem of defining, 27–28, 32
Compressibility of strings, 38
Computational complexity
definition, 28, 100
and depth and crypticity, 100–101
Computers. *See also* Algorithmic information content; Machines that learn
decision tree for, 72–73
internal model, 72–73
parallel processing, 32, 36, 308
quantum, 101
Comte, Auguste, 108
Condensed matter physics, 113
Conduitt, 85–86
Consciousness, 155, 156–157, 317
Conservation International, 334–335
Conservation laws, 82, 158
conservation defined, 158
of isotopic spin, 262–263
in quantum mechanics, 178
Context dependence, 28, 33
Continental drift, 285
Cooperation
need for, 304–305
of schemata, 242–244
Corliss, R. William, 288
Cosmological arrow of time, 227
Coulomb's law, 82
CPT, 195
Creative thinking, 261–273
brainstorming, 267
fitness landscape for, 266–267
incubation period, 264–267
measurement of, 269–270
problem formulation, true boundaries of problem, 270–273
shared experiences in, 264–265
speeding up process of, 265–266, 268
in theoretical science, 261–264
transfer of thinking skills, 268
Cretaceous extinction, 99, 239, 330
Crick, Francis, 66
Crude complexity, 23, 84
conciseness and, 34
defined, 34, 36
information and, 23–41
Crutchfield, James, 57
Crypticity
defined, 101, 105
and theorizing, 105–106
CSICOP. *See* Committee for Scientific Investigation of Claims of the Paranormal
Cultural diversity
enlightenment vs., 342
preservation of, 338–341, 361–362
Cultural DNA, 292–294, 296, 338, 343
Cultural evolution, 292, 304, 329
Cybernetics, 72–73

Darwin, Charles, 118, 293
Origin of Species, 300
Dawkins, Richard, 292
The Blind Watchmaker, 317
DeBono, Edward, 267–268
Decaying exponential, 132
Decision tree, 72–73
Decoherence
entanglement and mechanisms of, 147, 148–149
for object in orbit, 148–149, 151
Decohering coarse-grained histories, 146–147
in quasiclassical domain, 149, 150
tree-like structure of, 149–150
Delbrück, Max, 118
Deming, W. Edwards, 298
Depth, 100–105
and algorithmic information content, 103–105
and biochemistry, 114–115
a deeper look at, 102–103
defined, 101, 103

Depth (*cont.*)
hypothetical example, 101–102
and simplicity, 101–102
Deterministic chaos, in financial markets, 47–48
de Vries, 300
Digital organisms, 314–315
Dionysians, xiii
Dirac, Paul Adrien Maurice, 109, 138, 179
Direct adaptation, 70–74
Directed evolution, 298–300
Diversities under threat, 329–343
conservation of biological diversity, 331–338
enlightenment vs. cultural diversity, 341
importance of tropics, 331–332
information (misinformation?) explosion, 342–343
local people and conservation, 335–337, 367
man-made extinction, 332–333, 348
preservation of cultural diversity, 338–341, 361–362
rapid assessment, 333–335
reconciliation, 343
science and tropical conservation, 331–333, 365
spectrum of conservation practices, 337–338
universalizing vs. particularizing, 341
universal popular culture, 342
DNA, 66, 69
Donelaitis, 338
Drug resistance in bacteria, 63–74
evolution of, 65–69
incorrect theory of, 70–71
resistant mutant, 67–68
Durrell, Gerald, 303

E. coli. See Escherichia coli
Eastern Ecuador, 3
ECHO, 317
Ecological communities, 237–238, 256, 304
computer simulation of, 317–318
diversity of, 246–248, 332
Economics
evolutionary approach to, 320–324
power laws in, 319
for a sustainable world, 353–356
Ecosystems, complex and simple, 28–29
Effective action, 209–211
Effective complexity, 50, 100–102, 162, 228
biochemistry and, 114–115
complex adaptive systems and, 54–56, 59
defined, 56
intermediate AIC and, 58–60
internal, 56
potential complexity and, 70
regularities and, 54–56, 162
of universe, 134
Ehrlich, Paul, 351
Einstein, Albert, 39, 99. *See also* General-relativistic gravitational theory
dream of unified field theory, 126–127
explanation of Brownian motion, 152
mass-energy relation, 192
rejection of quantum mechanics, 127, 168–169
special relativity, 262
Ekert, Artur, 173
Eldredge, Niles, 239
Electromagnetic force, 182, 189
Electromagnetism, theory of, 81–84. *See also* Maxwell's equations
relativistic quantum-mechanical, 109
Electron, 177–179, 190
and electron neutrino, 186–189
flavor of, 186
fundamental physics of, 109–110
Electron neutrino, 186–189
Elementary particle physics, 6, 8, 112
laws of, 9
multiplicity of particles explained, 129, 196–198

Emergent structures, 99–100
Emmons, Louise, 335
Empirical theory—Zipf's law, 92–97
Energy state, 110
Englert, François, 193
Entanglement, 147
Entropy, 101
  of algorithmic complexity, 226–227
  and coarse graining, 225–226
  decrease of, 221
  erasure and shredding, 224–225
  as ignorance, 219–220
  increase of, 221–222, 223
  and information, 219, 223–224
  as measure of disorder, 218, 220
  new contribution to, 223–224
  and second law, 217–218, 220, 222–223, 325
EPRB experiment, 171
  and hidden variable alternative, 172
  potential applications of, 174–176
Equivalence classes, 145–146
Equivalence principle, 87
Erwin, Terry, 29
Escherichia coli, 66–68
Ethnobotany, 336, 339–340
Everett, Hugh, III, 137–138, 150
Evolution. *See* Biological evolution
Expansion of early universe, 140, 148, 149, 154, 195, 206, 220
Expert systems, 72
Extremophiles, 236–238, 248

Families of fermions, 190–192
Faraday, Michael, 81, 83
Faraday's law, 82
Fermi, Enrico, 149, 205, 262, 287
Fermions, 177
  defined, 124
  families of, 190–192
Feynman, Richard, 138, 178, 189, 208
Feynman diagram, 189
Fields, 124, 127
Fierz, Markus, 287
Financial markets, deterministic chaos in, 47–48
Fine-grained histories of universe, 143–144, 145
First law of thermodynamics, 217–218, 325
Fisher, Sir Ronald, 77
Fish falls, 287–288
Fission-track dating, 154
Fitness
  and genetic algorithms, 312, 317
  in neural net computation, 308–310
Fitness, biological concept of, 248–256, 295, 317
Fitness landscapes, 249–250
  basin of attraction, 266–267
  in computer learning, 310
  for creative ideas, 266–268
  deep basin, 267, 312
Fitness of sex, 252–255
Fitzpatrick, John, 252
Flapdoodle, 167
Flavor forces, 189
Flavors of quarks, 181
Fluctuations, 151–152, 157
Followed quantities, branch dependence of, 160
Fontana, Walter, 315
Foster, Robin, 335
Fowler, William, 187
Fractals, power laws and, 97
Franklin, Benjamin, 177
Franklin, Rosalind, 66
Free will, 156–158
Friedman, Jerome, 185
Frost, Robert, "The Road Not Taken," 149, 157
Frozen accidents, 133–134, 227–230, 295
Fundamental science. *See* Hierarchy of the sciences

Gateway events, 240–241, 246, 259–260
Gauge theories, 192

Gell-Mann, Arthur (father), 15–16, 88, 340
Gell-Mann, Ben (brother, now Ben Gelman), 12, 14–15
Gell-Mann, Margaret (first wife), 7, 78
General-relativistic gravitational theory, 16–17, 85–88, 126–127, 208, 320
 confirmed by observations, 86, 152
 equation, for gravitational field, 87–88, 208
 and equivalence principle, 87
 and geometry of space-time, 87
 Newton's theory vs., 86–87
 propagation of gravitational influence, 87
 simplicity of, 88
Genetic algorithm software
 applications of, 312
 and classifier system, 310–311
 description of, 310–311
Genome, 64, 69–70, 239–241, 251, 300
Genotype, 65, 67–69, 239, 253
Gentry, Alwyn, 335
GeV (Giga-electron-volt), 192–193
Gilbert, William, 81
Glashow, Sheldon, 189, 192
Gluinos, 205
Gluons
 flavor-blind, 182
 sensitive to color, 182–185
"Goblin worlds," 165
Gödel, Kurt, 39
Goldbach's conjecture, 40, 101–102
Goldberger, Marvin "Murph," 21, 191
Gould, Stephen Jay, 239
"Grand unified theory," 126, 206
Gravitation
 Einstein's general-relativistic theory of, 16–17, 85–88
 law of gravity, 16
 universal, 85–88
Gravitational constant, 201–202
Gravitational wave astronomy, 216
Gravitino, 205
Graviton, 124, 128, 205
Green, Michael, 128
Greenberg, Joseph, 294
Griffiths, Robert, 138, 140
Grossman, Marcel, 208
Guatemala, 4
Guralnik, Gerald, 193
Gutenberg, Beno, 99

h bar (Planck's constant/2π), 201–202, 207
Hagen, C. R., 193
Haldane, J. B. S., 77
Half-life, 132
Hamilton, William, 253
Hartle, James, 137–138, 155, 163, 367
 on initial state of universe, 131, 140, 206, 209, 217
 "The Wave Function of the Universe," 138
Harvard-MIT theoretical seminar, 75–76
Havel, Václav, 361
Hawking, Stephen, 153, 215, 227, 367
 on initial state of universe, 131, 140, 206, 209, 217
 "The Wave Function of the Universe," 138
Heisenberg, Werner, 202
Heisenberg uncertainty principle, 24, 139, 143, 163, 178
Helmholtz, Hermann von, 264
Henderson, Hazel, 292
Herder, 338
Heterotic superstring theory, 127–129, 195, 199, 203
 spatial aspect of, 210
Hidden variables, quantum mechanics and, 169–172
Hierarchical learning, 57
Hierarchy of the sciences, 107–120
 biochemistry, 114–115
 biology, information for reduction of, 113–114
 bridges or staircases and reduction, 111–113

Hierarchy of the sciences (*cont.*)
 chemistry and fundamental physics of electron, 109–111
 chemistry at its own level, 111
 criteria for being fundamental, 108–109
 life, between order and disorder, 115–116
 mathematics, special character of, 108–109
 psychology and neurobiology, mind and brain, 116–118, 156–157
 quantum electrodynamics, 109–112
 "reductionism," 118–119
Higgs, Peter, 193
Higgson (Higgs boson), 193–194, 196, 205
Hinshelwood, Sir Cyril, 70–71
Histories
 alternative, at race track, 141–142
 alternative, in quantum mechanics, 142–143
 alternative, of universe, 138, 140–141, 159, 210
 coarse-grained. *See* Coarse-grained histories
 combined, 141–142
 defined, 140–141
 fine-grained, 143–144, 145
 quantum mechanics in terms of, 138
 quasiclassical, 154
Holdren, John, 351
Holland, John, 310–311, 317, 322
Homeopathic version of sympathetic magic, 89–90
Hopfield, John, 309
Horowitz, Mardi, 301
Hufstedler, Shirley, 21
Hunt, Morton, *The Universe Within,* 265

IGUS (information gathering and utilizing system), 155–156, 164–165, 212–213
Immune system, 19–21
Imprinting, 301–302
Inca Empire, 3
Inclusive fitness, 250–252
Incompressible string, 38, 44, 105
Indeterminacy, 139, 276
 from quantum mechanics and chaos, 24–27
Indifference, condition of, 220, 227
Individual fitness, 251–252
Individuality, notion of, 6–8, 231
 universality vs., 9–10
Information. *See also* Algorithmic information content
 cost of, 325
 and crude complexity, 23–41
 defined, 37–38
 in diversity, 330
 and entropy, 219–220, 223–224
 transition, for sustainable world, 362–366
Information gathering and utilizing system (IGUS), 155–156, 164–165, 212
Information theory, 34, 36–37
Initial condition of universe, 137, 140, 206, 209, 227
 and arrow(s) of time, 129–131, 195, 215, 220, 227
 candidate for, 131
 and causality, 216–217
 role of, 213
Integrated pest management, 352
Interaction strengths, 309–310
Interference terms, 143–148
Intergenerational equity, 356–357
Internal effective complexity, 56
Irreversibility, 225, 325, 354
Ishi (last Yahi), 339
Isotopic spin, 262–263

Jaguarundi, sighting of, 6–9
Johst, Hanns, *Schlageter,* 153

Joos, Erich, 138, 149
Joyce, James, *Finnegans Wake,* 180
Judson, Olivia, 254

Kapitsa, Pyotr L., 286
Kauffman, Stuart, 319
Kendall, Henry, 185
Khwarizmi, Muhammad ibn Musa, al, 35, 243
Kibble, Thomas, 193
Kin selection, 251
Kirschvink, Joseph, 286
Kolmogorov, Andrei N., 35–36, 223
Kroeber, Alfred, 339
Kroeber, Theodora, 339
Kugelblitz (ball lighting), 285
Kuhn, Thomas, *The Structure of Scientific Revolutions,* 86–87, 136

Landauer, Rolf, 223
Langton, Christopher, 299, 319
Language acquisition, 294–295, 302
 by a child, 51–61
 grammar as partial schema, 53–54
Language evolution, 13–14, 150, 294–295, 296
Laser, 124
Learning. *See also* Machines that learn
 and creative thinking, 261–273
 with genes or brain, 19, 60–61
 hierarchical learning, 57
 supervised learning, 308
Least action, principle of, 209
Lederberg, Joshua, 66
Lee, T. D., 189
Length, fundamental unit of, 203
Lewis, Harold W., 286–287
Life
 distinguishing features of, 64
 between order and disorder, 115–116
 on other planets, 114–115, 257, 277
Limits to Growth (Club of Rome), 362

Lintgen, Arthur, 290
Lloyd, Seth, 369
Lorenz, Edward N., 26
Lorenz, Konrad, *King Solomon's Ring,* 301–302

MacCready, Paul, 269–270
Machines that learn, 307–325
 evolutionary approach to economics, 320–325
 genetic algorithms, 310–312
 interaction strengths, 309–310
 neural net computation, 307–310
 reading English aloud, 308–309
 simulation of biological evolution, 314–318
 simulation of collectivities of adaptive agents, 318–320
 simulation of complex adaptive systems, 312–314
MacNeill, Jim, 355
Macrostates, 218–219
Magical thinking, 275
Magnetism, study of, 81–82. *See also* Electromagnetism
Maladaptive schemata, 291–305, 343
 external selection pressures, 296–297
 leading to extinction, 293–294, 330
 persistence of, 301–305
 and time scales, 303–305
 and windows of maturation, 301–303
Maloney, Russell, *Inflexible Logic,* 49
Mandelbrot, Benoit
 The Fractal Geometry of Nature, 97
 work on power laws, 94–96
"Many worlds," 138, 150
Marcum, Jess, 45
Mars, positions of, 149–151
Mass-rest energy equivalence, 192–193
Massachusetts Institute of Technology (MIT), 75–76
Mathematics
 continuum, 320

Mathematics (*cont.*)
  discrete, 320
  pure vs. applied, 108
  rule- and agent-based, 320–321
  special character of, 108–109
Matter-antimatter symmetry, violation of, 195–196
Maximal quasiclassical domain, 159, 162–165
Maxwell, James Clerk, 81, 84, 222
Maxwell's demon, 222–223
Maxwell's equations, 81–82, 126–127, 320
  confirmation of, 84
  consequences of, 82–84
  displacement-current term in, 83
  and known speed of light, 83
  symmetries of, 262
  wave solutions for, 82–84
Mayan language, 4
Mayr, Ernst, 12–13, 288
Meme, 292
Mendel, Gregor, 300
Microlending, 356, 365
Microstates, 218–219
Mind and brain. *See* Brain and mind
MIT. *See* Massachusetts Institute of Technology
Mixed quantum state, 140
Monte Carlo method
  for doing sums, 46
  random numbers and, 45–46
Moral equivalent of belief, 279–281
Morgan, T. H., 118
Morowitz, Harold, 213, 240–241
Multiple universes, 210–212
Munn, Charles A., III, 258–259
Muon, 190
Muon neutrino, 191
Mutation of gene, 67–69, 259–260, 300
Mutual information, 58

Napo River, 3
Natural selection, 64
Neddermeyer, Seth, 190
NETalk, 308–309
Neural net computation, 307–310
Neurophysiology, 116–118
Neutrino, 186–192
Neutrino astronomy, 216
Neutron, 181, 185
  discovery of, 180
Neveu, André, 128
Newton, Isaac
  equations of motion, 151
  gravitational constant, 201–202
  law of gravitation, 85–88, 136
  legend of falling apple, 85–86
Nietzsche, xiii
Nobel, Alfred, 107
Nobel prizes, 107, 179, 185, 189, 193
"No-boundary boundary condition," 131
Nonlocal, 173
Nuclear force, 185
Nuclear particle states, 186

Odysseans, xiii
Omnès, Roland, 138, 140
Oppenheimer, Robert, 286
Optimization, 312, 315
Origin of life on Earth, 16, 20, 115, 134, 213, 220, 240, 314
Ørsted, Hans Christian, 81

Packard, Norman, 48, 319
Palmer, Richard, 322
Paña Cocha, 4
Paradigm shifts, 86–87
Parallel processing computers, 32, 36, 308
Parameter, defined, 95
Pareto, Vilfredo, 319
Parker, Theodore A., III, 334–335
Parthenogenesis, 252–253
Particle-antiparticle symmetry, 179–180
Pattern recognition, 89, 277, 296
  in the arts, 279
Pauli exclusion principle, 124, 177

Perkins, David, 269
Person schema, 301
Peterson, Ivars, *Newton's Clock,* 25
Phenomenological theory, 93
Phenotype, 65, 69, 239, 251
Photino, 205
Photons, 124, 177–180
  virtual exchange of, 178–179, 182, 200
Piranhas, 4
Planck, Max, 168, 201
Planck mass, 199, 200, 202, 204, 205–206
Planck's constant, 201–202, 207
Planetary bargain (debt swaps), 337–338, 358
Plasticity, 301–302
Pliny the Elder, 348
Plotkin, Mark J., *Tales of a Shaman's Apprentice,* 339
Podolsky, B., 168, 171
Poincaré, Henri, 265
  Science and Method, 25–26
The poor, struggles of, 351–352, 358
Popper, Karl, 78
Population biology, mathematical theory in, 77
Positron, 124
  discovery of, 179
Potential complexity, 70
Power laws, 94–97
  application to sand piles, 97–99, 319
  connection to fractals, 97
  in economics, 319
  for events in nature, 99
  for resource distribution, 319–320
  self-organized criticality, 97–100, 319
  size and incidence of earthquakes, 99
Power of theory, 89–106
  depth and crypticity, 100–106
  empirical theory—Zipf's law, 92–97
  scale independence, 97–100
  "theoretical," two uses of word, 90–91
  theorizing about place names, 91–92
"Princeton string quartet," 128
Probabilities, 140–142. *See also* Alternative histories
  exact probabilities and quoted odds, 147–148
  in quantum mechanics, 24, 131–133, 135, 140, 144
Project 2050, 346
Protons, 180–183, 185
Pseudorandomness, 46–47, 48, 73
Psychoanalysis, as theory of human behavior, 78–80
Psychological arrow of time, 227
Psychology
  physicochemical nature of, 116
  as study of mind, 117
Punctuated equilibrium, 238–240, 260, 314
Pure quantum state, 140

QCD. *See* Quantum chromodynamics
QED. *See* Quantum electrodynamics
Quantum (quanta)
  defined, 124
  positively and negatively charged, 189
  Z^0, neutral, 189, 191
Quantum accidents, 133–134
Quantum approach, 123
Quantum chromodynamics, 183–185, 190, 192
  vs. quantum electrodynamics, 184–185
  simplicity revealed by, 185–186
  spin of quanta in, 196
Quantum computers, 101
Quantum cosmology, 138–139, 210–211, 212
Quantum cryptography, 173–175
Quantum electrodynamics, 109–112, 177–179
  approximations for chemical processes, 110, 111
  vs. quantum chromodynamics, 184–185

Quantum field theory
assumptions, 177
particle-antiparticle symmetry, 179–180
QED as example of, 177–179
of quarks and gluons, 183
Quantum flavor dynamics, 189, 190, 192, 194
spin of quanta in, 196
Quantum mechanics
alternative histories in, 142–143
approximate quantum mechanics of measured systems, 136–137
and classical approximation, 25, 135–165
vs. classical physics, 6, 135
collapse of wave function, 156
contemporary view of, 135–165
discovery of, 123
and flapdoodle, 167–175
as framework for physical theory, 6
indeterminacy from, 24–27
modern interpretation of, 6, 137–139, 164
probabilistic nature of, 24, 131–133, 135, 140, 144
protean character of, 162–163
successes of, 6, 169–170
to treat individuality, 6, 8, 9
Quantum states, 124, 218
of universe, 139–141
Quantum universe, 24
quantum state of universe, 139–141
simplicity and randomness in, 123–134
Quark-gluon interaction, 182–183, 186
Quarks
bottom, 191
charmed, 191
colors of, 181
confined in neutron and proton, 181–182, 185
experimental proof of, 185
flavors of, 181
forces between, 182
making neutron or proton out of, 181
mathematical vs. real, 182
naming of, 11, 180
prediction of, 11
strange, 191
top, 191
Quasiclassical domain, 135–138, 149–150, 152
coarse graining for inertia and, 153–154
of familiar experience, 158–160
individual objects in, 160–162
maximal, 159, 162–165
measurements, 154–155
possibility of inequivalent domains, 163
Schrödinger's cat, 152–153
Quichua, 3

Radioactive decay, 133, 136, 152–154
Ramond, Pierre, 128
RAND Corporation, 41, 43–45
Randi, James, 290
Random bit string, 48
Random features, in data stream, 105
Randomness, 43–50, 230
AIC as measure of, 41
effective complexity, 50
meanings of "random," 44–45, 48
and Monte Carlo method, 45–46
pseudorandomness, 46–47, 48
in quantum universe, 123–134
separated from regularity, 56
Shakespeare and proverbial monkeys, 48–50
and stochastic processes, 44, 48
Random numbers, 44, 48
computer-generated, 46
definition, 46
and Monte Carlo method, 45–46
Random process, 48

Random string, 38, 41, 104
Random walk, 47, 244–245. *See also* Stochastic processes
Rapid Assessment Program, 334–335
Ray, Thomas, 314–317
Reduction, 112–116, 119. *See also* Hierarchy of the sciences
"Reductionism," 116, 118–119
Regularities
  in data stream, 23, 55, 56, 58, 105, 276
  and effective complexity, 54–56, 162
  from frozen accidents, 133–134
  identifying classes of, 57–58
  invention of, superstition, 276–278, 283
  separated from randomness, 56
Relativistic quantum-mechanical theory of electron and electromagnetism, 109. *See also* Quantum electrodynamics
Renormalizability, 200, 205
Richardson, L. F., 64
Richter, Charles F., 99
Rising exponential, 132
Robots, 72–74
Rosen, Harold, 128
Rosen, N., 168, 171
Rosenberg, C. R., 308
Rubbia, Carlo, 189

Sakharov, Andrei, 195
Sakiestewa, Ramona, 11
Salam, Abdus, 107, 189
Santa Fe Institute, 48, 243, 295, 310
  economics program, 320–324
  founding of, xii, 9, 119
  Project 2050, 346
  as rebellion against reductionism, 119
  research on simplicity and complexity, 17, 99–100
  Science Board Symposium, 299
Scale independence, of power laws, 97–98
Scaling laws, 94–96, 98. *See also* Power laws
Schema (schemata). *See also* Adaptive schemata; Maladaptive schemata
  in adaptive process, 17–18
  competition of schemata, 24, 87, 329
  cooperation of schemata, 242–244
  representation by interaction strengths, 309
  theories as schemata, 75
Scherk, Joël, 128
Schrödinger equation, 320
  with Coulomb forces, 110
Schrödinger's cat, 152–153
Schultes, Richard E., 339
Schumaker, John F., *Wings of Illusion,* 281
Schwartz, Peter, *The Art of the Long View,* 363
Schwarz, John, 128
Scientific enterprise, 75–88
  competition of schemata, 87
  falsifiability and suspense, 78–80
  formulation of a theory, 77
  role of theory in science, 75–76
  selection pressures on, 80–81
  simplicity of unifying theories, 84–85, 88
  theories that unify and synthesize, 81–84
  theory and observation, 75–78
  theory tested against observation, 77–78, 296
  universal gravitation, theories of, 85–88
Search for Extraterrestrial Intelligence (SETI), 213, 257
Second law of thermodynamics
  application of, 221–222
  biological evolution and, 235–236
  entropy and, 217–218, 220, 222–223, 325
  irreversibility and, 325, 354
Segregation distortion, 251

Sejnowski, Terrence, 308
Selection in biological evolution, 235–260
  co-evolving species, 237–238
  cooperation of schemata, 242–244
  death, reproduction, population, 255–256, 315
  deception among birds, 258–259
  filling of niches, 256–258
  fitness, 248–255
  gateway events, 240–241, 246
  and increased complexity, 241–242, 245
  punctuated equilibrium, 238–240, 260, 314
  selfish gene, 251
  small steps and large changes, 259–260
Selection pressures, 24, 296, 304, 340
  exerted by human beings, 298–301
  favoring drug-resistant bacteria, 68
  nonscientific kinds of, 296–297
  in organizations, 298
  on scientific enterprise, 80–81
Selectrons, 205
Self-awareness, 156–158, 317
Self-consistency, principle of, 129, 209
Selfish gene, 251
Selfish scheme, 277, 296
Self-organized criticality
  application to sand piles, 97–99, 319
  power laws and, 97–100, 319
SETI (Search for Extraterrestrial Intelligence), 213, 257
Sexual reproduction, 250, 252–255
  advantages of, 253–254
Shannon, Claude, 37, 223
Simon, Herbert, 61
Simplicity
  defining of, 27–28
  depth and, 101–102
  in quantum universe, 123–134
  revealed by quantum chromodynamics, 185–186
  of unifying theories, 84–85, 88
Sims, Karl, 299–300, 318
Skeptical Inquirer, 281–282, 289
Skeptics' movement, 281–282
  alleged phenomena challenging science, 288–289
  and claims of the paranormal, 281–283
  and science, 284–288
Smith, John Maynard, 255
Social Darwinism, 366
Sociobiology, 360–361
Solomonoff, Ray, 35–36, 223
Solvay Conference (Brussels), 168
Southwick, Marcia (wife), 85
Specialization vs. integration, 14–15, 345–346
Special relativity, 262
Speciation, 12
  co-evolving species, 237–238
  definition of species, 13
  diversity of species, 17, 331
Speed of light, 83, 201–202, 216
Sperry, Roger, 117
Speth, J. Gustave, 349
Spin angular momentum (spin), 76, 171, 196
Spontaneous symmetry breaking, 193–194, 195–196, 229
Squarks, 205
SSC. *See* Superconducting supercollider
Standard model, 177–198
  Anderson-Higgs mechanism, 193–194
  defects in, 125–126
  as generalization of QED, 180
  multiplicity of elementary particles, 196–198
  nonzero-masses (or energies), 192–193, 194
  renormalizability of, 200, 205
  spontaneous symmetry violation, 193–194, 196
  and violation of time symmetry, 195
  zero-mass approximation, 192, 193
Starobinsky, A. A., 165

Stochastic processes, 44, 47, 48. *See also* Randomness
price fluctuations as, 47–48
Strange particles, 262–263
Strange quarks, 191, 263
Strong interaction, 186, 262
"Student method" of betting, 45
Summing over, 146–148, 151, 158
Superconducting supercollider, 125, 204
Supergap, 204
Supernova, 221
Superpartners and new accelerators, 204–205
Superstandard model, 205–206
Superstition, 263
from fear, 276–277, 283, 296
from invented regularities, 276–278, 283
mental aberration and suggestibility, 284
vs. moral equivalent of belief, 279–281
myth in art and society, 278–280, 343
and skepticism, 275–290
Superstring theory, 127–129, 199–213
apparent multiplicity of solutions, 206–207, 210
approach to Planck mass, 205–206
basic units of energy, 201–202
branching tree of solutions, 210
comparison with observation, 200–201
heterotic form, 128–129, 195, 199, 203
low-mass sector of, 199–200
meaning of "superstring," 203–204
particle masses and basic unit, 202–203
predictions of, 205–206
Sustainable world, 345–366
demographic transition to, 350–351
and destruction from human activity, 348–349
economic transition to, 353–356
ideological transition to, 359–362
informational transition to, 362–366
institutional transition to, 357–359
intergenerational equity, 356–357
meaning of "sustainable," 347–348
population and natural resources, 349
social transition to, 355–357
specialization and integration needed, 345–346
sustainable quality, 355–357, 359–361, 363–364
technological transition to, 351–353
Swift, Jonathan, 64
Sympathetic magic, 89–90, 275, 277, 340
Sze, Arthur, x, 11
Szilard, Leo, 21, 223

Tauon, 191
Tauon neutrino, 191
Taylor, Richard E., 185
Theory. *See* Power of theory; Scientific enterprise
Theory of everything (TOE), 129
Thermonuclear reactions, 113
Thurber, James, 33
TIERRA, 314–318, 321
Tikal, 4
Time's arrows. *See* Arrows of time
Time symmetry, violation of, 195
TOE (Theory of everything), 129
Top quarks, 191
Total quality management (TQM), 298
Transitions, 345
Travels in Central and South America, 3–8, 237, 334
Tree of branching histories, 149–150, 210, 229
observation as pruning of, 154–156
Trivers, Robert, 251
True probabilities, 146–147
"Truly selfish gene," 251, 297

Ulam, Stanislaw, 202
Uncertainty principle, 139, 143, 163, 178
Uncomputability of AIC, 38–41, 56
Undecidability, 39–40
Undirected graph, 30
Unified quantum field theory, 127–129, 199. *See also* Superstring theory
    incorporation of Einsteinian gravitation, 200–201
Universal constants of nature, 201
Universal laws of physics, 8–9, 11
    law of gravity. *See* Gravitation
    laws of electromagnetism, 84
Unrolling the loop, 315
Updike, John, "Cosmic Gall," 186–187

van der Meer, Simon, 189
Virtual exchange of photons, 178–179, 182, 200
Viscaíno, Sebastián, 92
Volta, Alessandro, 81

Wage distribution, law of, 319
Wallas, Graham, 265
Watson, James, 66
Weak force, 262
    neutral, 189
    reactions resulting from, 187–188
Weinberg, Steven, 189
Weisskopf, Victor, 72
Wells, H. G., 14
Wheeler, John A., 137
Wiener, Norbert, 72
Wiesenfeld, Kurt, 97
Wilkins, Maurice, 66
Windows of maturation, 301–303
"Winos," 205
Wolfram, Stephen, 77
Woolfenden, Glen, 252
World Commission on Environment and Development, 355
World Resources Institute, 346, 349
Wouthuysen, 287
Wright, Sewall, 77
Wright, Will, 363

Yale, 15, 75
Yang, C. N., 189
Yang-Mills theories, 192

Zeh, Dieter, 138, 149
Zel'dovich, Ya. B., 195
Zipf, George Kingsley, 94–95
Zipf's law, 92–96, 97
Żurek, Wojciech ("Wojtek"), 138, 149, 224
Zwiebach, Barton, 208